Effective Fortran 77 for Engineers and Scientists

David T. Barnard
David B. Skillicorn

Queen's University

wcb
Wm. C. Brown Publishers
Dubuque, Iowa

Copyright © 1988 by Wm. C. Brown Publishers. All rights reserved

Library of Congress Catalog Card Number: 87-13382

ISBN 0-697-06754-8

No part of this publication may be reproduced, stored in a retrieval system, or transmitted, in any form or by any means, electronic or mechanical, photocopying, recording, or otherwise, without the prior written permission of the publisher.

Printed in the United States of America

10 9 8 7 6 5 4 3 2 1

Contents

Preface   xiii

1 Working with Algorithms   1

    1.1   Algorithms   2
    1.2   Languages   4
    1.3   Reasoning about Programs   9
           Programming Example   13
           Design, Testing, and Debugging   14
           Chapter Summary   15
           Define These Concepts and Terms   15
           Exercises   16

2 Introduction to Computers   19

    2.1   Data and Programs   20
    2.2   Manipulating Data   20
    2.3   Storage Devices   22
    2.4   A Model for Execution   24
    2.5   Input and Output   25
    2.6   General Purpose Machines   26
           Design, Testing, and Debugging   27
           Chapter Summary   27
           Define These Concepts and Terms   28
           Exercises   28

3 Working with Computers 31

- 3.1 The "Life Cycle" of Software 32
- 3.2 Tools 39
- 3.3 A Sample Program 45
  - *Design, Testing, and Debugging 47*
  - *Style and Presentation 47*
  - *Chapter Summary 48*
  - *Define These Concepts and Terms 48*
  - *Exercises 49*

4 A Piece at a Time 51

- 4.1 A Problem-Solving Methodology 52
- 4.2 Sample Algorithm Development 53
- 4.3 Turning Algorithms into Programs 55
- 4.4 Subprograms 57
- 4.5 Textual Placement of Subprograms 59
- 4.6 Simple Output 59
  - *Programming Example 64*
  - *Design, Testing, and Debugging 66*
  - *Style and Presentation 66*
  - *Fortran Statement Summary 66*
  - *Chapter Summary 67*
  - *Define These Concepts and Terms 68*
  - *Exercises 68*

5 Simple Calculating 73

- 5.1 Using Memory 74
- 5.2 Integers 76
- 5.3 Real Numbers 76
- 5.4 Operators 78
- 5.5 Constants 80
- 5.6 Operator Precedence 82
- 5.7 The Assignment Statement 86
- 5.8 Inputting Information 87
- 5.9 Variables and Subprograms 89
  - *Programming Example 93*
  - *Design, Testing, and Debugging 95*
  - *Style and Presentation 96*
  - *Fortran Statement Summary 96*

Chapter Summary 98
Define These Concepts and Terms 99
Exercises 99

6 Control Statements 103

6.1 Definite Iteration 104
6.2 Indefinite Iteration 108
6.3 Selection 118
6.4 Logical Data 123
6.5 Numerical Integration 125
6.6 More General Selection 129
6.7 Getting Answers from Subprograms 129
Programming Example 131
Design, Testing, and Debugging 133
Style and Presentation 134
Fortran Statement Summary 134
Chapter Summary 136
Define These Concepts and Terms 137
Exercises 137

7 Functions 141

7.1 Motivation for Functions 142
7.2 Constructing a Function 142
7.3 Invoking a Function 143
7.4 Intrinsic Functions 145
7.5 Using Functions 146
7.6 Tracing Programs 148
Programming Example 151
Design, Testing, and Debugging 153
Style and Presentation 153
Fortran Statement Summary 153
Chapter Summary 154
Define These Concepts and Terms 154
Exercises 155

Engineering Problem 1
Electron Emission 157

Engineering Problem 2
Heat Transfer through Windows 159

8 Mathematics for Engineering 161

8.1 More Numerical Integration 162
8.2 Simpson's Rule for Integration 166
8.3 Evaluating a Series 168
8.4 Solving Differential Equations 171
8.5 Passing Functions as Arguments 176
8.6 Root Finding 178
Programming Example 186
Design, Testing, and Debugging 189
Style and Presentation 189
Chapter Summary 189
Define These Concepts and Terms 190
Exercises 190

Engineering Problem 3
Escape Velocity of a Space Vehicle 192

Engineering Problem 4
Failure of Electronic Components 195

Engineering Problem 5
Time between Failures 197

Engineering Problem 6
Finding a Set of Sample Points 199

Engineering Problem 7
Calculating the Force of Impact 201

9 Input, Output, and Formatting 203

9.1 Simple Output 204
9.2 Formatting of Output 204
9.3 Simple Input 216
9.4 Structure in the Input Stream 217
9.5 Files and the File System 218
9.6 Plotting 224
Programming Example 228
Design, Testing, and Debugging 232
Style and Presentation 232
Fortran Statement Summary 232
Chapter Summary 233
Define These Concepts and Terms 234
Exercises 234

Engineering Problem 8

Path of an Electron Beam 237

10 Arrays 239

10.1 Structured Data 240
10.2 Using Arrays 241
10.3 Recalculating the Standard Deviation 244
10.4 Other Uses of Arrays 246
10.5 Indexes 250
10.6 Other Types of Indexes 251
10.7 Arrays as Vectors 252
10.8 Polynomials 260
10.9 Linear Search 264
10.10 Binary Search 268
10.11 Sorting 271
10.12 An Example: The Sieve of Eratosthenes 275
10.13 Using the Common Statement 278
Programming Example 280
Design, Testing, and Debugging 282
Style and Presentation 283
Fortran Statement Summary 283
Chapter Summary 284
Define These Concepts and Terms 285
Exercises 285

Engineering Problem 9
Building a Custom Resistor   287

Engineering Problem 10
Motion of Particles   289

Engineering Problem 11
Separation of Components   291

11   Computing with Characters   295

- 11.1 An Algebra of Strings   296
- 11.2 Converting Numeric Representations   298
- 11.3 More Program Tracing   302
- 11.4 Text Editing   307
- 11.5 Text Formatting Techniques   317
- 11.6 A Framework for Formatting   324
  - *Programming Example*   331
  - *Design, Testing, and Debugging*   333
  - *Style and Presentation*   333
  - *Fortran Statement Summary*   333
  - *Chapter Summary*   334
  - *Define These Concepts and Terms*   334
  - *Exercises*   334

Engineering Problem 12
File Compression   339

12   Numerical Linear Algebra   341

- 12.1 2-Dimensional Arrays   342
- 12.2 IO with 2-Dimensional Arrays   344
- 12.3 Matrix Operations   346
- 12.4 Gaussian Elimination   350
- 12.5 Inverting a Matrix   357

12.6 LU Decomposition 361
12.7 Eigenvalues and Eigenvectors 367
12.8 Plotting Functions of Two Variables 371
12.9 Linear Regression 375
Programming Example 379
Design, Testing, and Debugging 382
Style and Presentation 383
Chapter Summary 383
Define These Concepts and Terms 384
Exercises 384

Engineering Problem 13
Closing a Traverse in Surveying 386

Engineering Problem 14
Text Compression 389

Engineering Problem 15
Calculating the Huffman Encoding 391

13 Models, Simulation, and Games 395

13.1 Models and the World 396
13.2 Static Simulation 397
13.3 Randomness 399
13.4 Generating Pseudo-Random Numbers 400
13.5 Monte Carlo Methods 403
13.6 Dynamic Simulations 407
13.7 Playing Games 410
Programming Example 422
Design, Testing, and Debugging 427
Style and Presentation 427
Chapter Summary 427
Define These Concepts and Terms 428
Exercises 428

Engineering Problem 16
Detecting Transmission Errors 430

Engineering Problem 17
Placing Chips onto Circuit Boards 432

Engineering Problem 18
Simulated Annealing 435

Engineering Problem 19
Neutron Scattering 437

14 Computing with Other Numeric Types 439

14.1 Double Precision Variables 440
14.2 Complex Numbers 441
Programming Example 443
Design, Testing, and Debugging 444
Style and Presentation 444
Fortran Statement Summary 444
Chapter Summary 445
Define These Concepts and Terms 445
Exercises 445

Engineering Problem 20
Fractal Geometry 446

Appendix A: Coding Conventions 452
Appendix B: Fortran Summary 454

Appendix C: Other Fortran Features   459
Appendix D: Intrinsic Functions   473
Appendix E: Accessing External Files   475
Appendix F: Building Your Own While Statement   477
Appendix G: Selected Solutions   478
Glossary   488
Index   495

Preface

To Instructors

Intention

Effective Fortran77 for Engineers and Scientists is designed for a first programming course for engineers and scientists. It assumes no background in programming in any language. The book covers all that a beginning programmer needs to know; it presents both the general techniques of programming and specific details of Fortran. A student who uses this book for a semester or two can go on to become a seasoned programmer by mastering a few additional language features (Appendix C), and by dealing with additional real world problems.

Effective Fortran77 for Engineers and Scientists contains a wealth of programming examples drawn from various parts of engineering. Thus it is also suitable for an engineering applications course, which is commonly taken after an introductory programming course. For example, it could be used for a numerical applications course following the traditional introduction to programming.

Features

The book has these distinguishing features.

- Modern ordering of topics. Subprograms are introduced early, allowing the problem-solving methodology used to be translated directly into program structures.
- Engineering problem sections. These specifically discuss problems that students will encounter in their first and subsequent years.
- Introduction to engineering and numerical applications. This gives students an understanding of techniques and algorithms used in packages and tools they will encounter in subsequent years.

- Machine independence. This will allow students to program engineering applications. Some details required for using popular mainframe computers and personal computers are included.

Approach

Programming is a skill that needs to be carefully taught if students are to become competent programmers. It requires more than simply a listing of program syntax elements bolstered by occasional examples. As with mastering any skill, some theoretical knowledge must be learned and some techniques must be practiced.

A great deal is now known about how to teach programming. Two ideas are central: that it is natural for humans to think about solutions to problems a piece at a time, and that much of the skill of an experienced programmer comes from having a library of program fragments, or paradigms, together with rules for combining these paradigms to make new programs.

The methodology of stepwise refinement has been developed to deal with the first idea. Students are exhorted in most introductory courses to break their programs into manageable parts that are organized hierarchically. However, most books don't adequately present the language mechanisms for doing so until students have already written programs of moderate size. Having learned how to handle such pieces of code, most students are reluctant to use the overhead of modular decomposition techniques until they encounter large programs (if they do in introductory courses). The result is that the exhortations have very little effect on actual programming practice. In this book we begin by teaching the language constructs that support stepwise refinement even before we discuss assignment. This forces students to appreciate the flow of control involved in procedure invocation and return. Students tend to assume that material taught late in courses is harder than material taught earlier. Presenting flow of control early provides more time for it to sink in and helps it to become second nature for students by the middle of the course.

To support the second idea, that students learn to program by example, we have included many programs that are extensively discussed in the text. We find that one of the major difficulties faced by students in introductory programming courses is that they don't know how to begin when faced with a problem. The solution to this is to make students aware that the major reason that experienced programmers (and professors) know how to start when they encounter a problem is that they have probably seen something like it before. We suggest a continual emphasis on reuse of concepts and program fragments to reinforce this notion.

Thus, students need to understand that the different programs they see are to be understood as paradigms which they should add to their repertoire. Once they really understand how a program works, they are in a good position to modify it to create a new program.

Organization

The ordering of topics in this book is designed to give students the tools they need to develop good programming habits when they need them. We begin with a discussion of the fundamental concept of an algorithm and start to develop a technique for working from a problem to an algorithm. We then discuss the machine model that is assumed by Fortran and some of the operating system and other tools that are available. We assume that at this stage in a typical course, students are encountering a real machine, some for the first time, and Chapter 2 attempts to fit the pieces together.

We then introduce the top-down design methodology and illustrate how to handle subprograms in Fortran. The only executable statement introduced at this time is the output statement, so students have nothing to confuse with the central idea of flow of control.

In Chapter 5 we introduce expressions and cover the basics of calculation. Variables are also introduced for the first time.

Chapter 6 introduces the control statements for iteration and selection. We cover iteration first because it seems natural to treat the idea of repeating a sequence of steps before considering variations in the steps. These are immediately illustrated with some useful engineering applications.

We then cover functions, using the background of subprogram invocation and parameter passing that should have been mastered by this point.

We do not think that engineers graduating in the nineties will do much programming themselves, and they will surely not be writing large programs. Therefore we have not attempted to introduce the numerical applications with full rigor, since these will typically be encountered as parts of larger packages. Instead, our approach has been to provide some insight into the algorithms used in packages, believing that having some idea of how a tool works helps in its use. Examples that are clearly targeted at real engineering applications also help to motivate students. Chapter 8 introduces some common numerical applications that are used in the remainder of the book. In Chapter 9 we deal with enough details of the IO system to allow writing programs using formatted data and files. In Chapter 10 we introduce arrays in one dimension and illustrate some simple applications. Chapter 11 covers the handling of character data and strings. In Chapter 12 we cover arrays in higher dimensions and show their applications, particularly in linear algebra. Much of this material provides a different viewpoint on the linear algebra that most freshman engineers encounter, reinforcing both the mathematics and the programming. Chapter 13 deals with simulation and games, an interesting and useful set of applications, with emphasis on random numbers. Chapter 14 shows how to compute with more sophisticated types. This includes complex numbers and double precision numbers.

Several sections appear at the end of each chapter.

- A **detailed programming example** is presented that builds on ideas outlined in the chapter, and puts them together into a complete programmed solution.

These examples include details of presentation and documentation that space prevents us from using in the body of the book. They provide guidelines for students on the presentation of their own programs.
- Material on the **design** of programs deals with the practical issues of testing programs and making them easy to debug.
- Material on programming **style** increases the students' awareness of good programming habits.
- A list of the **language features** presented in the chapter, together with an example of each, serves as a convenient reference.
- A **summary** recapitulates the main points of the chapter.
- A list of **key words**, which is suitable for self-test and review, is given.
- Chapters conclude with **exercises** which vary from easy reviews of chapter contents up to substantial programming projects. Many of these exercises also make suitable programming assignments. Solutions to exercises numbered with a power of two are provided at the back of the book.

Course Syllabus

The material in this book can be covered in two semesters.

A one-semester introductory course should cover chapters 1 through 7, together with 9 and 10.

A one-semester applications course following an introductory programming course should cover Chapter 8 and chapters 12 through 14.

If a two-semester course is being given, it is probably best to cover chapters 1 through 14 in order.

Engineering Orientation

In addition to the engineering emphasis in the examples used throughout the book, we have made an attempt to include problems that are interesting both from an engineering point of view and inherently as programs. These include some of the exercises at the end of each chapter and the engineering problem sections.

The engineering problem sections are substantial problems chosen from topics that many students will encounter in their first- and second-year engineering courses. They are often simplified and can be handled with common sense; deep understanding is not required. We have tried to make them self-contained. We use these problems as programming assignments in our courses. They should take an average student about two weeks to complete. It was not our intention that students should attempt all of these problems. They are intended to provide a great deal of choice so that student interests can be matched where possible.

Language Topics

We have not attempted to provide an exhaustive coverage of all language features in the body of the book; many of those we have omitted are used safely only by experienced programmers. The omitted features are included in Appendix C for recognition purposes.

Mathematical Sophistication

We have assumed throughout that students will be encountering introductory calculus and linear algebra concurrently with the material that this book covers. Our approach has therefore been to provide a brief, almost incidental, review of the mathematical content of the programming topics covered. We are interested more in providing an interesting and practical sidelight on the mathematics than covering it with full rigor.

The more important mathematical background required is integration, manipulation of series, ordinary differential equations, vector spaces and vector manipulation, matrices, and eigenvectors. Most of this material is covered in a typical engineering freshman year. It may occasionally be necessary to supplement the mathematical treatment of this book if relevant topics have not yet been encountered in mathematics courses.

To Students

Computers pervade our society; we encounter them in both our work and our recreation. In particular, computers have become a fundamental tool for engineers. They are used for financial applications in engineering companies (traditional data processing); for preparing project proposals, reports, and other documents; for mathematical and graphical modeling of structures being designed or analyzed; and for many other things.

To be effective as an engineer you will have to know about the capabilities of computers—what they can do and how they do it. Sometimes you will need to write programs to do specialized things that have not been done before, but most times you use a computer you will probably use programs that someone else has written. We will introduce you to some of the general capabilities of computers, but most of the book is taken up with teaching programming. This is a useful skill in its own right, and it will help you understand more deeply what computers do, and do well.

As with all skills, with computer programming there is a certain amount of material that you have to know and some techniques that have to be learned. It often happens when learning a new skill that it seems overwhelmingly difficult at first, but after a while this difficulty disappears and you can't understand why it seemed so hard. Think about learning to swim or to ice-skate. Programming is just like this—it may seem difficult at first, but it gets easier. In fact, we find that this is particularly true of programming; students often have a "Eureka" experience, when the whole thing suddenly becomes clear and they don't really understand why it seemed so confusing. So if you find things difficult at first, stick with it and see if you can get over the initial hurdles.

Programming involves finding problem solutions that can be expressed as a sequence of steps to be carried out. We are primarily interested in sequences of steps that can be carried out by a computer, although the skills that you will learn can be applied in many other areas as well. Unfortunately, our own minds work against us as we try to do this. One of the ways our brains make it possible for us to think is by handling many routine actions for us in a subconscious way, in other words, by making them into habits. For example, we (mostly) manage to walk without devoting much thought to it. Sadly, when we come to think about the exact steps needed to do things that we "know" quite well how to do, we find that we can't actually describe them properly. We have knowledge at one level, but we don't have it at another. We are often faced with this kind of problem in programming. For example, we can all select the largest number from a list held up in front of us. But we have a lot of difficulty describing the steps we went through to make the selection (try it).

There are two main methods that are used to make the task of programming easier. The first is to break solutions up into pieces, each of which is small enough to be thought about as a unit. Psychologists tell us, and personal experience confirms, that the capacity of the human mind to remember details is severely limited. Good problem-solving techniques take this limitation into account. Part of the skill to be learned here is how to break the solution up in a sensible way and keep the relationship of the pieces straight.

The second method is to build a repertoire of program pieces that do particular things and understand how these pieces may be combined to form new programs. All experienced programmers work this way—like chess players, they don't remember all possible combinations, but they do remember meaningful chunks. You will learn standard ways of handling common programming problems from the examples in this book.

We have used examples and problems drawn from engineering throughout the book. You will have seen some of these in your other courses. You will see others in your engineering courses in later years. If you have already seen some of the problems in your other courses, then looking at them from a computational point of view should help shed some new light on them. We have made them self-contained enough so that if you haven't already encountered them, you should be able to see why they are interesting problems for engineers.

You may not do much actual programming as an engineer. But you will certainly use programs written by others and it's a good idea to understand what goes on under the covers. It is also important to recognize that writing large programs is qualitatively different than writing small programs. To say this another way, writing a large program does not simply take longer; there are actually new design and management problems introduced when programs become very large. This is difficult to appreciate from a first course in programming because the programs you write and modify are of necessity quite small.

You may also have to read and understand programs written by others. We have provided an appendix in which we describe some language features that we don't recommend using (these features sometimes cause undesirable problems with programs), so you can recognize them if you encounter them. Since the development of the Fortran77 standard, there has been an unfortunate situation arising from the difference between older versions of Fortran and this new standard. We stick to the standard throughout the book when we are describing language features, or else we explicitly indicate otherwise in those few places where we step outside it. But some implementations of the language are not as restrictive as the standard is; some of the things we indicate as restrictions may not be true of the system you use.

We begin in Chapter 1 with a description of algorithms, one of the underlying concepts of computer science. In Chapter 2 we introduce some details about the hardware technology that is used in computers. Chapters 3 and 4 are concerned with problem-solving techniques appropriate for program development. The remainder of the book presents details of the programming language Fortran, together with instruction in program development. Particularly interesting statements in programs are set in color.

The later chapters end with substantial programs developed as solutions to presented problems. We use these to bring together some of the ideas introduced in the chapters and show you in more detail how to build complete programs. In particular, we include more program documentation than we have room for in the chapters themselves. These examples show you how to write your own programs well.

The programming example is followed by two sections that discuss issues in program design and program writing style. Most of these things are hints that will help you to develop good programs that will be easy to read and understand.

Each chapter also contains a summary and examples of use of the language features introduced, a summary of the important ideas in the chapter, and a set of important words that you can use either as a quick review or as a test of your understanding of the material.

Finally, chapters conclude with exercises. These exercises are usually short and often do not deal with real problems. Solutions to exercises numbered with a power of two can be found at the back of the book. We supplement these exercises with several sections presenting engineering problems in some detail. These tend to be somewhat simplified settings of practical problems that are encountered in professional engineering situations.

Acknowledgments

Our wives, Marcia Barnard and June Skillicorn, and DTB's children, Stephanie and Benjamin, have been more supportive than we could have expected them to be; there were too many days and evenings when we worked on this book or the course associated with it, rather than being with them. That is over now — at least for a while.

Mike Jenkins and Howard Staveley used an early version of the book in CISC131* in 1987, and provided many helpful suggestions. Our students in that course also used the book and their comments were both extremely straightforward and extremely helpful. Brent Nordin patiently read early drafts.

Engineering problems were suggested to us by several colleagues, including Bob Crawford, Peter Douglas, Sid Penstone, and Francis MacLachlan.

Olga MacMillan supplied the recipe for raisin biscuits.

This book was written and formatted on a computer system. Mike Smith, of Queen's University Computing and Communications Services, went far beyond the call of duty and the normal commitments of friendship to help us with many aspects of the typography. Rick Pisani and Rose Chan, of that same group, were very helpful in our (sometimes frustrating) struggles to make things come out as we wanted them. The shortcomings that remain are ours, but the quality we did achieve is due to their efforts. The book was typeset at Typesetting Systems, Inc., in Kingston, Ontario.

The Department of Computing and Information Science has been a comfortable place to work on this material, since the Department, as does Queen's University as a whole, takes a great deal of care over its undergraduate teaching commitments.

Working with Algorithms

1.1 Algorithms
1.2 Languages
1.3 Reasoning about Programs
      Reasoning about Termination
      Reasoning about Execution Time
      Reasoning about Correctness

1.1 Algorithms

Many of the things that we do in the normal course of life could be described by a precise set of steps to be followed. Some examples are the routine you follow between awakening and actually getting out of the house each morning, the instructions for cooking a cheese omelet, the procedure for doing the laundry, the directions to follow to get from your house to City Hall, and so on.

Although these are not usually expressed formally, they certainly could be. In each case the formal expression would include a list of conditions that must be true (for example, you're lying in bed), a list of "material" to work with (for example, three eggs, 1/4 pound of cheese, etc.), and a list of things to do, in order.

Of course, many of the most interesting aspects of life can*not* be written down in this manner, including developing and nurturing personal relationships, making judgments about nonquantitative problems, and establishing personal career goals. These are processes that are not understood well enough to codify with a specific set of instructions. Perhaps they never will be.

These two different kinds of activities are sometimes referred to as *algorithmic* and *heuristic*, respectively. An **algorithm** is a precise set of steps to follow to solve a problem—that is, to transform a given situation into a desired one. The "given situation" is characterized by a set of conditions (such as the availability of certain materials—called the inputs to the algorithm) and the "desired situation" is also characterized by a set of conditions that must hold (such as the availability of certain new materials—the outputs). If you were making an omelet, the inputs would include eggs, cheese, a skillet, and a fork, and the output would be an omelet. An algorithmic activity is one for which there is an algorithm.

We use the word **heuristic** in two senses. First, it can refer to a "rule of thumb" or general principle that is too general to be encoded as a set of specific actions. For example, "be loyal to your country" is a heuristic in this sense—it's a general principle but we don't immediately know how to apply it in any particular circumstance. Second, the word can refer to a specific set of rules (really an algorithm) that only provide an "approximate solution" to a problem. For example, when packing the trunk of a car, a good approach is "get the biggest items in first and fit the smaller items around them," but doing this doesn't always maximize the amount we can fit in. For some types of problems we cannot expect an algorithmic solution, but for problems like packing the trunk there definitely *is* an algorithm (for example, try every possible packing order and location for each item), but it's so inefficient that it's not practically useful. It would be more accurate to describe the first approach to packing the car as an "approximate algorithm" and the second as an "exact algorithm."

This book only considers problems with algorithmic solutions. Many problems of real interest do have algorithmic solutions and for many of these it is possible to use a computer to do some or all of the work required. This book teaches how computers can be used to implement solutions to problems of this kind.

We can distinguish two classes of things that we need to be concerned about when we design algorithms. The first class is the objects (such as eggs or a skillet) that are manipulated by the algorithm; the second class is the operations (such as break an egg, stir, or serve) that we can perform on the things in the first class. Both of these classes can be further divided into simple (or *primitive*) things, and complex (or *structured*) things. For example, if we are interested in algorithms used in cooking, then primitive materials are things like a measuring spoon and an egg; structured materials are things like a *set* of measuring spoons and a *rack* of spice bottles; primitive operations could be "measure a teaspoon of cinnamon" and "separate the white from the yolk of an egg"; complex operations could be "for each of six eggs, separate the white from the yolk" or "sift the flour twice." If we're completing a tax return, primitive materials could be things like a statement of income issued by an employer or a receipt from a charitable organization; structured materials could be a *set* of dependents or a *list* of charitable donations; primitive operations could be "enter a number in a specific position in the form" and "add up the numbers in two specific positions in the form"; complex operations might be "complete the separate form for self-employed earnings" or "add up all the deductions for dependents." Most of these things depend on the kind of problems we are concerned with (the problem domain).

This isn't strictly true, though, of structured operations. They are (more or less) independent of the problem domain. To say it another way, there are a few structuring mechanisms for operations that will allow us to solve many problems in many different areas. Structuring mechanisms for operations include:

- **Sequencing** — do a set of operations in order;
- **Selection** — do one operation chosen from a specified set of operations;
- **Iteration** — do an operation repeatedly; and
- **Abstraction** — do a separately defined named operation.

Abstraction is one of the most important and powerful notions in computer science. The ability to specify *how* to compute something in one place and subsequently use it as though it were a primitive part of the computing system is the key to building successively more interesting and complex software systems. In fact, the human brain also uses abstraction and that's one of the reasons that we can think about complex things without having to think about breathing and keeping our heart beating.

There are two types of iteration — *definite* and *indefinite*. For **definite iteration** it must be possible to determine how many times the operation is going to be repeated before the iteration begins. For example, if an algorithm contains the structured operation "repeat 3 times: add one egg" that's definite iteration. **Indefinite iteration** is more general and does not require that we know at the beginning how many times the action will be repeated. For example, if an algorithm contains the structured operation "turn a page of the telephone directory until you come to the names beginning with *K*" that's indefinite iteration.

Here are some examples of structured operations. In the instructions for completing a tax return we might encounter statements of this kind.

- *Sequencing:* First enter your total income, and then enter your deductions.
- *Selection:* If you made charitable donations, deduct the amount from your total income.
- *Iteration:* For each of your dependents, enter the person's name and age, together with the amount to be deducted.
- *Abstraction:* Complete the supplementary form that shows moving expenses.

In a cookbook we might encounter statements like these.

- *Sequencing:* Mix the dry ingredients together, and then add the milk.
- *Selection:* If you're making the raisin muffins, add the raisins now.
- *Iteration:* Sift the flour twice.
- *Abstraction:* Serve with whipped butter (see page 135 for making whipped butter).

Algorithms for execution on computers use various kinds of data for basic materials (for example, integers such as 3 and 127, and characters such as "*a*" and "*x*") and various kinds of structured data objects (for example, arrays such as the vector (15, 5, 7), and strings such as "Hello, world."). The basic operations are data-type specific. This means that operations only make sense for specific kinds of things. This shouldn't be too surprising; you can "select a station" on a television set or a radio, but not on a microwave oven. Similarly, you can express a number as a percentage by dividing by 100, but it doesn't make sense to do the same thing with the phrase "Hello, world." Some of the basic operations that computers can do are addition, subtraction, multiplication, comparing two values, and moving values from one place to another. The structured operations that computers can carry out include variants of the sequencing, selection, iteration, and abstraction mechanisms we've described.

1.2   Languages

It's not enough to know the materials and operations required for a particular algorithm. We must use some notation for writing down the steps that are required and the objects to be manipulated. Thus we need a language to express algorithms. Algorithms can be expressed in many different languages. Often there is a known convention for a particular problem domain. Here are several examples.

- The procedure for calculating an individual's annual income tax is encoded in the tax return form, together with a manual written in English that explains some of the more confusing terms.
- Here's a recipe for making raisin biscuits—try them, you'll like them.
    3 cups flour
    3 tsp. baking powder
    1 cup brown sugar

    1 cup shortening
    1 cup raisins
    1/2 tsp. salt
    1 egg white, slightly beaten, and fill cup with milk

Combine in a bowl the flour, baking powder, brown sugar, raisins, and salt. Mix together to coat raisins and cut in shortening. Add the egg white and milk and mix just enough to form into a ball. Roll on a lightly floured board, not too thin, and brush the top with egg yolk. Cut any shape you wish with a knife. Bake at 400 degrees for 15 to 20 minutes or until done.

- A parent might try to tell a young child how to prepare her breakfast in the following words: Go to the cupboard and get a yellow bowl and put it on the counter. Then get whatever cereal you want from the pantry and put some in the bowl. Put the cereal back. Then get the milk pitcher from the refrigerator. If it's not too full, then pour the milk. If it is too full, come back and get me to help you. Then put the milk pitcher back in the refrigerator, and get your spoon from the drawer.

These are all algorithms, but the language used varies depending on the problem domain and the audience. No doubt you can think of many other examples. The forms of expression range from natural language (English) more or less unconstrained, to forms that are explicitly tailored for a class of problem (the income tax return).

One of the major constraints on deciding how to express an algorithm is the need to make the algorithm unambiguous. If the person or machine executing the algorithm has common sense, then we can tolerate a loose expression of the algorithm because we expect the user to resolve any ambiguities sensibly. However, it's difficult to write algorithms in English unambiguously (which is why legal documents are inscrutable to all but those trained in their use, and frequently even to them). Therefore algorithms are often written in a restricted language where the precise meaning can be made more apparent.

To use computers to solve the kinds of problems encountered in this book, it's necessary to devise an algorithm and express it in an appropriate language. In this setting, *appropriate* means *executable by a computer*. These languages are called **programming languages** and the process of writing algorithms in programming languages is called *programming*. There are many such languages, and one way to characterize them is to consider the level of detail at which the operations must be expressed. In the next few paragraphs we discuss several classes of languages, beginning with very detailed ones.

As you may know, computers are usually built from electronic circuitry with fundamental components that can represent and recognize only two states, often written 0 and 1. Chapter 2 will discuss computer hardware in more detail, and it suffices for the moment to know that all the basic circuitry *understands* is binary numbers (that is, base two, using the digits 0 and 1), so all programs and data must ultimately be written using these same symbols. This notation is called **machine language** since it's what the machine directly understands. It's a very low-level expression of what we want done, by which we mean that it includes all the grimy details of what operations are to be done, where the operands are to be found, how they're to be manipulated, and so on. Each model of computer interprets binary sequences in different ways, and thus defines its own machine language. Human users do not program directly in machine language—it's much too complex, tedious, and error-prone. Here is an example of what some operations written in machine language might look like.

01001101
10000000
01110111

A more easily used, higher-level language can be defined by making up short names for instructions (such as Add, Store) and for operands, that is, quantities and locations to be manipulated. This is called an **assembly language** because it requires a piece of software called an *assembler* to translate the symbols into the binary sequences that the machine understands. Figure 1.1 shows this process. This translation is typically a very simple one—for instruction symbols it's just substitution, and for operand symbols it's only a little more complex. Since the translation is so simple, the symbolic form still corresponds very closely to what the machine understands, so it is, again, specific to a machine model—the assembly language for an IBM System/370 will not run on an IBM Personal Computer. Most human users of computers never program in assembly language. As systems increase in sophistication, less and less software is written in assembly language because it is difficult and expensive to write and cannot be used on other machines (that is, it's not *portable*). Here is an example of some statements written in an assembly language:

```
LD          R1,A
ADD         R1,B
STO         R1,C
```

High-level languages are used to avoid these problems. These languages allow programs to be executed on more than one machine model and written in a notation more comfortable for human users. The lower-level languages described earlier are closely matched to machines but are, in general, ill-suited for humans. There are many high-level languages, and they vary in many ways, which we'll not attempt to point out. Some of the oldest such languages still in wide use are Fortran and Cobol; others include PL/1, Basic, Pascal, Ada, C, Lisp, Snobol, and APL. Each of these languages can be used on many different kinds of computers. To make this possible a program called a *compiler* is used to translate the high-level language to something (assembly or machine language) that is already understood by the machine. Figure 1.2 shows this process. Because the machine language varies from machine to machine, each machine must have its own compiler for each language that is to be used. The translation from a high-level language to machine language is more complex than the translation from assembly language, and the details vary with the language and with the machine on which it is to execute. Most human programmers use a high-level language. In your career as a practising engineer, when you program (at least for the next several years) it will probably be in a high-level language. Here are some examples of statements to compute the sum of the numbers in a list in several of these high-level languages:

Fortran
```
      Sum = 0.0
      do 100 I=1 to 10
         Sum = Sum + A(I)
100   continue
```

Basic
```
10 let Sum = 0.0
20 for I=1 to 10
30    let Sum = Sum + A(I)
40 next I
```

APL
```
Sum <— +/A
```

After the introductory material, this book will deal exclusively with programming in the high-level language Fortran. The name Fortran comes from FORmula TRANslation and indicates the mathematical orientation of the designers of the language. (The word is often written in all uppercase symbols, but since it has effectively become a proper name over the past three decades, we'll write it with only an initial uppercase letter.) It was originally designed in the 1950s and there have been several versions. Fortran I was not widely used, but Fortran II was a popular tool for several years. It was eventually succeeded by Fortran IV, which

Figure 1.1
Processing an Assembly Language

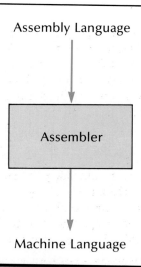

Working with Algorithms

was the "official" version until 1977, when the specification for Fortran77 was produced. Fortran77 has not totally displaced Fortran IV even yet, since there are many programs already written in Fortran IV and many programmers who are comfortable writing it. The major changes were the introduction of a richer set of control structures and better handling of character strings. Even though it's old by standards of the computer industry, Fortran is still used in much engineering and scientific programming.

There is still another level that is usually recognized, sometimes called **very high-level languages**. These languages are diverse and difficult to characterize. Some are used to provide easier access to databases. For example, a user of such a language might say "show me the names of all staff members who will retire in the next five years" and have a program process the request and produce the result; since this is an alternative to writing a program in Fortran to do the same thing, it can be said to be programming. Some other "languages" are used for other specific types of computation. For example, a *spreadsheet* is an array of labels, constant values, and formulas for deriving other values; when the user changes a value (for example, overhead percentage on sales), the system automatically calculates the derived values (for example, expected profit); since this is an alternative to writing a program in Fortran, it can also be said to be programming. Figure 1.3 illustrates this concept. You will doubtless use spreadsheets during your undergraduate and professional experiences, and possibly other very high-level languages as well.

One of the observable trends in computing through the three decades (more or less) that machines have been commercially available has been the development of increasingly sophisticated and higher level languages, and a corresponding growth in the number of programmers. This will continue, with specialization and the level of abstraction (that is, suppression of details) becoming more and more apparent. Some think the ultimate achievement would be programming in a natural language, but this won't be possible for a considerable time, if ever.

Figure 1.2
Processing a
High-Level
Language

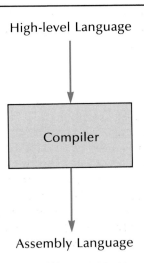

The high-level languages vary in the data types and operations that are provided, and in the specific syntax that is used. In this book we will present the details of the Fortran language.

1.3 Reasoning about Programs

Algorithms expressed in high-level programming languages are precisely defined objects. They can be understood by another computer program—the compiler—and thus one could expect them also to be precisely understood by human readers. In fact, they can be. But it requires training to understand utterances (in this case, programs) that are unambiguous and have very little of the redundancy or "noise" (words not carrying specific content) of most human communication. You've probably been infuriated by someone who interpreted everything literally—or perhaps you've read Asimov's stories about robots. There are several questions we might ask about an algorithm.

- Will it terminate? Perhaps termination depends on the conditions that hold when the algorithm starts. If so, under what conditions will it terminate? For example, an algorithm to compute your personal taxes will always terminate—sometimes with depressing results! But an algorithm that tells you to "stir a powder into ten times its volume of water until dissolved" will terminate if the powder is sugar, but not if it is sand.
- How long will it take? The answer should be in terms of the size of the problem. For example, many sorting algorithms take four times as long to sort $2n$ numbers into order as they do for n numbers.
- Is it correct? If the starting conditions are satisfied and the algorithm terminates, will it have accomplished what it is "supposed to" accomplish? For example, if the input is a list of numbers and the program terminates, will the list be sorted?

There are other questions that can be asked about algorithms (for example, are all the materials used), but we will only address these three in the remainder of this chapter.

Reasoning about Termination

Determining if an algorithm terminates can be a difficult and subtle process. We will make some simplifying assumptions. We'll assume that all the basic operations provided in the language terminate in a finite time; in other words, we're assuming that expressions like $127 + 3579$ will produce an answer in a finite (and short) time. This seems sensible (and it *is* sensible), and now the only algorithm components of potential danger are the iteration constructs. Definite iteration, too, is not a problem if we assume that the number of iterations (remember, it's always known in advance) is finite—and, of course, this is a safe assumption in a computer with finite storage.

This leaves only indefinite iteration to be considered. It is easy to write algorithms that have indefinite iterations that do not terminate. For example, an algorithm containing the operation "while soup is not salty enough stir in more sugar" is doomed to the abyss of nontermination. Intuitively it's easy to see that if the operation to be repeated doesn't manipulate the material that is examined in the condition controlling the iteration, there is no hope of termination. However, manipulating the material examined in the condition is not sufficient to guarantee termination. The operation "while the floor is dirty move the dirt to the other side of the room" will also not terminate.

To prove **program termination**, it must be possible to show that each iteration makes some measurable progress toward the goal expressed in the controlling condition. In most cases in this book, and also in most cases you will encounter in your own work, it is straightforward to show this.

Consider dealing an unknown number of cards to a specified number of players, until the deck is exhausted. This is indefinite iteration (we don't know how many cards there are), but it's easy to see that the process will terminate because the deck is of finite size and it gets smaller as each card is dealt. Now consider a more complicated iteration. Suppose two players are engaged in a card game; each begins with k cards. The play consists of each player turning up the top card in his pile; the player whose card has the higher denomination takes both cards and adds them at the bottom of his pile; if the denominations are the same, each player turns up the next card and the comparison is repeated. The game terminates when one player holds all the cards. It is much more difficult to argue about the termination of this indefinite iteration; a considerable knowledge of the problem domain is required to make a convincing case one way or the other. Similarly, consider playing a game of chess; the obvious structure for a description of this game involves players alternating in making a move until one king is checkmated or a draw is declared. But recognizing a draw position requires a high level of intelligence and knowledge of the game. These examples illustrate that, although

Figure 1.3
A Spreadsheet for Grades

	a	b	c	d
1	Student	Scores		Grade
2		Midterm	Exam	
3				
4	Kairi	75	86	81
5	Karababa	85	80	83
6	Kavanagh	80	90	85
7				
8	Averages	80	85	83

The entries in the Grade column and the Averages row will be formulas that cause the required values to be computed. Changing one of the values in the Exam column would result in automatically recalculating the Grade entry in that row, as well as the Averages entries in those two columns.

in many cases termination arguments are simple, they can also be very complex and subtle.

It will be our practice to make a termination argument about indefinite iterations in the book. These will be informal, for the most part, but—we hope—sufficient to be convincing. If we wanted to, we could formalize these arguments. We hope you will become comfortable with these arguments and develop the habit of making them yourself for the algorithms that you devise.

Abstractions can cause difficulties in arguments about termination if they invoke each other in complicated ways. We'll ignore this nasty set of possibilities. None of the programs in this book are complicated in this respect.

Reasoning about Execution Time

Every operation in an algorithm takes some time, and to get an accurate estimate of the time required to carry out the entire algorithm one would have to estimate how many operations would be carried out and how much time each would take. However, by convention, most of this detail is ignored.

We ignore the time taken by each specific operation and concentrate on how the execution time is related to the size of the problem. And further, again by convention, the size of the problem is determined by the amount of input (more or less—this is a simplification but it's not too deceiving).

Let's consider some examples to make this idea clearer. Suppose we were given a pile of cards with an integer on each card. If the algorithm we are considering is counting the number of cards, then we need to handle each card exactly once. If there are n cards and it takes time k to handle each one, then going through the pile will take time kn. There will be some extra time at the end to announce your answer, say m time units. Thus the total time for this algorithm is $kn + m$.

Now if you ask a friend to carry out the same algorithm and she can look at each card more quickly, she will take a shorter time because k will be smaller for her. However, we ignore these kinds of variations in considering the time taken for an algorithm for the following reason: suppose that we plotted the times taken by you and your friend to count the number of cards for different numbers of cards. Both graphs would be straight lines (with different slopes). It is the concept that the execution time is a straight line as a function of the number of things being manipulated that is interesting. We say that this algorithm is *order n* (since a straight line is a linear function of n) and write it as $O(n)$.

Now consider the same pile of cards and the following problem. Suppose that the cards have been laid out in a long line. Rearrange them so that they are laid out in reverse order in their original positions. Now we have to handle each of the cards twice because we have to move them from their original positions to new ones and then put them back in line in reverse order. If it takes time k to do the first move and time l to do the second, then the total time for this operation is $(k + l)n$. Now although this algorithm is (probably) about twice as slow as the previous one, it is still an $O(n)$ algorithm because $(k + l)n$ is still a linear function of n.

Next consider an algorithm to sort the pile of cards into order. One way to do this is to look through the pile for the largest number and move it to the bottom. Then we look for the second largest and move it to the second place from the bottom, and so on. We first have to determine the time required for the operation "look for the ith largest." When we are looking for the largest we have to look at all of the cards, so the expense of that operation is clearly $O(n)$. When we have only a few cards left, the list to be checked is much shorter. If we consider the average length of the list being checked to be $n/2$, then each operation "look for the ith largest" is $O(n)$ because $n/2$ is $1/2(n)$. Now we have to repeat this operation n times so the cost of sorting is $O(n^2)$. If we drew the graph of the amount of time required against the number of cards to be sorted, it would be parabolic in shape. If sorting 10 numbers requires 10 time units, then sorting 20 numbers requires about 40 time units.

Because we ignore the constant terms in estimating the execution time of algorithms, the only features of algorithms we need to consider when doing this reasoning are the iterative constructs, such as the construct required to "look for the ith largest" above.

In some of the examples in the book we will derive an expression like the previous ones for the execution time, which is sometimes called the **complexity**, of the algorithm involved. We hope you'll become comfortable with the techniques we use.

Reasoning about Correctness

If we could attach a formal meaning to each of the basic operations in a language and devise combining rules showing the meanings of each of the structured operations in terms of the meanings of the basic operations, then we could devise a formal meaning for entire programs. Fortunately, it is possible to do this for all the constructs in commonly used languages.

Here's an example. Suppose the condition before an algorithm fragment starts is $x > 0$, and the condition afterwards is to be $x > 1$. The algorithm fragment *replace x by x+1* is sufficient to make this happen.

To work formally with the assignment of a value to a variable, we need a general rule, and this specific example must be covered by that rule. If we write R for the condition that is to hold after the assignment *replace x by E*, then the condition that must hold before the assignment is $R(x/E)$, which means "R with every occurrence of x replaced by E." In the example R is $x > 1$, the assignment is *replace x by x+1* and $R(x/E)$ is $R(x/x + 1)$ or $x + 1 > 1$, which can be manipulated by standard algebra to become $x > 0$ as expected. The general rule is written

{R(x/E)} replace x by E {R}

The rules for sequencing, selection, repetition, and abstraction are more complex, as one would expect, but it is certainly possible to work with them for small algorithm fragments. However, for algorithms that are reasonably large, the expressions (for the conditions that hold at various points in the algorithm) become large and complex because there are many data items to be considered. Manipulating these expressions is tedious and error-prone, so that producing a complete proof of an algorithm is at least as difficult as devising the algorithm in the first place. The only hope for making the approach practical for large programs is to have automated assistance, that is, a program that assists the user in proving algorithms.

Proving that programs do what they are supposed to is important, since **program correctness** is clearly critical. In the remainder of the book we will occasionally argue informally about the correctness of algorithms or programs. It is always possible, if it's required, to formalize such arguments. In the examples, we'll introduce other rules as they're required. Perhaps the greatest advantage of this material is that it can help cultivate a more critical awareness of what is required when algorithms are devised, even if a proof is not constructed.

Programming Example

Problem Statement

Write an algorithm to take a deck of cards and reorder it so that the cards are in the opposite order, that is, the bottom card of the deck is now on top and the previous top card is on the bottom.

Inputs

A deck of cards in a pile.

Outputs

A deck of cards in a pile with opposite order to the input deck.

Discussion

This can be done by moving one card at a time from the top of the starting pile to the top of the final pile. We assume that our table is big enough to hold two piles of cards simultaneously.

Program

```
repeat
    take top card from initial pile
    place card on top of new pile
until "initial pile is empty"
```

Testing

It is easy to see that this program works for the ordinary case that you might imagine. However, the program fails if the initial pile is empty, since we are supposed to take the top card from an empty pile.

Discussion

It might seem silly to consider the case when the initial pile is empty, because any human would know what to do if that occurred. However, algorithms that are executed by nonhumans cannot rely on common sense. You will find that one important skill to develop when designing algorithms is the ability to think clearly and without unarticulated assumptions.

Design, Testing, and Debugging

- Always carry out a termination argument for any algorithm you design.
- Think carefully about the boundary cases that may occur in any algorithm you design—those situations that are at the limits of the problem or those that seem improbable but might still occur. Does your algorithm still perform correctly in these situations?
- Think about the time an algorithm will take as a function of the number of its inputs. Is the time reasonable for the scale of problem you expect to solve? Try some realistic problem sizes. For example, the algorithm we gave for sorting a deck of cards works reasonably well for 52 cards. What happens if there are 500 cards? How long would it take you to carry out this algorithm?
- Work to get the algorithm right before translating it into a programming language. It's much easier to alter the algorithm than to alter the program, so it pays to invest time in the initial design.

Chapter Summary

- An algorithm is a precise, unambiguous specification of the set of steps to follow to solve a problem.
- Algorithms can be built from simple and structured operations. The structured operations of sequencing, selection, repetition, and abstraction are sufficient to describe all algorithms. The simple operations used depend on the problem domain.
- The objects manipulated by algorithms can be either simple or structured. For computers, the objects manipulated are pieces of data, either numeric or non-numeric.
- An algorithm is expressed in a language. Computer algorithms are expressed in programming languages. Writing algorithms in a programming language is called programming.
- Programming languages exist at different levels. Machine language is the set of instructions that a computer understands directly. An assembly language is a symbolic form of machine language, more suitable for humans to use. High-level languages allow humans to express algorithms without worrying about the details of the machine language equivalent. Very high-level languages allow humans to express the goals of their programs rather than the methods to be used.
- It is possible to reason about programs. Properties of programs that are interesting are: whether or not they terminate, how long they take to execute as a function of the amount of information they handle, and whether they work correctly (that is, as required).
- The number of steps that an algorithm or program takes, as a function of the amount of input it uses, is called the order of the algorithm. For example, if an algorithm takes a number of steps that increases as the square of its input, then we say it has order n^2, which we write as $O(n^2)$.

Define These Concepts and Terms

Algorithm
Heuristic
Sequencing
Selection
Definite iteration
Indefinite iteration
Abstraction
Programming language

Machine language
Assembly language
High-level language
Very high-level language
Program termination
Complexity
Program correctness

Exercises

1. Which of the following are algorithms? For those that are, under what conditions will they terminate? For those that are not, explain why not.
   a. Given a number, take 2 away from it repeatedly until it is zero.
   b. Keep crossing the largest remaining number off a list until there are no numbers left.
   c. Look between pages 157 and 158 of a book.
2. What is the complexity of each of the following operations?
   a. Look at a deck of cards to see if the jack of hearts is present.
   b. Arrange a deck of cards by suit.
3. Your opponent thinks of a number between 1 and n. You try and guess the number and, for each guess, your opponent tells you whether you guessed too low or too high. How many guesses will it take you, expressed in terms of n?
4. Write instructions for playing a record on your stereo. Give enough details of what to do and what to avoid that you would be comfortable letting your twelve-year-old neighbor carry out the instructions.
5. Examine the owner's instructions for some major appliance in your house (perhaps a microwave or electric oven). Are the instructions detailed enough to allow someone who has never seen such a device to operate it safely?
6. Think about what you would do if you came back to your house on a cold day in January and found there was no heat. Did you come up with a heuristic or an algorithmic description of activities? (California residents substitute flood for furnace failure.)
7. What are the primitive materials and operations involved in doing your laundry?
8. Make up algorithms for each of the following:
   a. making a phone call
   b. starting a car
   c. checking a book out of the library

   Write down arguments about correctness and termination for each one. Verify your algorithms by getting someone to follow them. Do they manage successfully without resorting to common sense?
9. What activities do you carry out in a normal week that have specialized "languages" or vocabularies to describe them? Think of sports, hobbies, jobs, cooking, and so on.
10. Give examples of the use of sequencing, selection, repetition, and abstraction in the owner's manual of a major appliance in your home.
11. A certain card game is played with a standard deck as follows. The cards are shuffled and held in a pack face down. The following steps are repeated until there are no cards left in the pack: the top four cards are turned face up, making four "piles"; any of the second, third, and fourth cards that are of the same denomination (for example, kings) as the first are placed on top of the

first card; if there is only one pile remaining, it is removed; otherwise the piles are picked up from left to right and placed at the bottom of the pack. Will this game always terminate? Give a convincing argument if you think that it will. Give an example of a pack of cards that cannot be made smaller if you think that it will not.

12. Take an ordinary deck of cards and remove one suit (for example, all the clubs). Shuffle these cards. Time yourself to see how long it takes you to sort them into ascending order by denomination. Repeat this ten times and compute your average time. Now repeat this entire process with two suits, with three suits, and with the entire deck. How does your sorting time increase as the number of cards increases? Is it linear—that is, does it take twice as long to sort twice as many cards—or worse than linear?

13. The following algorithm attempts to find the greatest common divisor of two integers. Does the algorithm always terminate? Is it correct?

```
while x is not equal to y
    if x is greater than y
        set x to x - y
    else
        set y to y - x
```

The algorithm should leave the two values the same, and equal to the greatest common divisor of the initial values.

14. Is the game tic-tac-toe guaranteed to terminate?
15. Is the board game of snakes and ladders guaranteed to terminate?

Introduction to Computers

2.1 Data and Programs
2.2 Manipulating Data
2.3 Storage Devices
2.4 A Model for Execution
2.5 Input and Output
2.6 General Purpose Machines

In this chapter we describe some characteristics of modern computers. There are various ways in which computers can be organized. These organizations are called **computer architectures**. Almost certainly the computers you are likely to encounter—especially as you use this book—will have a great deal of similarity and we will restrict ourselves to one particular architecture in this description. We include this material because it will help you understand what the computer is doing as it executes your programs, and how algorithms and languages are reflected in hardware.

2.1 Data and Programs

There were two domains mentioned in Chapter 1 when discussing algorithms: objects and operations. When we consider algorithms to be executed on computers, the object domain is *data* and the operation domain is *programs*. A program is something that is executable—it can be loaded into the machine and given control, so that the operations to be performed by the machine are determined by the program. The data are the items of information that are available to be operated on.

Since both programs and data are encoded as strings of binary digits in a computer, the difference between them is not inherent—it's simply a matter of interpretation. By way of analogy, consider a box of filing cards, each containing a recipe. If you want to make chili, you extract the appropriate card and it becomes your algorithm or program—it tells you what to do. But if you want to find all the recipes that use tomato paste, the chili recipe card is just a datum to be examined.

Similarly, programs can be data items for other programs in a computer system. We already gave an example of this in Chapter 1 when we said a compiler (a particular program) translates a user's program from a high-level language into a form the machine can understand. As another example, suppose that you write a program and that the program uses some data kept in a file. If you want to give the program and data to someone else, then you have to make copies of each. All computers have some standard program (called a **utility program**) that makes a copy of a file. The utility program is itself a program (that makes copies of other programs). It does not particularly care what it is that it makes copies of—all files look like strings of 1s and 0s. So both your program file and your data file are data to the copying program.

2.2 Manipulating Data

We have already said that each computer has a language that it directly implements. This language consists of primitive operations on primitive data items. The task of a compiler is to transform the high-level language program into an equivalent machine language program, using the primitives that are built into the machine. In this section we discuss what facilities are available in typical machines you might use.

Every computer you will encounter will have **integers** built in as a primitive data type. The usual algebraic operations (add, subtract, multiply, divide) are available. Sometimes other operations (change sign, modulus) might be supplied, but basic number theory tells us that the previously named ones are more than enough to perform any computation. The major difference between integers as implemented in computers and integers as we all know them from school algebra is that those in the machine are of bounded size—in other words, they cannot be arbitrarily large or small, but are confined to some predetermined range. If you've used a calculator, this won't be a surprise. If you haven't, think of doing simple arithmetic on telephone numbers, which have exactly seven digits. You can doubtless add one to your telephone number (unless it's 999-9999) and still have something that at least looks like a telephone number (it's still seven digits long). But if you multiply your telephone number by itself you'll definitely get a result that has more than seven digits—and you wouldn't be able to write it down in the space reserved for a telephone number on any form you might have to fill out. The space on a form (that is, providing room for exactly seven digits) is analogous to what happens inside a computer; some fixed amount of space is reserved for each integer. This means it's possible to have expressions that cannot be correctly executed. Multiplying your telephone number by itself is simply computing x^2, which is always computable in algebra but not always on a computer. This is our first example of the limitations that the finiteness of computers imposes on computation; we'll encounter several more.

Most computers, especially large ones, also have **floating point** numbers built in. This is the primitive data type that is used to implement the numbers that are called **real** in many programming languages. However, it's important to keep in mind that there are differences between these numbers and the real numbers we're familiar with from algebra. These differences arise, as in the case of integers, because of the finite amount of storage allowed for each number. This manifests itself in two ways. Rather than storing an arbitrary number of digits for a real number, computers use a variant of what is called *scientific notation*. Here a value is expressed as a *mantissa* and an *exponent*; the exponent is used to raise some *base* (in normal science this is 10, the base of our decimal numbering system) to a power that is multiplied by the mantissa. Thus an approximation to π could be represented as 3.14159 or 0.314159×10^1 or 0.0314159×10^2 or 31.4159×10^{-1}. There are limitations on both parts of the representation—there is a maximum number of digits that can be stored in the fractional part, and there is a maximum magnitude for the exponent. Thus we can only store approximations to the irrational numbers (such as π, which we represented with six significant figures above as an example), and also some rational numbers. This last fact shouldn't be surprising; some rational numbers can be precisely represented as decimal expansions (for example, 1/2 is .5) and some cannot (for example, 1/3 is .333...). Computers use binary rather than decimal representations, so there are some values that can be accurately represented in one system but not the other; however, the principle is the same. These numbers that the machine provides are called *floating point* rather than real numbers to make clear that there is a difference. We will return to the implications of these differences several times in the book. The usual algebraic operations are available on floating point numbers, but since the operands are approximations, the results will also be approximations.

Machines also have **logical** or Boolean values built in. It may seem unusual to you to speak of logical data in the same sense as integer and floating point, but we can "compute" with them. They are used to represent the truth value of propositions such as $x > 0$. The operations available on truth values (logical values or *Boolean* values) are *and*, *or*, and *not*. For example, *true and false* has the value *false*. This kind of "calculation" (often called Boolean algebra) might be familiar from secondary school.

The final machine data type we will mention is **character**. Machines can read characters from an input device, and they can write characters to an output device. Reading a character really means just moving it from a device (such as a keyboard) into the machine, and writing a character really means just moving it from the machine to a device (such as a display screen). Characters can be manipulated by putting them together in different ways.

When we say these types of data are *built in* to the machine, what we really mean is that the operations to manipulate these data items are available in the basic circuitry. But we have to take care how we use these operations, since a value does not identify itself as an integer or a floating point number—any more than programs and data identify themselves to a file copying utility (as discussed in the previous section). A particular sequence of binary digits can be interpreted as a character, an integer, or a floating point number, and so on. If we wrote in assembly language, we would have to take care that we did not, for example, attempt to add two characters together. Fortunately, when we write in a high-level language, the compiler ensures that the values are always interpreted appropriately so that only sensible operations are applied to them.

Programming languages can have other primitive data types that are not directly representable in the machine and can also have structured types. It is part of the compiler's job to decide how to use the raw capabilities of the machine to provide these other data types.

We've already said that all values in the machines we're interested in are stored as sequences of binary digits (zeros and ones), which are also called **bits** (*b*inary dig*its*). Another unit of storage is the **byte**, which is the number of bits required to store a character, such as "G" or "%". On most machines a byte is 8 bits, so it can really store 2^8 or 256 different values; this is more than the number of characters needed for any language like English or even Fortran, but it's a convenient number for machine designers to use.

2.3   Storage Devices

Computers work with binary digital values. The previous section described some of the ways in which these values can be manipulated, and we now describe how they are stored.

A computer has a hierarchy of **storage devices**. The higher-level devices are smaller, faster, and more expensive than the lower-level devices. Some of the levels often present are:

- high-speed registers inside the computational unit,
- a main memory cache,
- main memory (sometimes called *core* since it used to be built of little iron doughnuts or cores, but it's always some form of electronic memory now),
- a disk cache,
- disk drives, and
- tapes.

In the next few paragraphs we will describe these in a bit more detail.

Storage devices vary in capacity. The high-speed registers can store a few bytes; the main memory cache can usually store a few thousand bytes or tens of thousands of bytes; main memory and the disk cache can usually store hundreds of thousands to millions of bytes; and disks and tapes can store billions of bytes.

Programs tend to exhibit what is called *locality*. This simply means that references to data are not randomly distributed across all the data values, but instead tend to cluster during a period of time. Intuitively this isn't too surprising; for example, if we had a program to print a transcript for each student in a class, the data would be processed in a predictable, nonrandom manner. As we processed the file of information about students, rather than jumping at random from one character in the file to another, we would look at the entire record for one student, then at the entire record for the "next" student (possibly in alphabetic order), and so on. Because of this locality phenomenon, it makes sense to provide cache mechanisms that temporarily move related data items into higher-speed storage. The easiest way to explain this is by analogy: the set of file folders a person keeps on his desk is a cache between the working surface (those open in front of him) and all the files in his filing cabinet; they are kept together in a pile because a file tends to be used several times once it's being used at all, so it's kept where it can be accessed quickly. For example, you might have a file with information for your income tax calculations; when you get near the deadline for submission, you take it out and perhaps open it several times, doing some part of the calculation each time, before completing the tax return form and putting the file back in your filing cabinet. Caches in computer systems are just like the small (it is to be hoped) pile of folders on the desk that can be accessed more quickly and easily than the larger number of folders in the filing cabinet. Smaller machines, especially microcomputers, do not usually have caches and have to access the material directly in the larger storage unit.

When your income tax form is in a file folder in your filing cabinet, you cannot read from the form or write on it; all you can do is move the file from the cabinet to your desk (or to the top of the filing cabinet, or some other convenient place, but we'll assume the only working surface available is your desk). Once you've got it there, you can open the file and actually work with the form. To say this another way, the only operations available on the large, slow "storage device" (your filing cabinet) are moving files into it and moving files out of it—you cannot directly manipulate data stored there. Computers work in an analogous way with their storage devices. Data cannot be directly manipulated on tapes or disks. They must first be moved into the machine's main memory. Before programs can be executed, they must also be moved into the main memory of the machine.

Programs and data are usually stored on disks—rigid disks on a large machine and either rigid or floppy disks on a microcomputer. Other media (punched cards, cassette tapes, paper tapes) are possible but rare. Tapes (either reels or cartridges) are used for large amounts of data, and as a way to archive both data and programs that are not frequently used.

Moving data between levels of the hierarchy sometimes occurs implicitly. For example, when a programmer specifies that a variable in a program is to have its value incremented, the movement from main memory through the cache (if there is one) into a register and then back again is implicit. However, if a program must read a list of names from a file on a disk, it must explicitly ask for them to be moved into the main memory. We'll see how these things work for Fortran in subsequent chapters.

2.4 A Model for Execution

How does a machine actually carry out the operations indicated in a program? The essential idea is that the hardware itself has a number of instructions built in, as previously described. These are encoded according to some convention specified by the manufacturer of the machine but we can ignore it. The basic action of the hardware when it is carrying out an instruction is:

- fetch the next instruction from main memory (this might be the binary value 10010011);
- decode it—that is, determine which of the built-in primitives it stands for (such as "add two integers");
- fetch the operands (the objects that it manipulates) from main memory (such as the integers 37 and 15);
- perform the operation (actually carry out the addition, producing the result 52);
- store the results produced (if there are any) in main memory (in the example, put the 52 in some memory location); and
- determine the next instruction to be executed.

These steps are repeated indefinitely. This repeated set of steps is called the **instruction execution sequence**.

Of course, there are several other details that must be dealt with. How does the process get started? This is the "bootstrap problem": when the machine is turned on, how does it get its first instruction? The usual solution is to have the circuitry look in a specific place (perhaps on a disk) to find a program to run—the so-called bootstrap program (referring to pulling one's self up by one's bootstraps).

The bootstrap program reads in the operating system (see Chapter 3) and gives it control of the machine (in other words, lets it supply the next instruction to be executed). A second problem is terminating a program. This is handled by letting the program return control to the operating system.

This all takes place in the **central processing unit** (CPU). In most computers, the central processing unit has two major responsibilities. The **arithmetic and logical unit** (ALU) is the circuitry that carries out the instruction execution we've described. Specifically, it would carry out the addition of two integers or select a character out of a string. The second responsibility is to control all the activities of the computing system. This is done by the **control unit**, which initiates and controls communication among the system components. This includes reading a character typed on a keyboard and displaying a character on a screen.

We can think about the various components of a computer and their interactions using a simple picture. The memory of a computer is rather like a blackboard, divided up into squares. One section of the blackboard contains a list of instructions to carry out, while another contains data values. The control unit goes down the list of instructions and carries out each one in turn. Calculations are done by extracting data from the squares on the blackboard and sending them to the ALU to be manipulated. Results are then written into other squares on the blackboard. Input and output are rather like receiving slips of paper under the door and copying them to the blackboard (for input), or copying values from the blackboard onto pieces of paper and passing them out through the window (for output). This picture is, of course, simplified, but it gives you an idea of what goes on inside a computer.

2.5 Input and Output

Input and output (sometimes referred to as I/O or IO) are performed by devices attached to the system under control of the CPU. Devices available for connection to computing systems include terminals (keyboard with screen), printers (based on many different technologies, including impact, laser, ink jet), plotters, graphical displays, sound synthesizers, optical character readers, paper tape readers and punches, and card readers and punches.

The **IO devices** can be attached in various ways. In simple systems, such as many microcomputers, the devices are directly connected to the CPU and main memory. In more complex, larger systems, the devices are usually attached to *controllers* or *channels* that are directly attached to the system. These intermediary devices take over some of the control responsibilities from the CPU, thus allowing it to spend more of its time executing the user's instructions. As an analogy, think of a lawyer working on behalf of clients, with access to a large collection of files. She might go to the filing room herself to retrieve material, and this is similar to simple computer systems where the CPU controls the devices directly. Or she might send a clerk to retrieve material for her, and this is similar to more complex computer systems, where a separate processor (controller or channel) is used to control the device while the CPU does other work.

2.6 General Purpose Machines

The type of machine we have been discussing is illustrated in Figure 2.1. There are other ways to build machines, with more components, or with other methods of interconnecting them. However, at the level of a language like Fortran, the structural variations are not visible to the programmer. Consequently we can safely ignore such considerations for the remainder of the book.

The machines we have been describing are *general purpose* because anything that is computable can be computed on them (according to theories we don't want to explain here), *stored program* because the instructions are put into the memory before the execution begins (unlike, say, a simple hand-held calculator where the instructions are entered and immediately executed), *electronic*, **digital computers**. It is possible to compute with nonelectronic mechanisms (such as fluid flows in pipes), but we're not interested in them.

These machines are digital because the values are stored using digital representations. It's also possible to store values in other ways; machines that do so are called **analog computers** because some physical property varies in an observable manner proportional to the value being represented. As examples, the deflection of the needle in the speedometer of your car can represent the speed of the vehicle, and the amount of expansion of the liquid in a thermometer represents a temperature. Properties of electrical currents can represent values, and one can build electronic analog computers. However, we will not be concerned with them in the remainder of the book.

This entire discussion of computers is here only to help you understand a bit about what goes on "under the covers" inside the system you will be using. It isn't essential to understand this material in detail to learn how to program in Fortran. However, it is sometimes helpful to take some of the mystery out of the process. You may find it useful to return to reread some of this material from time to time

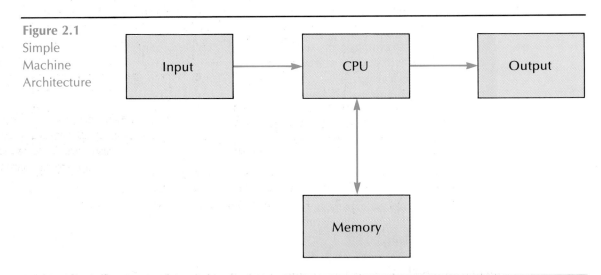

Figure 2.1
Simple Machine Architecture

as you gain a deeper understanding of programming and want to know more about what's going on at lower levels in the system.

Design, Testing, and Debugging

- Remember that the representation used for real numbers (floating point) only allows approximations to most values. Thus calculations involving floating point numbers may not always behave as the rules of mathematics might lead you to expect.
- Never test to see if two floating point values are equal—because of the imprecise representation, they might not actually be when you (and the laws of mathematics) expect them to be.

Chapter Summary

- Computers have some primitive objects that they directly manipulate, including some types of data that they can represent and operate on. For most computers these primitive types of data are integers, floating point values, logical values, and characters.
- Primitive data types are all stored in the memory of a computer as sequences of bits, that is, locations that can be in one of two states. Bits are accessed in groups, called bytes.
- Computers have a hierarchy of storage devices attached to them. Storage devices that can be quickly accessed are expensive and hence relatively small. Slower speed devices are cheaper and have much larger capacities. Most computers would have: registers, a cache, main memory, a disk cache, disks, and some form of long-term storage such as tapes.
- Computers follow an endless loop of tiny operations consisting of the following steps: fetch the next instruction, decode it, fetch the operands that the instruction is to manipulate, execute the instruction, store the result, and determine the next instruction.
- Most computers contain the following units: the central processing unit, which oversees the execution of instructions; the arithmetic/logic unit, which carries out calculations on data; the memory devices; and some input and output devices.
- The computers that we consider are called digital computers because they operate with components that can only be in discrete states. Analog computers, which can represent continuous quantities, are also possible.

Define These Concepts and Terms

Computer architecture
Utility program
Integer data
Floating point or real data
Logical data
Character data
Bit
Byte

Storage devices
Instruction execution sequence
Central processing unit
Arithmetic/logic unit
I/O devices
Digital computers
Analog computers

Exercises

1. Find out the manufacturer's name for the central processing unit of the computer on which you will be programming.
2. Find out how much main memory your computer has.
3. Find out how much disk storage capacity your system has.
4. Calculate about how many characters there are in the average novel. What about this book? How many characters are there in your local library?
5. How many bits would be required to represent telephone numbers if every human in the world had his or her own personal telephone? Ignore business telephones. Assume there are about five billion humans alive at the present time.
6. Is your telephone a general purpose, stored program, electronic digital computer? Is your microwave oven such a device? Does your microwave oven *contain* such a device?
7. Write a list of all decimal numbers that can be represented using a one decimal digit fraction and one decimal digit exponent. Draw a number line and mark on it the values that can be represented exactly.
8. A value written in binary notation can be converted to the corresponding decimal notation in the following way: the rightmost bit corresponds to the units column, the column to the left of that to the 2s, the column left of that to the 4s, and so on. So the value of the binary string 110101 is

$$1 \times 2^5 + 1 \times 2^4 + 0 \times 2^3 + 1 \times 2^2 + 0 \times 2^1 + 1 \times 2^0$$

which equals 53 in decimal. Calculate the decimal representations of the following binary numbers.
   a. 1110001
   b. 1010101
   c. 10111
   d. 111111

9. You're familiar with writing values between 0 and 1 in decimal notation as a string of digits following a decimal point. We can do the same thing in binary. The number 0.1011 is a binary fraction made up of binary digits following a binary point. The decimal value corresponding to this binary fraction can be calculated by realizing that the first position after the binary point represents the number of halves, the second digit the number of quarters, and so on. So 0.1011 in binary is equivalent to the decimal fraction

$$\frac{1}{2} + \frac{0}{4} + \frac{1}{8} + \frac{1}{16}$$

or 0.6875. Convert the following binary fractions to their decimal form:
   a. 0.111101
   b. 0.00101
   c. 0.11
   d. 0.01010101

10. Decimal fractions can also be converted to the equivalent binary form. We show how to do this by considering the decimal fraction 0.6875 from the previous question. At each step we multiply the value by 2 and consider whether the result is greater than 1 or not. The fractional part is used in the next iteration. Thus we get

$$0.6875 \times 2 = 1.375$$
$$0.375 \times 2 = 0.750$$
$$0.750 \times 2 = 1.5$$
$$0.5 \times 2 = 1.0$$

The binary fraction is read off from the units column of the products as 1011 so that 0.6875 is 0.1011 in binary. Calculate the binary fractions corresponding to:
   a. 0.1
   b. 0.375
   c. 0.25

11. Using the technique of the previous problem, find another terminating decimal fraction that does not terminate when expressed in binary.

12. Computing doubtless pervades your college or university. To get an idea of the extent and importance of computer use, try to determine the following from available documents or someone who can make an informed estimate: the number of microcomputers on your campus, the number of staff members of your central computing support organization, and the percentage of the university's budget allocated to computing-related expenditures.

13. Communication speeds inside computer systems vary from about one thousand characters per second between a CPU and a standard terminal to millions of characters per second between a CPU and a disk. How fast can you communicate (in characters per second) in legible handwriting? How fast can you communicate in speech?

14. What are the access rates for the storage devices of the system that you will be using to program?
15. How many bytes are required to store a copy of the Fortran compiler you will use on a disk on your system?

Working with Computers

3.1 The "Life Cycle" of Software
　　Problem Statement
　　Developing a Solution
　　Coding
　　Internal Documentation
　　Testing and Debugging
　　External Documentation
　　Maintenance
3.2 Tools
　　File System
　　Editor
　　Compiler
　　Debugger
　　Printing
　　Text Formatting
3.3 A Sample Program

Solving a problem using a computer involves several steps. The computer can be useful not only in executing the final solution program but also in several of the steps along the way. In this chapter we discuss the steps involved and some of the tools that can be used, and then show an example of a complete Fortran program. Some of the material in this chapter may be difficult to grasp on first reading; if so, don't worry. It will doubtless make more sense as you learn more about programming in Fortran, and you may want to reread parts of it.

3.1 The "Life Cycle" of Software

Computer programs are sometimes written to be used only once for a specific computation and then discarded, but this is very rare. Most programs will be used many times, will be used by different people than those who wrote them, and will evolve as the circumstances in which they are used evolve (since most work environments are not static). Think about programs that store and manipulate your grades in your institution's computing system; these programs have a lifetime of several years, are used by many administrative staff members who have no programming expertise, and will evolve over the years as rules about course selections, prerequisites, and so on, are changed. Because of this dynamic nature, we often speak of the "life cycle" of computer software.

This cycle involves several steps. These include understanding the problem (sometimes called analysis), specifying the algorithm, developing the program to execute the algorithm, preparing the data, testing and debugging the program, and using the program in its intended setting. At each stage it's necessary to prepare written descriptions of the program and its use; this is called documentation. Documentation has two audiences. Those who will be using the program need to know how it is used, what information it needs, and what answers it will produce. Those who are responsible for making sure the program works properly and for updating it if necessary will also need documentation. These stages are illustrated in Figure 3.1.

Each of the stages in the life cycle can feed back to previous stages. For example, experience with the program will often produce new requirements, resulting in necessary modifications, further testing, revision of documentation, and so on. We will describe these steps in more detail in the following subsections.

Problem Statement

The first step in developing a program is understanding the problem to be solved. This sounds trite, but it is a surprisingly difficult task for large applications. Think about the various activities that go on at the bank you use, and how challenging it would be for you to understand those mysterious processes well enough to design and implement a computerized information system to support them. There are bits of wisdom not written down anywhere that must be found, as well as the bewildering array of explicit procedures that exist. Further, most users do not simply want to automate what they currently do; they want to change (improve?) procedures on the way, make procedures more responsive to clients, reduce costs, and so on. Even in much simpler cases there can be difficulties in terminology and implicit assumptions, both of which make understanding the problem requirements a difficult task.

Of course the potential available in the tools being used also affects the problem statement. If you want to build a deck on the back of your house and the only tools available are a hammer and a hand saw, you would doubtless have different designs in mind than if you had a power saw. Many people looking for computer solutions to information processing problems don't have a good grasp of what is possible, or of the particular detailed design decisions that will be made, so they don't know how explicit to make the requirements statement.

You will encounter these difficulties when you develop programs, even as solutions to the assignments and exercises in this book. In these cases you should simply make a reasonable assumption and proceed. Of course, when you are the eventual user—you're building to your own specifications—you're aware of the potential designs available to you, and the problem statement and the solution often develop hand in hand. In production environments this is the ideal situation to achieve as well, and it is approximated by having users and designers (called systems analysts) work together to produce problem statements.

Developing a Solution

As we described in Chapter 1, it is certainly possible to make a completely formal statement of the requirements for a program. However, in most cases you will not

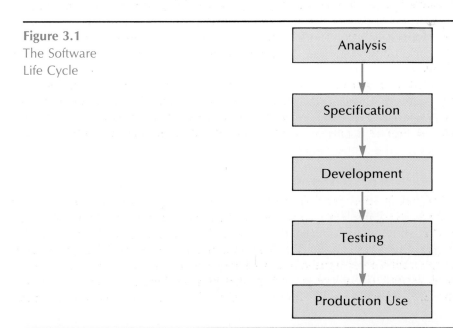

Figure 3.1
The Software Life Cycle

be given a formal requirements specification, and it's not necessary to produce one yourself.

Once you've got a reasonably precise definition of the problem, you can start devising an algorithm to solve it. A good approach is to start with a general statement of what you're trying to do, and successively choose pieces to make more specific. This is sometimes called **top-down** development because you start with a high-level abstract description and work "down" from that to more concrete descriptions — eventually ending up with an executable program. The process of making a general statement more specific is called **refinement**. A good rule of thumb for choosing the next part for refinement is to work on the hardest part first. There are many decisions to be made when developing a program (how to store the data, how to encode various things, and so on) and if you do the hard parts first, you might be able to make them easier by making appropriate choices for some of these things. When you get to developing the algorithm and program for the easy parts it's unlikely you'll have made them much harder by your earlier choices.

Don't be surprised if you sometimes have to back up in the development process and change some earlier decisions about how to structure the program or the data. Programmers, especially good ones, do this all the time. When you present the program to someone else, and when we present programs in this book, it's normal not to show the mistakes and blind alleys tried and abandoned and instead to show only the beautiful final product. Most of science is like that — most of the theorems we learned in mathematics weren't immediately conceived in their final beauty by their creators, but were the results of painstaking efforts with (usually) mistakes along the way; but we don't see that at the end of the process. So don't be discouraged if you can't form beautiful programs without making mistakes or changing your mind about some decisions, because that's how good work gets done, even if many people don't seem to admit it explicitly.

Sometimes it is a good idea to separate the design process involved in constructing an effective algorithm from the details of the particular computer language that is going to eventually describe it. Algorithm development is often done using **pseudocode**, which could be described as a high-level language for humans, part way between English and a computer language. Pseudocode should have about the same level of detail as the computer language that you are aiming for, but neglect the fine details of syntax that a human can infer. We won't use a pseudocode in this book; when we develop algorithms we'll use English and Fortran.

Most solutions are created using the ideas from simpler pieces of program that you've written or patterns of behavior that you've included in other programs. These are called **paradigms**. Paradigms are sometimes coded in self-contained pieces of program that do particular operations. These pieces can be put together in a number of sensible ways: they can be joined together sequentially or they can be merged. Much of the skill of an advanced programmer comes from having a wealth of paradigms in mind from which solutions to new problems can be constructed. Many of the examples that we will use in this book provide paradigms for common operations. You should learn to use them to construct your own solutions—not by memorizing them but by understanding them.

Coding

After the algorithm has been developed, it has to be translated into a high-level computer language. This step is a routine one, but can be frustrating to a beginner. Most computer languages require the steps of the algorithm to be described in notation that may seem slightly unnatural, especially at first. This usually involves placing language statements only in certain parts of lines and using punctuation in seemingly odd ways. In Fortran, as we shall see, almost all lines begin with six spaces, and each language statement must begin on a new line.

After the program has been written, it can be translated to machine language by a compiler. Usually, at this step, you discover that you have made some trivial errors in expressing what you meant in the computer language. The compiler will tell you about parts of the program that it couldn't understand and you can then correct them.

In the following example we illustrate some typical compiler output for a program that contains a syntax error.

```
   program Syntax
   ineteger N
        ^
Error: Undefined Type Declaration
   read *, N
   print *, N
   stop
   end
```

The compiler produces an error message for each error it detects and prints the message directly after the line on which the error occurred. Most compilers, as in this example, try to also indicate where on the line the error was found (in this case by using the ^ (caret) symbol under the offending word or symbol). The error in this program is caused by misspelling the word "integer." (Don't worry about the other lines in the output for now.) Sometimes, particularly when you are first writing programs, you will express something so badly that the compiler will be unable to interpret it very well. When this happens, it may produce error messages that are essentially false, that is, they arise from earlier errors that have confused the compiler. If you can't understand an error message, and it comes as part of a long sequence of such messages, then it will often go away after you fix the mistakes you do understand.

Sometimes it may happen that you have made a genuine mistake in expressing yourself in your program, but the error message does not seem to relate to the mistake you actually made. When this happens, it is usually because you, being much better at handling language than a compiler, can see the "real" mistake,

Working with Computers

whereas the compiler sees a different, syntactic mistake. For example, consider the following program:

```
program Show
integer A
print *, A(1)
stop
end
```

The intention in this program was to declare something called an array. The correct version of the program would contain the line

```
integer A(10)
```

However, the line

```
print *, A(1)
```

can have two possible meanings. With the declaration above it is an instruction to output something from memory. Because the declaration is wrong (it isn't what we meant), the other meaning, a function invocation, is assumed. This program produces an error message at line 3 saying

```
Error: Function A has not been defined
```

This message does not really help us to discover our "real" mistake.

Eventually you will correct all the detected errors, your program will be understood by the compiler, and you can execute it. At this point you can begin testing your program to ensure that it works correctly.

Internal Documentation

Programs should be easy to read. Because of the need to update programs to perform new functions, humans need to be able to look at a program and tell what it does. Even the original author needs to be able to look at the program during the testing step if some test shows that it is not performing according to its specifications. There are a number of ways to enhance readability.

- Even though Fortran does not require it, program text can be arranged on the page to reveal its logical structure. This is often done by indenting some of the program statements to show that those parts are dependent on other parts. Programs *should* be indented to reveal their logical structure. We'll do this consistently throughout the book. You should adopt some convention for your own use, either ours or some other, and stick to it. Ours is described in Appendix A, which summarizes the form we'll develop as we present the de-

tails of the language throughout the book. Use it for reference, but it's not necessary to look at it now.
- Names chosen by the programmer should reflect their logical use; Volts is probably better than V in a program dealing with characteristics of electric circuits.
- Fortran allows the programmer to insert **comments** in the body of the program. It is not necessary to describe everything in detail, because effective use of the previously described techniques can make most things in a program clear. However, it is probably a good idea to use commentary for
  — providing a header for the program saying who wrote it, and so on,
  — describing the purpose of the program,
  — describing each separate part of the program and its parameters,
  — describing difficult algorithms and giving literature citations for them if necessary, and
  — describing particularly complex statements or expressions.

We don't do this in the examples in this book because of space limitations, but we've tried to give you some indication of what's appropriate in the end of chapter programming examples.

Testing and Debugging

While it is attractive to think of writing programs that are correct as they are originally conceived, that is an accomplishment beyond the abilities of most — and you should not expect it in your first programming course. This doesn't mean it's not a standard to aim for. However, your programs will usually contain errors. These are called *bugs* as though they were intruders that had sneaked in unawares and had to be eliminated. (A better name would be "programmer errors.") Getting the bugs out is called *debugging*.

Once you get beyond the stage of writing programs in which the language syntax is frequently wrong, most of your debugging will be directed towards finding semantic errors. These are places where your program attempts to do something that is illegal, at least under certain circumstances, or where the program as you have written it doesn't describe the computation you wanted done.

The way to find these kinds of errors is by *testing* your program. Testing is difficult and complex, because it's not easy to know exactly what to test. Most programs will accept many different sets of input data, and it's impossible to test every set. Even a program that simply reads two integers and adds them together has more specific sets of data than any conscientious person could hope to test exhaustively.

Obviously we must exploit our knowledge of the program to test representative sets of data. If there are boundary conditions, it's worthwhile testing them all. Some examples of boundary conditions follow.

- If a program is to work with a list of data items, try the minimum number it should accept—possibly zero items.
- If a program is to work with a list of data items, try the maximum number of items it should accept.
- If a program performs a computation with extreme or singular points in its range, try them. For example, you could try to force a divisor to be zero and see if the program checks for that.

Another approach is to try and make sure that every piece of code in the program is executed at least once. Of course this can't be done with a single test run, but by using different sets of data we can ensure that each loop is executed at least once, that every possible choice in each selection statement is executed, and so on.

In some programs it's not enough to do these things because combinations of conditions may provoke otherwise unseen behavior. A good rule of thumb is to *test every significantly different internal state of the program*. What constitutes significant difference is something only an intelligent reader of a program can determine. Throughout the book we'll indicate bits of testing strategy for various programs so you can start to learn how to do it for yourself.

External Documentation

Most programs require some external **documentation**. External here means "outside the program listing" and usually is considered to be for two audiences.

The users of the program need to know how to run it, what data it operates on, what options are available, and what techniques are used (if there are standard techniques available, then they'll know how to expect the program to behave). Every program you write needs some user documentation unless it's only for your own use. Certainly every program you'll write as part of your work in this course will require this. It can be quite short, and should never get beyond a page or two for the things you'll do in the course.

The programmer(s) who will have to work on the program need some more detailed information. The next section talks about maintenance, which is a fact of life for most programs. Sometimes the original programmer will be the one doing maintenance and sometimes not. Even if you are the one returning after a period to modify a program you wrote, you may find it surprisingly difficult to remember the details of your design decisions. It is thus important to produce a document that supplements the internal documentation (obviously available to someone working on the program) with whatever detailed information should be presented about why design decisions were made, what choices were considered and rejected and why (if there's any possibility of someone reversing the decisions later on), how this program relates to other programs (if it's part of a collection of programs), and so on. It's difficult to say exactly what should be in such a document because it varies considerably depending on the complexity of the program and the sophistication of both the techniques used and the programmers who may eventually have to work on it.

For most programs you will write in your first programming course, the system documentation will be simple or nonexistent. For larger programs you might encounter in your professional engineering career, the system documentation will be extensive.

Maintenance

There are two reasons why programs have to be changed. First, programs often contain errors that do not become apparent for some time after they have been introduced into regular use. The error may only appear when certain values of data are supplied to the program, or under unusual sets of circumstances arising from interaction with other programs. When these errors are detected they have to be fixed because the program does not meet its specifications. For example, consider a payroll program for casual employees; after several months of normal use it is discovered that if an employee works few enough hours, then deducting contributions to benefit plans from the employee's salary results in a negative amount being printed on a check.

Second, programs that are used have their requirements changed over time. The environment for which they were built may change, or users may see other things that can be done with an enhanced program. For example, in a payroll system there are periodic changes required to reflect changes in the legislation that governs income tax rates and the deductions that employers must make.

In either of these cases more work is required on the program, and possibly on the documentation. This kind of work should be expected—many programming groups spend between 25 and 50 percent of their time doing **maintenance** (fixing things that are broken when the program does not meet its specification) and enhancement (changing the program to meet changing requirements). Once the changes become extensive enough, it's often better to rewrite the program entirely rather than continue with incremental changes. Thus the final stage of a program's life cycle is when it is discarded.

We have now discussed the various stages in a program's life cycle, and we turn to the computer-based tools available to assist us in working with programs.

3.2 Tools

There are a number of programs and features of a computer system that can be used to good advantage in the various phases of the software life cycle.

File System

The results of various stages in the life cycle can be maintained on your computer. These are kept in a **file system**. A file on a computer system is much like a file in a conventional filing cabinet. It contains pieces of information (data) that are related in some way. You might keep a file of text containing the statement of requirements for a program, another containing the program itself (perhaps several files containing several versions of the program if it's evolving over time), another for the user documentation, and so on.

Each file has a name by which the computer system knows it, and usually other attributes as well. These might include information about who is allowed to access it (the owner, immediate colleagues, the entire group of users of the system) and in what ways (a file can be read, written, or added to, but not otherwise changed, or a program can be executed, and so on).

Many file systems are hierarchic. This really means they contain a distinguished type of file called a **directory**. The directory contains information about other files and directories. This allows the file system to be organized as a hierarchy. There is a directory at the "top." It contains references to files or other directories; these other directories may contain references to files and other directories, and so on.

This is rather like posting a list on the outside of a filing cabinet that lists the files that are kept in it. The list could itself be thought of as a file. It would then be a directory.

If we were working with a roomful of filing cabinets, we would probably use a hierarchic directory. It might work something like this. A sheet of paper on the wall could tell us which bank of filing cabinets to consult for information about staff, which to consult for corporate financial records, and so on. A list near the cabinets containing staff records could tell us which cabinet to look in for names beginning with the letter "T," and, finally, a list posted on that particular filing cabinet could tell us if there was a file for "Turcke, David J." This is illustrated in Figure 3.2.

In such a system, one usually thinks of being "located" at a particular directory in the hierarchy. Files can be named in absolute terms (that is, from the top of the entire system's hierarchy) or in relative terms (that is, from the current location). By judiciously managing the placing of files in hierarchies and the naming of directories, one can make the large amounts of information stored in many computer systems much easier to access.

Unfortunately, not all systems have this hierarchic structure. They are sometimes called *flat* file systems since all the files are visible at once—as if a hierarchical system had been squashed into one level.

Editor

An **editor** is a program that edits (changes) files. There are more editors than any single person can ever hope to learn to use. Most computer systems have several editors available, and, perhaps because they're the most frequently used programs, people tend to become emotionally attached to particular ones. Try to avoid that temptation. Your personal productivity is improved as you become familiar with the tools you use, so you shouldn't arbitrarily jump from one editor to another.

On the other hand, it is often useful to have different tools suited to particular tasks, all of which you can use effectively. Even if you stick with one editor for a period of time, you'll eventually find another that's "better" in some ways, and you'll be well advised to change. We'll describe some of the features of the kinds of editors you're likely to encounter, and leave details to system documentation you can obtain with your microcomputer or from your computing center.

The essential idea of an editor is that the user can specify various changes to be made to the file the editor is working on. The basic editing operations are "add a character" (so we could transform "bead" to "bread") and "delete a character" (so we could transform "stump" to "sump"). In fact, with these two operations, any file can be transformed into any other file, if necessary, by first deleting all the existing characters and then adding in all the desired ones; these two operations are all that are needed. However, all editors have more complicated operations built from these basic operations in various ways, including things like adding several lines, deleting a word, interchanging adjacent characters, moving a block of characters from one location in the file to another, undoing the last change, including text from another file, and so on.

An *interactive* editor accepts commands that the user types at a terminal. Usually these editors are *full-screen* so that they maintain an image of a portion of the file on the terminal screen. The changes made to the file are continually displayed and the user can move through the file as required. Sometimes editors are *line-oriented* in that they do not show the contents of the file unless the user asks explicitly to see certain lines, which are then displayed.

It's also possible to have a *batch* editor. In this case the specifications for changing the file, that is, the editing commands, are put in a file by the user (using some other editor) and then used as a set of instructions, or script, for the editor to change a specific file. These can be useful for programs that produce data in

Figure 3.2
A Hierarchical File System

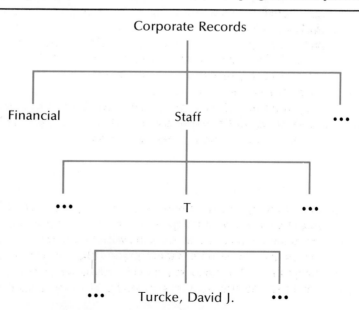

Working with Computers 41

a format that is incompatible with a later program that requires the same data. We would then build a script for the batch editor, and, whenever we need to do so, modify the output from the first program to be suitable for the second.

Compiler

We pointed out earlier that a **compiler** is a program that translates a high-level language into machine language. Once a program in Fortran, or any other high-level language, has been written using an editor, it must be compiled using a compiler.

Most compilers you will use require that a program be complete and correct before it can be translated and executed. If the program contains syntax errors, then the compiler will indicate these but will not go on to produce a translation of the program.

The compiler is run as a totally separate step after the program has been entered using an editor. Compilers typically have options that allow you to see, for example, the machine language generated, the addresses in memory used for all the variables, or the size of each piece of the program you've written after translation. We won't try to give details of particular compilers; you'll have to check the documentation available with the system you're using.

Compilers do two things: they understand or analyze a program, and then they translate it. The understanding involves both the form and the meaning of the program. Thus, a compiler can tell if the program is well-formed according to the rules of the language; if not, it will usually respond with some type of **syntax error** message. A compiler can also tell if the meaning of a program is reasonable; if not, it will respond with a **semantic** (that is, meaning-related) **error** message. The English statement "Pigs flies." is syntactically incorrect because the subject is plural and the verb is singular. The English statement "Pigs fly." is syntactically correct but semantically erroneous because, except in imaginary worlds, pigs don't fly.

Even when the compiler cannot find syntactic or semantic errors, the program may not be well-formed. For example, a program that attempts to divide one integer by another makes sense unless the divisor is zero, and then it will generate an error when it's being executed—a *run time error*. A compiler could not possibly always detect such an error because the divisor might be a value entered by a user when the program was executing. The compiler cannot tell in advance what value the user will enter, and hence whether or not it will be zero. In a case like this the program specification should state what input values are allowed.

When the compiler terminates, it leaves a translated version of the program in a new file on your system. On some systems this file may be directly executed (perhaps by typing the name of the file). On other systems, another process, called **linking**, may have to be done first. Linking involves putting your program together with other program pieces that may have been compiled at other times. This is almost essential when programs become large or when many people are working on the same program. It also makes it possible to develop libraries of program pieces that can be used in many programs. The linking process puts all the program pieces together to make one executable program.

Sometimes compilers and editors are more cooperatively bound together. It is then possible to move quickly from the editor to the compiler without writing the program out into the file system, for example. It might also be possible to have the compiler automatically invoke the editor when an error is detected in the program and position the editor on the erroneous line.

Even more closely cooperative environments, called *program synthesizers* or *structure-oriented* systems, are possible. In these cases the editor has knowledge of the structure of programs and uses that knowledge while the program is being produced. The programmer might be allowed to select one of the possible forms for a statement in the language from a menu, rather than being allowed to type in any arbitrary text and then subsequently have a compiler attempt to decide what it is. In this way most syntax errors are avoided since the editor does not allow the programmer to produce incorrect programs. And since the two functions are bound together closely, the analysis of the program's meaning can take place incrementally, whenever the program is changed, rather than just at the end of the program's development. Of course, there is still a need for translation of the program into machine language, so not all of the compiler's traditional work is made redundant by this approach.

Debugger

When a program is being tested, the programmer often wants to observe various aspects of its behavior, including

- the flow of control (what routines and statements are executed and in what order),
- the values of variables at specified points in the computation, and
- the amount of time spent in various parts of the program.

While it is certainly possible to modify the program itself to produce some of this information, this is not a satisfactory approach since it takes a considerable amount of time and effort. Many computer systems provide software to help with this task.

A *symbolic debugger* is a program that can superintend the execution of another program. Typically these programs observe and understand the flow of control and the values of variables (and possibly other things as well). They can report these things to the programmer, and allow the programmer to halt execution and change the values of variables or explicitly branch to particular parts of the program.

Sometimes a debugger is matched to a specific compiler for a specific programming language. In such systems the compiler puts information into the translated program to allow the debugger to do its job. In other computer systems the debugger is a more general program that can work with output from a variety of compilers.

In the program synthesizer systems described earlier, the debugger is often integrated into the compiling-and-editing environment, thus further enhancing the attractiveness of this approach.

You will have to consult the documentation for the system you are using to see if a debugger is available for your use. Simple microcomputer systems often have no debugger. If your system has one, use it.

There are also some techniques to be learned that make debugging easier. The process of **debugging** always involves reasoning back from an observed malfunction of the program to the error in the program that caused it. One approach is called the "wolf fence." The idea is to separate those parts of the program in which the error does not appear from those in which it might.

For example, we might discover that an output of the program has a spurious value. We can conclude that the error cannot possibly have occurred before the first time that the variable was given a value. Therefore, we can remove that part of the program from consideration. By observing the value at several places in the program's execution, using a debugger or simply displaying the value, we can narrow down the part of the program in which the problem must lie. A relatively small number of checkpoints is usually enough to narrow down the location of the error to a few statements.

Don't be disheartened if you find debugging difficult at first. Resist the temptation to blame errors on the compiler or operating system—they almost never are the real cause of problems.

Printing

You will often want to print files you've produced, such as a program, documentation, or data. If you simply want to see the contents of the file without any fancy processing, you'll use a command available on your computer system to do that. The appropriate command might be "*print* <fileSpecifier>" or something similar.

If you're working on a small system (such as a personal microcomputer), there will be at most one printer attached to it and the simple command should be sufficient. If you're working on a large shared computer system, there may be many printers. In this case you may have to specify a specific printer as an option to the *print* command, or in some other command that establishes environment settings. You'll have to consult your system documentation to see how to do this.

Text Formatting

The computer system you're using will probably have some means of producing documents such as business letters, research papers, and books. There are two common ways to do this.

The first is to use a **word processor**. This is a program that combines the full-screen editing capabilities discussed earlier with some formatting capabilities such as generating running headings, emboldening section headings, generating a table of contents, justifying (producing a solid right margin for) paragraphs of text, and so on.

These are sometimes called *what you see is what you get* or *wysiwyg* (pronounced whizzy-wig) systems, because the user sees on the screen what will eventually be produced as output on a printer. Actually in most cases these are *what you see is almost what you get* systems, because most printers that are used for output have more flexibility than the screens we use. For example, most printers (especially laser printers) provide several different kinds of lettering (called fonts), often including proportional fonts (where small letters like "i" are narrower than wide letters like "m", as in this book and unlike most typewriters), and these cannot be represented on most computer screens. This will doubtless change over the next few years as displays become more sophisticated and are capable of representing arbitrary fonts.

The second way to produce formatted documents is to use an editor and a **text formatter**. The editor is used to build a file containing the text required in the document, together with some *markup*. The markup comprises instructions to the formatter telling it what the pieces of the document are (section, paragraph, figure, and so on) or how to format them (skip a line, indent a half inch, or keep the next five lines together on a page). This file is then processed by a formatter that lays out the text on the page as indicated by the markup.

Some people prefer the wysiwyg approach even though it may have the limitation of using the more restricted representations that are possible on many screens. Some prefer the edit-and-format approach even though it has the limitation of not showing immediately on the screen what the final product will be. The best, and newest, approach is to have a system with both sets of advantages — maintaining a marked-up version of the text, but allowing the user to edit a formatted version on the screen. You should try the various options available on your computing system and use one you find sufficient for your needs for documenting your programs.

3.3   A Sample Program

We now present a program that calculates the areas of triangles, so that you have a chance to see what a real program looks like. You can probably guess what most of the statements do. We will introduce all of these statements in the next few chapters and you may want to refer back to this program to see them in use.

```
*       Program 3.1
*       Calculate the areas of triangles
*       given their bases and heights, in
*       meters. Print areas in square mms.
        program Area
        real Base, Height
        real Area1, Area2, Calc
        integer I
        do 100 I = 1, 5
            print *, ' Enter base and height in meters'
            read *, Base, Height
            Area1 = Calc(Base, Height)
            call Convrt(Area1, Area2)
            print *, ' Area is ', Area2, ' square millimeters'
100     continue
        stop
        end

        real function Calc(B, H)
        real B, H
        Calc = 0.5 * B * H
        return
        end

        subroutine Convrt(Meters, MilliM)
        real Meters, MilliM, Factor
        parameter (Factor = 1.0e6)
        MilliM = Meters * Factor
        return
        end
```

The output from this program is shown next. The values marked in bold are the things that a user would enter. The remaining values are produced by the program.

```
Enter base and height in meters
1  2
Area is           1000000.0000000 square millimeters
Enter base and height in meters
3  4
Area is           6.0000000E+06 square millimeters
Enter base and height in meters
5  6
Area is           1.5000000E+07 square millimeters
```

```
Enter base and height in meters
7 8
Area is            2.8000000E+07 square millimeters
Enter base and height in meters
9 10
Area is            4.5000000E+07 square millimeters
```

Design, Testing, and Debugging

- Always understand the problem properly before you spend time designing the solution. Solving the wrong problem is a waste of your time.
- Learn to develop your solutions methodically. You're less likely to miss something important or to be overwhelmed by the details. Use the top-down approach we've described in this chapter.
- Write documentation even if no one but you will ever look at or use the program. Humans forget details very quickly, and even you will forget important details about your program from one day to the next.
- Write the documentation at the same time as you write the program. Don't postpone it until the program is written and working. Often the exercise of writing down what the program is supposed to do reveals aspects that you would not otherwise have thought of.
- Test your programs thoroughly. Never assume that because your program works for a few cases, it is error-free. Consciously plan the situations that you will use as tests to make sure that you have covered the bases.
- Understand that almost all programs evolve over their lifetimes. Try not to design programs that will require major changes to provide small changes in functionality.
- Learn about the support tools that your system provides. They can be big time savers for you and can increase your personal productivity dramatically. Read the documentation that goes with your file system, editor, and compiler.
- If your system has a debugger, use it. It will speed up the rate at which you detect problems in your programs.

Style and Presentation

- When you write documentation, try to keep in mind the two distinct audiences who will read it—the users and the maintainers. Remember especially that users do not know anything about the internal working of programs. You can't express things to users in internal terminology.
- Read the documentation that comes with your system with a critical eye. Notice its flaws and resolve never to duplicate them.

Chapter Summary

- The general word for computer programs is software.
- We say that a piece of software goes through a life cycle. This begins with a response to a problem. A problem statement is first developed and agreed upon. An algorithm to solve the problem is then developed. This algorithm is then translated to a program, usually written in some high-level language. The program contains internal documentation that describes its purpose. The program is tested to ensure that it is working as specified, and any faults are removed by debugging. External documentation for users and maintainers is written. The program then goes into production use. As it is used, faults are discovered, or enhancements wanted. Program maintenance is the ongoing alteration of the program to achieve this. Finally, the program's useful life comes to an end.
- A file system is the part of an operating system that maintains sets of information for users.
- An editor allows you to create a new file or to modify the contents. Editors can be line-oriented, allowing you to manipulate only a single line at a time, or full-screen, allowing you to manipulate a "window" on the terminal screen.
- A compiler translates a program written in a high-level language into machine language. It checks that the program conforms to the rules of the high-level language and checks, where possible, for inconsistencies in the program. Error messages are produced to show where there are mistakes in the program.
- A debugger is a program that can assist you in finding places where your program is not doing what you thought it would do. It provides the ability to halt the program part way, to examine the values of memory locations, and to determine how much time is spent executing various parts of the program.
- A text-formatting program allows you to produce documents that are laid out well on the printed page. There are two kinds: what you see is what you get, and markup.

Define These Concepts and Terms

Top-down development	Editor
Refinement	Compiler
Pseudocode	Syntax error
Paradigms	Semantic error
Comments	Linking
Documentation	Debugging
Maintenance	Word processing
File systems	Text formatter
Directory	Run time error

Exercises

1. Investigate the editor (or one of the editors) on the system you will be using. Which of the following operations can it do?
   a. Reverse two adjacent characters (very useful because it fixes a common typing mistake).
   b. Include the contents of some other file inside the file you are editing.
   c. Search for a particular word or phrase.
   d. Show a particular line of the file (say line 56).
2. What languages (other than Fortran) are available on your system?
3. Is there document-processing software available on your system? Is it a wysiwyg or markup system?
4. Type in the sample program from the last section of this chapter using your text editor. Compile the program and execute it. Do you get the same output? Don't be surprised if you get some error messages from the compiler at first. It's hard to type programs without making mistakes. Your system may require that all the characters be entered in uppercase (capitals).
5. What command do you use to list your files on the computer system you work with? Is the file system of this computer system hierarchically structured? If it is, draw a sketch of the arrangement of your own files. Is there a limit to the amount of space available to you? If so, what is it?
6. Does your computer system have a debugger for use with Fortran programs? If so, obtain documentation so that you can use it in the remainder of this course.
7. Use your document-processing software to create and print a description of what you expect to obtain from a university education and the role of computing courses in this. Make the document double-spaced, longer than one page but no more than two, and center your name and the date in boldface at the top of the first page.
8. Use your editor and compiler to enter and execute the following program. Type carefully to avoid diagnostic messages.

   ```
   program First
   integer I, J
   I = 10
   J = 2
   print *, I / J
   stop
   end
   ```

9. Use your editor to modify the program in Exercise 8 by replacing the word "print" with the word "priny" (which could happen if you pressed the wrong key). Rerun the program and see what happens.
10. Use your editor to modify the program in Exercise 8 by replacing the number 2 with the number 0. Rerun the program and see what happens.
11. Use your editor to modify the program in Exercise 8 by replacing the number 10 with the number 9. Rerun the program and see what happens. This will be explained in a later chapter.
12. Use your editor to modify the program in Exercise 8 by replacing the word **integer** with the word **real**. Rerun the program and see what happens. This will be explained in a later chapter.
13. Write specifications for an algorithm that is to sort a list of integers. Try and make it comprehensible to someone who isn't familiar with computers and algorithms. This illustrates what can happen when defining a problem: the person wanting the solution may have a vocabulary ("sort") that the person developing the solution may not share.
14. Find out the details of the command that will print a file on the computer system you will be using. Use the command to print one of the programs you have written.
15. As software is modified and enhanced, it is periodically redistributed to its users, labeled with a "release" or "version" number. Determine the official name and the release or version number of the compiler you will be using.

A Piece at a Time

4.1 A Problem-Solving Methodology
4.2 Sample Algorithm Development
4.3 Turning Algorithms into Programs
4.4 Subprograms
4.5 Textual Placement of Subprograms
4.6 Simple Output

4.1 A Problem-Solving Methodology

We saw in the previous chapter the steps that are involved in taking a real-world problem, developing an algorithm to solve it, implementing the algorithm in a programming language, testing it, and documenting it. However, knowing the steps to be followed doesn't automatically mean that we can follow them. In this chapter we look at a methodology for dealing with problems for which we want algorithmic solutions.

Having a methodology means that there is an approach that we can follow to help us to design and implement algorithms. The first steps are usually the most difficult — once we have designed the algorithm, it's usually a great deal easier to translate it into a programming language and get it executing.

We have to be aware of how humans solve problems in general so that we can tell if a methodology is going to work. Humans find it hard to remember many small items of information even over short periods of time. When we come to think about an algorithm that is at all complicated, it seems as if there are thousands of details to be remembered about what steps to take and in what order to take them. So our methodology must have a way of helping us to remember all of these details and get them arranged correctly in the final algorithm.

To handle this human limitation we try to break up the solution into separate pieces, each of which can be considered more or less independently of the others. By doing this we need to think only about the details of the piece we're working on as we construct it. If we have done the division into pieces cleverly, then we will only have a small number of things to think about when we put all of the pieces together.

Now this is not as easy as it sounds, and perhaps it doesn't sound all that easy either. That's where the second part of our methodology comes in. Before we look at the specifics of our solution, we try to come up with a general outline. Once we have the general outline, we successively refine our ideas as we see more and more of the details and, as we do so, the natural groupings of steps become more obvious.

It may happen that, as we go through this process of designing an algorithm, we discover that we can't do what we intended with the steps we have designed. Then we have to go back a bit and look at the problem more generally until we see a way around the problem, and then continue breaking our solution down into small pieces.

After you have had some practice doing this, it will come a lot more easily. However, algorithm design is a definite skill to be learned. Our brains are designed to keep from us the details of many of the things we do by handling them automatically — we call these automatic operations habits. There are many things that we know how to do, but find very difficult to express algorithmically in detail. For example, we know how to walk, but would find it very difficult to explain to someone else exactly when each muscle involved should contract. Something of the same difficulty arises in constructing algorithms in other problem domains.

We know, at some level, what needs to be done, but find it very difficult to express the steps in enough detail that they qualify as an algorithm. Think about how you might arrange a list of names in alphabetical order; now think about telling your three-year-old niece how to do it. This problem is compounded by the fact that when we describe algorithms to other humans we assume some common sense on their part, so that imperfect algorithms do actually work. This is not the case for computer-executed algorithms because computers have no common sense and will do exactly what they are told, no matter how ridiculous. A great deal of the skill of programming comes from being able to think algorithmically.

As with human habits, after a while you will acquire a stock of standard ways to carry out operations that are needed regularly in programs. These paradigms should quickly become second nature to you so that, when you encounter a new problem, you can concentrate on what is new about it. So in some ways, learning to program is like driving in a strange city—at first you have to look at everything and you frequently get lost. After a while you learn some common ways of getting around and you stop noticing the details. Eventually, you can find your way around almost without thinking about it.

4.2 Sample Algorithm Development

We can illustrate the technique we have been describing outside of a computer context. Suppose we want to paint a room that has plaster or wallboard walls. We'll assume that the color has been chosen and that all the required material is on hand. The most abstract description of the algorithm is

paint a room

but this is essentially only a name or label for the process, since it certainly doesn't tell anything about *how* to get the job done.

If you've painted a room before, you might start with a refinement like this:

to paint a room:

prepare the room
apply the paint
clean up

Preparation is probably the least pleasant part of this job, and in many ways the most difficult, so we might as well try to refine that part of the algorithm first. There are several things to be done. Assuming that only the walls are to be dealt with (and not the floor or the ceiling), they have to be cleared of obstacles and repaired if necessary; then the floor should be protected against accidental spraying and spilling.

to prepare the room:

take down drapes, ornaments, pictures, shelving
remove electrical switch and outlet covers
move these items and furniture away from the walls,
    to provide working room all along the walls
put drop cloths or runners on the floor by the walls
if there are holes in the walls that will not be reused,
    or if the walls have been damaged,
      then overfill the holes with wallboard compound
when the wallboard compound is dry, sand it smooth
repeat the last two steps as required
protect baseboards, door frames, etc.,
    that border the surfaces to be painted,
    with masking tape

If all has gone well (it never does), we're now ready to apply the paint. There are two stages to this: "cutting in" with a brush and applying the paint to the large areas with a roller. We'll need two roller coats.

to apply the paint:

put some paint in a small pail
using a brush, paint along each border of each wall
   (next to floor, ceiling, doors, windows, corners)
   being sure to use sufficient paint to cover well
   (this gets done only once)
put some paint in a roller tray
using a roller, paint each wall,
   overlapping with the border applied by the brush
repeat the last step for a second coat on each wall

There are tricks of the trade that one should know for using a brush, using a roller, and pouring paint from one container into another; we will not give those here, even though they're essential to a successful, acceptably clean execution of the job.

There's one other complicating factor that we have not described here: for almost any room, the roller tray won't hold enough paint to do the job. This means that periodically, when the tray is empty or almost empty, we have to interrupt the "normal" sequence of putting the roller in the tray and then spreading paint on the wall, to add paint to the tray. This is easy enough to say in an imprecise way, as we've just said it, but writing down explicit instructions for handling conditions that can arise like this part way through a task can get to be quite tricky—some programming languages have special ways of talking about this type of processing, but unfortunately Fortran does not; we'll just have to be as careful as we can when developing algorithms.

Leaving aside these complications, then, we're ready to refine the last part of the algorithm. If an oil-based paint was used, a solvent like varsol or turpentine is needed in the cleanup; for a water-based paint (a latex) the only solvent required is water. If you're basically lazy (like at least one of us), you might decide to throw away your equipment rather than cleaning it. Of course, if you are using good equipment this is silly, but if the job was small and you used cheap or old equipment, it's an approach to consider—cleaning paint out of brushes and rollers isn't much fun.

to clean up:

put away remaining paint
roll up runners and drop cloths
clean roller trays
if they're worth the effort, clean brushes and rollers
clean yourself
replace furniture
replace drapes, ornaments, shelves, and so on
agree with yourself that you'll never do this again

This relatively simple task uses sequencing (prepare the room, *then* apply the paint), selection (if there are holes in the walls, fix them), repetition (repeat as required), and abstraction (cleanup is defined one place and used another) in its algorithmic expression. We've also made use of the refinement technique to develop a solution by working from a very abstract description (paint a room) to a quite specific one. We have not, though, given enough detail for someone with no painting experience to get the job done. This is not at all unusual—algorithms, like other communications, have to be constructed for a particular audience.

4.3 Turning Algorithms into Programs

We've demonstrated how to construct the skeleton of an algorithmic solution to a problem, but we need to talk about how to turn such a skeleton into a program if it's for an algorithm that a computer can execute.

At this point we begin to introduce the details of Fortran, and this will continue throughout the remainder of the book. We will do it partly by showing *examples* of features and usage, and partly by showing *general patterns* for the features; we'll use whatever technique seems most appropriate for the different pieces of the language. Appendix B contains a summary of the Fortran details presented in the book and can be used as a reference. You can look up particular things in the index and thus find the place where we introduce the material if you want to review the presentation.

Every program has two parts: the **program name**, and the list of steps that the program is going to carry out (the *executable statements*). The first piece of information is in the program statement, which looks like this

> **program** *name*

The word **program** must be typed exactly as it is shown—it is a special word and serves as a marker—in this case showing that this is a program statement. Where the word *name* appears you can use any name, preferably one that describes what the program is about. You can think of it as the title for the program. All Fortran names must be formed according to the following rules:

- They must be no longer than six characters.
- They must begin with a letter of the alphabet. We, by convention, use a capital letter.
- They can contain letters and numbers.

Some compilers also allow a dollar sign, "$", to appear in names, but this is non-standard and you should not do so. Thus the following are all valid names that may appear in Fortran programs:

> Volts, FMax, Length, Sum1, N1234

whereas the following names are all illegal:

> Weight1, Amperes, Current, 2Way

The **program** statement is preceded by six blanks on the line. This may seem artificial to you, especially if you've programmed in other languages such as Basic. Like several other features of Fortran that we'll see later, this is a legacy from the early days when Fortran was designed. In fact, the **line layout** of programs is in four pieces: the first five character positions (sometimes called columns) are set aside for numeric statement numbers (we'll see later how to use them); position (column) 6 is used for a continuation character when a statement of the program is too long to fit on a physical line and must be continued; columns 7 through 72 are used for the program statements themselves; and, finally, columns 73 through 80 are reserved for sequence numbers (since not all statements have to be numbered, this gives room to number each physical line in the program). These restrictions are a result of the 80 column punched card that was the standard input medium when Fortran was designed. Even though cards are almost never used now, the restrictions they imposed have become part of the Fortran language standard. Some Fortran compilers may not enforce these rules.

Each Fortran statement appears on a separate line of the program. If a statement is too long to fit on a single line, then it may be extended in a continuation line. Such a line has a non-blank character in column 6. In our programs we have chosen to use the ampersand, "&", as the **continuation character**. If the Fortran statement is too long to fit on the original line and a continuation line, then it can be extended with more continuation lines. Some compilers will limit the number of continuation lines, probably to about ten. Some older styles of programming use other characters such as "+" for the continuation character. We discourage this as it is easy to confuse it with a character in the statement itself.

The rest of the program follows the program statement. The **executable statements** (those that tell the computer to do something) are written in the order in which we want them to be done. The whole list of statements is terminated by the word **end**. A complete program looks like this:

```
      program Outlin
*     beginning of executable statements
         Stmt1
         Stmt2
         ...
         StmtN
      stop
      end
```

The second physical line of this program begins with the character "*" in column 1. This indicates that the statement is a *comment*. It is used to communicate to human readers, but is otherwise ignored by the computer system. Comment lines may be used anywhere in a program. The last executable statement of any program must be the **stop** statement, which is used to pass control back to the operating system.

In this book we have adopted a convention for printing programs that uses boldface for reserved words in Fortran, and a standard font for all other words and characters. When you prepare programs for your computer, ignore our boldface convention and type all words and characters in the normal way.

We have also adopted a further convention, which is that we put keywords in all lowercase characters, and use both upper- and lowercase characters in identifiers created by the programmer. We capitalize the initial letter of each word or word fragment in our identifiers. Some systems will require keywords to be in uppercase; some systems may require identifiers to be in uppercase as well; some systems may accept the forms we have used but produce warning messages. Adopt a convention you find easy to work with; if you like ours, try an example program to see if your system will accept it.

4.4 Subprograms

Programming languages provide a language structure to implement the idea of a piece of an algorithm that can be thought about on its own. A **subprogram**, as the name suggests, is a little program in itself, but it doesn't have an independent existence and is subject to the control of another program. In terms of the earlier discussion, the subprogram is the implementation of the refinement of a more abstract notion.

A subprogram consists of a sequence of steps that are usually logically associated with each other. The steps are executed when the subprogram is invoked (started) by some other program. When the subprogram has finished going through all of its steps, it stops, and the program that invoked it continues executing. Subprograms naturally match the pieces that we have used to build and describe algorithms.

There are two kinds of subprograms in Fortran, called **subroutines** and **functions**. We will talk about simple subroutines in this chapter; more complicated versions, as well as the discussion of functions, will come later.

A subroutine is like a program except that it begins with a subroutine statement, instead of a program statement. A subroutine looks like this:

> **subroutine** name
> Stmt1
> Stmt2
> ...
> StmtN
> **return**
> **end**

We must include the word **subroutine** exactly, but we can use any name. The executable part consists of a sequence of statements, with a **return** statement as the last one. An **end** follows the last statement.

The subroutine is invoked by writing a **call** statement, followed by the name. A program that invoked a subprogram named Invert would look like this:

> **program** Xample
> **call** Invert
> **stop**
> **end**

When the main program is executed and the invocation is encountered, the main program stops and the statements making up the subprogram are executed sequentially all the way through. When the last statement has executed, the main program resumes executing at the statement after the invocation. The main program behaves as if the statements of the subprogram have been inserted into the main program at the place of invocation.

Because we can always tell where we are when a program is executing (that is, which statement is providing the instructions the CPU is executing; in other words, which statement is controlling the CPU), we speak of the **flow of control** through the program. The normal flow of control is to execute statements sequentially, one after the other. An invocation of a subprogram alters this flow of control because the statements of the subprogram are executed before the next statement of the invoking program. Subprograms can, in turn, invoke other subprograms, altering their own flow of control.

4.5 Textual Placement of Subprograms

In the program of the previous section, we did not actually include the subprogram Invert. The text of subprograms is placed either before or after the text for the main program—we prefer after since we think it's easier to read. A typical program with two subroutines would therefore look like this:

```
program Xample
call First
call Second
stop
end

subroutine First
...
return
end

subroutine Second
...
return
end
```

In Fortran, subprograms can invoke other subprograms. The flow of control in this case is the same as if the subprogram had been invoked from the main program. When the first invokes the second, the statements of the second are executed. Control then passes back to the statement after the invocation in the first subprogram.

4.6 Simple Output

To show some complete programs, we need to be able to *do* something. The simplest thing to do in Fortran is produce output. The simplest output operation is the printing of some characters on a screen (or possibly a printer), or, on some computer systems, into a file on a disk. The Fortran statement that does this is the **print statement**. In its simplest form it looks like this

```
print *, 'text to be printed'
```

Whatever is placed between the quotation marks appears on the screen (or on the printer or in the file, depending on your system) when the statement is executed. In the case of the **print** statement just shown, the output would look like text to be printed

If there is no text given and no comma after the asterisk, a blank line is printed. We can use this statement inside subroutines and, by watching the output, we can tell which subroutines are executed and the order in which they are executed.

The **print** statement can actually be used to output several values on a single line. We could write

> **print** *, 'text', ' to be printed'

to produce the same output as in the previous example. Notice that the output strings will be printed without intervening blank spaces on the output device. If we want blanks between words in successive strings, then we must explicitly insert them; in this example we have explicitly inserted a blank before the word "to."

To show how this works, consider the following set of programs. The first program is simply a main program, without any subroutines, that prints a message.

```
*       Program 4.1
        program Simple
        print *, 'Message from the main program'
        stop
        end
```

The output from this program is

Message from the main program

on the standard output device. Now we can add a subroutine that prints another message.

```
*       Program 4.2
        program Simple
        print *, 'Message from the main program'
        call First
        print *, 'Another message from the main program'
        stop
        end

        subroutine First
        print *, 'Message from the subroutine'
        return
        end
```

The following output comes from this program.

Message from the main program
Message from the subroutine
Another message from the subroutine

Now we can make things more complicated by adding a second subroutine.

> *   Program 4.3
> **program** Simple
> **print** *, 'Message from the main program'
> **call** First
> **call** Second
> **print** *, 'Another message from the main program'
> **stop**
> **end**
>
> **subroutine** First
> **print** *, 'Message from the First subroutine'
> **return**
> **end**
>
> **subroutine** Second
> **print** *, 'Message from the Second subroutine'
> **return**
> **end**

The output of this program is

Message from the main program
Message from the First subroutine
Message from the Second subroutine
Another message from the main program

The following program uses the fact that subprograms can invoke other subprograms.

> *   Program 4.4
> **program** Simple
> **print** *, 'Message from the main program'
> **call** First
> **print** *, 'Back in the main program'
> **call** Second
> **print** *, 'Another message from the main program'
> **stop**
> **end**

```
      subroutine First
      print *, 'Message from the First subroutine'
      call Second
      return
      end

      subroutine Second
      print *, 'Message from the Second subroutine'
      return
      end
```

Its output is

Message from the main program
Message from the First subroutine
Message from the Second subroutine
Back in the main program
Message from the Second subroutine
Another message from the main program

Here is a more complicated program that prints out the words to an English folk song called the Dilly Song. Before looking at the output that it produces, try and work it out for yourself.

```
*     Program 4.5
*     Prints the first two verses of the Dilly Song
*
      program Dilly
      call Verse1
      print *
      call Verse2
      stop
      end

      subroutine Verse1
      print *, 'I''ll sing you one-oh.'
      call Green
      print *, 'What is your one-oh?'
      call Chor1
      return
      end
```

```
subroutine Verse2
print *, 'I''ll sing you two-oh.'
call Green
print *, 'What is your two-oh?'
call Chor2
call Chor1
return
end

subroutine Green
print *, 'Green grow the rushes-oh.'
return
end

subroutine Chor1
print *, 'One is one and all alone,'
print *, '    and evermore shall be so.'
return
end

subroutine Chor2
print *, 'Two, two, the lily white boys,'
print *, '    clothed all in green-oh.'
return
end
```

To determine the exact output of this program you have to follow the sequences of invocations. Here is the output the program prints. See if you can understand why.

I'll sing you one-oh.
Green grow the rushes-oh.
What is your one-oh?
One is one and all alone,
    and evermore shall be so.

I'll sing you two-oh.
Green grow the rushes-oh.
What is your two-oh?
Two, two, the lily white boys,
    clothed all in green-oh.
One is one and all alone,
    and evermore shall be so.

The output from each of the statements comes out exactly as it appears between the quotation marks except that the two quotation marks in a row in "I''ll" are collapsed into one apostrophe. Each statement begins its output on a new line. In the main program we have a statement with nothing to print. This prints out a string of nothing (on a new line) and so has the effect of creating a blank line in the output. This is how we separated the verses.

Programming Example

Problem Statement

Write a program to print out a grid made up of blank squares surrounded by stars on the output device.

Inputs

None.

Outputs

A grid of stars.

Discussion

We will construct the grid of stars from two different kinds of output lines. The first kind will print a line made up of stars. The other will print a star, followed by 5 blanks, followed by a star and so on. Subprograms will be used to print each line and also to print each empty square.

Program

```
*       Program 4.2
*       Prints a square grid of stars
*
        program PrGrid
        call Square
        call Square
        call Square
        call HLine
        stop
        end
```

```
subroutine Square
call HLine
call VLine
call VLine
call VLine
call VLine
return
end

subroutine Hline
print *,'********************************'
return
end

subroutine VLine
print *,'*         *         *         *         *         *'
return
end
```

Testing

The output from this program is

```
********************************
*     *     *     *     *     *
*     *     *     *     *     *
*     *     *     *     *     *
*     *     *     *     *     *
********************************
*     *     *     *     *     *
*     *     *     *     *     *
*     *     *     *     *     *
*     *     *     *     *     *
********************************
*     *     *     *     *     *
*     *     *     *     *     *
*     *     *     *     *     *
*     *     *     *     *     *
********************************
```

Discussion

This program is hard to modify if we decide that larger squares are more attractive. We will see in subsequent chapters that there are more general ways to write this type of program.

A Piece at a Time

Design, Testing, and Debugging

- Breaking algorithms into small pieces is an effective way of developing solutions to problems. Learn how to use it.
- When you have choices about where to textually place subprograms, consider the order that will seem most natural to someone reading through the program from beginning to end. Try to reduce the amount of looking ahead that is necessary; make the ordering reflect the logical structure of the invocations.
- Use output statements liberally, especially when you are developing a program. You can often spot a mistake in your program by the fact that some output that you weren't expecting appeared.

Style and Presentation

- Use meaningful names for your subprograms. Use the name to hint at what the subprogram does, so that someone reading your program doesn't necessarily have to look at the text of the subprogram to understand what's happening when it is invoked.
- Use output statements to generally improve the appearance of program output. Remember that program output is for humans and should be laid out so that humans find it easy to understand.

Fortran Statement Summary

Program

A program consists of any number of statements. It may be followed by any number of subprograms.

>   **program** *name.*
>   *statements*
>   **stop**
>   **end**
>   *subprograms*

Subroutine

Here is the simple form of the subroutine. We will see more complicated forms in subsequent chapters.

> **subroutine** *subroutineName*
> *statements*
> **return**
> **end**

> **subroutine** Examp
> **print** *, 'Hello World'
> **return**
> **end**

Call statement

This statement can be used to invoke a subroutine.

> **call** *subroutineName*
>
> **call** Examp

Print statement

This statement is used to produce output on some output device.

> **print** *, *outputList*
>
> **print** *, 'Program is running'

Chapter Summary

- Algorithms can be developed effectively using a technique called step-wise refinement. This involves breaking the solution into pieces at a high level, and then successively refining each of these pieces into smaller pieces.
- Good programmers have a mental library of paradigms, that is, pieces of algorithm that are often needed. They write large programs by combining these paradigms.
- Fortran programs consist of a program header, followed by a set of executable statements. These are terminated by the executable statement **stop** and the non-executable statement **end**.

- Comments can be inserted into a program anywhere. A line that begins with an asterisk or C is assumed by the compiler to be a comment.
- A subprogram is like a program except that it is only executed when invoked from somewhere else. A subprogram begins with a subprogram header, contains a sequence of executable statements, terminates with a **return** statement, and finishes with the non-executable **end** statement.
- A subprogram is invoked using the **call** statement. Subprograms can invoke other subprograms, but may not invoke themselves, directly or indirectly.
- Output from a program is produced by executing a **print** statement. The **print** statement is followed by a string which is displayed on the output device, usually a terminal screen.

Define These Concepts and Terms

Program name
Line layout
Executable statement
Subprogram
Subroutine

Return statement
Call statement
Flow of control
Print statement

Exercises

1. What is the output of this program?

    ```
    program  Output
    print *, '-----'
    call  X
    call  Y
    call  Z
    print *, '-----'
    stop
    end

    subroutine  X
    print *, 'XXXX'
    call  Y
    call  Z
    return
    end
    ```

```
subroutine Y
print *, 'YYYY'
return
end

subroutine Z
call Y
print *, 'ZZZZ'
return
end
```

2. What is the output of this program?

```
program Silly
print *, 'Z'
call Alpha
call Beta
call Alpha
stop
end

subroutine Alpha
print *, 'A'
return
end

subroutine Beta
call Alpha
print *, 'B'
call Gamma
return
end

subroutine Gamma
call Alpha
print *, 'C'
return
end
```

3. Write a program to print a version of "Old Macdonald" and be sure that each line of the song is printed from precisely one subprogram.
4. What would be the output of the Dilly Song program if every name using the digit 1 as a suffix had the 1 changed to a 2, and every name using the digit 2 as a suffix had the 2 changed to a 1?
5. Write an algorithm for preparing and serving a dinner with several courses. Use as many algorithm components as is helpful.

6. In the Dilly Song program, the subprogram Chor1, used to print the first chorus, is the only one that is used more than once. Rewrite the program so that the main program prints the verses and Chor1 is the only subprogram. Type in your program and execute it. Correct all errors until the output is as shown in the chapter.
7. Given that the next two verses of the Dilly Song are "Three, three, the rivals" and "Four for the gospel makers," modify the program so that it prints the first four verses correctly.
8. Using your editor and compiler, enter and run this program. What happens?

> **program** Bad
> **call** Alpha
> **stop**
> **end**
>
> **subroutine** Alpha
> **print** *, 'A'
> **call** Beta
> **return**
> **end**
>
> **subroutine** Beta
> **print** *, 'B'
> **call** Alpha
> **return**
> **end**

9. Write a program to print the following pattern.

```
*
 *
  *
   *
    *
   *
  *
 *
```

10. Modify the program of Exercise 9 to print the zigzag pattern three times in succession.

11. Write a program to print your age in large digits as in

```
222222      111
22222222    1111
22    22    11 11
      22       11
    22         11
  22           11
  22           11
  22222222  11111111
  22222222  11111111
```

12. Write a program to print a calendar for the current month, with each date in the corner of a large square in which you can write your schedule.
13. Modify the results of exercise 12 to produce a calendar for the next month.
14. Write a program to print an approximate representation of your institution's coat of arms or logo.
15. Write a program to print the following program.

program Null
stop
end

Simple Calculating

5.1 Using Memory
5.2 Integers
5.3 Real Numbers
5.4 Operators
5.5 Constants
5.6 Operator Precedence
5.7 The Assignment Statement
5.8 Inputting Information
5.9 Variables and Subprograms

5.1 Using Memory

When programs execute inside a computer, the machine's memory holds both the program (set of instructions) to be executed and the data on which the program operates. These data include values supplied to the program from the outside world (using an input device), intermediate values that it generates along the way, and results that will eventually be sent to the outside world using an output device.

The program is placed in memory when you type the command that causes it to be executed. Now we will see how we arrange for data values to be placed there. The computer's memory will only store zeros and ones, the so-called binary digits or *bits*. The things that we want to store in the computer's memory are numbers and characters. To do this we must use an encoding or representation that stores all the possible things we want to keep into memory as strings of zeros and ones. Most machines do not allow individual bits to be referred to directly. Instead, bits are collected into groupings of fixed size called **bytes**. Bytes are the smallest unit of memory that can be directly referenced. For obvious reasons, encodings are usually selected so that data values fit into an integral number of bytes. Fortunately, when we're programming we never have to be aware in detail of what these encodings look like, but it is important to remember that they exist. So, for the next few paragraphs we'll describe typical representations.

Let's look first at how we encode characters. The first thing to think about is how many characters we are likely to need: clearly we need the letters of the alphabet (upper- and lowercase), the digits 0 to 9, and some punctuation characters. This already gives 60 or 70 characters. There are some less obvious characters that are also useful—for example, the control characters (you get these by holding down a key on your keyboard named <ctrl> and pressing some other key simultaneously). In fact, we need a representation for every character that can be typed on a computer keyboard.

Some early computers had bytes containing 6 bits and could therefore hold up to 64 ($= 2^6$) different representations in a single byte. For practical reasons it is convenient to use an encoding in which a single character fits into a single byte. However, 64 characters don't leave enough possibilities to include the alphabet (twice), the numbers, and punctuation. As a result, the standard size for the byte soon became 8 bits, allowing 2^8, or 256, possible different symbols to be represented in each byte.

For historical reasons there was a divergence at the time when a standard encoding for characters was being agreed upon. Most of the world ended up using the **ASCII** (American Standard Code for Information Interchange) encoding, whereas the IBM Corporation used its own standard, **EBCDIC** (Extended Binary Coded Decimal Interchange Code). These encodings are completely different and even have different orderings of the representations of common symbols. Fortunately, they both have characters and digits ordered in the natural way.

There are also encodings for integers and real numbers, but they are more complicated and we will discuss them in the next two sections.

Now, in theory it would be possible to construct an encoding that would allow the representations for different kinds of data to be different. That would allow us to look at a particular string of zeros and ones and tell what kind of thing it represented and its value. However, such an encoding would be expensive because we would need to add extra bits to the representation to say that *this* string of bits represents an integer, while *that* one represents a character. Because of the expense, this isn't done in most machines and you can't tell what a bit string represents without knowing what kind of thing it describes.

When we write programs we need to allocate space in memory for all the values that we know the program is going to use. We could say to the compiler that we need to store two integers at locations 100 and 101 and a character at location 102. However, we don't usually have enough information (or care enough) to know which addresses we can use for our program, and it's a tedious and error-prone activity to keep track of which value is kept in what memory location.

Instead, the compiler allows us to say that we are going to require space for an integer and use a symbolic name for it throughout the rest of the program. For example, if we are calculating the mass of a body, we can use a symbolic name, such as Mass, that helps us to remember what that location is used for. Now we don't need to worry exactly where in memory the value is actually placed because we always refer to it by name rather than location. We also don't have to worry about how much space to allocate for it because the compiler understands that an integer value requires, say, 4 bytes and allocates that much space for it. Because we have told the compiler not only the symbolic name but also the type of value we are going to store there, it can also check that we are using the value properly.

Such a named location is called a **variable**; we can forget about its actual location and imagine it to be a named value that is kept for us. The step of informing the compiler about the names and types of variables that are used is called *variable declaration*.

If we declare a variable to be of **type** character and call it C, then the compiler will allocate a single byte for it. The compiler will also make sure that we don't try and do anything silly with the variable in the program by looking at all the places in the program where C is referred to. For example, if we tried to add something to C, then the compiler will produce an error message when it translates the program, telling us that it is illegal to do arithmetic with a character variable.

In Fortran, variables are declared near the beginning of the program and at the beginning of subprograms. The statements that declare the variables are called declarations. They are non-executable statements, and appear before any executable statements. Declarations of variables of different types is done using different declaration statements. Here is an example of the beginning of a program that declares some character variables.

    program Sample
    character*1 Ch, Letter

The declaration contains the keyword **character** followed by an asterisk and an integer (indicating how many characters are to be represented), and then the list of the names of the variables. All variables of the same type need not be declared on the same line. The following program beginning has the same meaning as the first one:

    program Sample
    character*1 Ch
    character*1 Letter

Successive declarations appear on successive lines. It is often a good idea to include comments with each declaration, explaining the purpose of each variable.

5.2 Integers

One of the common types of value for a program to manipulate is **integers** (whole numbers). We don't need to know exactly how integers are encoded within the computer, but it is useful to be aware of how much space in memory they occupy and therefore how many values can be represented. Most computers represent an integer using either 2 bytes or 4 bytes. If 2 bytes are used, then integer values from -32768 to $+32767$ can be represented. If 4 bytes are used, then integer values from -2147483648 to $+2147483647$ can be represented.

We declare integer variables in the variable declaration part in the following way:

    integer I, J, K
    integer First, Last

The names of the variables are preceded by the keyword **integer**. The form of the declaration is exactly the same as for character variables, and declarations of different types can be mixed together (each on its own line) in any way. We could have declared all of these integer variables in a single line like this:

    integer I, J, K, First, Last

and the effect would have been exactly the same. However, it never hurts to have more than one declaration for the same type of variable and it can often be useful because it allows variables to be separated into logical groupings.

5.3 Real Numbers

The encoding for real numbers is a bit more complicated than that for integers. This is because there are infinitely many reals as we move away from the origin, but there are also infinitely many reals in any nonempty interval of the number line. Because the memory of any computer is finite, we are going to have to settle for a finite set of real values that can be exactly represented. All the other reals are approximated by the nearest value that we can exactly represent. The numbers in the machine are actually **floating point** numbers, an approximation to real numbers, but not the mathematical entities you're familiar with. In fact they're a subset of the rational numbers.

The representation chosen for real numbers is that used for so-called scientific notation: each number is represented by a fixed number of significant figures and an exponent. A number represented using three significant figures would be written as 0.235×10^3 in the decimal number system. The memory space available to encode each real value is divided into four parts: space for the mantissa (the significant figures), space for the exponent (the power), and space for the signs of both the mantissa and the exponent. The sign of the mantissa is the sign of the real value. The sign of the exponent is the sign of the power and therefore describes whether the number is between -1 and $+1$ or outside that range. For example, 0.2×10^2 is 20 while 0.2×10^{-2} is 0.002.

Because the amount of space for the exponent is limited, we can only represent numbers within a limited distance of zero. For example, if we were only allowed a two-digit decimal exponent, then we could only represent values between 10^{99} and -10^{99}. The limited amount of space for the mantissa limits how accurately we can represent real values. The standard size used for representing real values is 4 bytes. This allows us to approximate values in the range 5.4×10^{-79} to 7.2×10^{75} on some machines (the asymmetric range has to do with details of the representation that do not need to concern us), and 10^{-38} to 10^{38} on others.

We can see how this kind of encoding works by considering the real values that can be represented using a decimal encoding with a single-digit mantissa and a single-digit exponent. First consider numbers of the form $0.n \times 10^1$. These represent the real numbers 0.0, 1.0, 2.0, 3.0, ..., 9.0. The representations $0.n \times 10^2$ encode the values 10.0, 20.0, 30.0, 40.0, ..., 90.0. As we increase the exponent we can eventually represent the numbers 100000000.0, 200000000.0, ..., 900000000.0. You see that large numbers cannot be approximated very accurately—for example, 850000000 has to be approximated by 900000000, an error of 50 million.

Now let's look at what happens as we make the exponent negative. The representations $0.n \times 10^0$ encode the real numbers 0.1, 0.2, ..., 0.9. The representations $0.n \times 10^{-1}$ encode 0.01, 0.02, 0.03, ..., 0.09. You see that we get better approximations as we get closer to zero because the numbers that we can exactly represent become more and more densely packed in the number line.

We use a negative mantissa to get negative numbers. The density of the numbers that can be represented exactly is the same in the negative direction as in the positive.

Exactly the same behavior occurs when we use a binary exponent representation. The intervals change, but our representation still exhibits the property that we can exactly represent only a small number of values, and these values are not evenly distributed. This property is important if we are going to write programs that deal with real numbers, in which the answers must be very accurate. This is usually the case for engineering problems. We must always be aware that our answers cannot be exact, not treat them as exact, and make sure that we check that the answers we get are as good approximations as possible.

We declare real variables using the following syntax:

    real X, Y, Z

This fragment declares the three variables X, Y, and Z to be of type real. They will each be allocated 4 bytes of memory on many machines and will be implemented using the machine's built-in floating point type. As before, these declarations could be divided up over different lines, like this

    real X, Y
    real Z

Declaration statements for different types can appear in any order, although it is good practice to place declarations for the same type on adjacent lines.

5.4 Operators

We have discussed declaring variables, and now we will use them in computations. Computations are described in Fortran by **expressions**. An expression is a combination of variables, operators, and constants. The **operators** describe what to do with the values of variables and constants to compute a new result value. Thus, an expression stands for a value, the value that is computed when the expression is calculated. Because it has a value, it also has a type. The value of an expression could be an integer, or it could be a floating point value, or a value of any other type.

For example, suppose that there is an integer value stored in the variable Number. Then an expression that computes the square of that value would be

Number * Number

The "*" is the multiplication symbol in Fortran and specifies that the two values (operands) described by the things on either side of it are to be multiplied together. When this expression is evaluated, the value stored in the variable Number is fetched from memory and sent to the ALU. The ALU performs a multiplication operation, using its circuitry, and produces the value corresponding to the square. What happens to the value then depends on the context in which the expression appears.

Fortran has ways to specify all of the standard arithmetic operations. For example, we can write

Number + Number

to describe an addition and

Number - Number

to describe a subtraction (this particular one is rather pointless since it gives a result of zero regardless of the value we started with).

Strictly speaking, we need to have different operators for each of the different types of variables. For example, we saw earlier how to write an expression that calculates the product of two integer values. If we have a real value in a variable X, then the expression to compute its square is

X * X

using the same operator ("*") as for the integers. However, a very different kind of operation is actually involved in doing the real multiplication than in doing the integer multiplication, because the representation of the values to be manipulated is different. The same symbol is used to represent both kinds of multiplication; the compiler is able to choose the correct machine language operator to use.

When we come to the operation of division, however, the type of the operands makes a big difference. This is because the division of one integer by another doesn't (in general) give a result that is an integer. An integer division of two integer variables Num1 and Num2 would be written

Num1 / Num2

and the result would be the result of the ordinary division, truncated to the next smallest integer. If Num1 contained the value 37 and Num2 contained the value 5, then the result of the division would be the integer 7 because 37/5 is 7.4 and the next smallest integer is 7. If the result were negative, the truncation gives the next largest integer. In either case, the resulting value is the integer obtained by discarding the decimal places of the result.

The exponentiation operator, written as "**", raises its first operand to the power of its second. The expression

Number ** 3

calculates the cube of the value in the variable Number. Notice that, although the operator consists of two characters, it is logically a single symbol.

We can build more complicated expressions by combining operators, variables, and constants. Here is an example of a more complicated expression:

Num1 * Num2 * 3

The value of this expression is calculated by taking the value stored in the variable Num1, multiplying it by the value in Num2, and multiplying the result of the first multiplication by 3.

Simple Calculating

We can put expressions into output statements. If Pi has a value, this statement prints out the quantity $4\pi/3$, which is the constant that is used to calculate the volume of a sphere [$volume = (4\pi/3) \times radius^3$].

print *, 4 * Pi / 3

Expressions may contain both integer and real operands. When this happens, the result of a computation has to be provided in one form or the other. For example,

3.1 * 2

contains a real number and an integer. In algebra this should produce the value 6.2, but another possibility would be to produce the value 6, which is the integer part of the result. Fortran in fact produces the real value 6.2, and this is an example of a general pattern of behavior: when there are both integer and real operands, the result will be real.

5.5 Constants

We often want to include constant values in our expressions. We can do this by simply writing the value directly as part of the expression. For example, if we want to compute half of a value stored in a real variable A, then we write

0.5 * A

Constants also have a type associated with them. A real constant such as 15.0 and an integer constant such as 15 are different in an important way. This difference isn't a mathematical one (for they represent the same mathematical value), but has to do with the different representations of values inside the computer. As you learn more about Fortran, you will find several places where integer constants are allowed but real constants are not.

It is straightforward to write expressions involving constants in this way but it is not necessarily a good idea to do so. The trouble comes when you go back to a program previously written by you or someone else and try to remember what the significance of the constants is. For example, the formula for calculating the energy equivalent of a mass M is

M * C * C

If C is the speed of light in a vacuum, then its value is 3×10^8 m/s. So the expression we want to calculate is

M * 3e08 * 3e08

and if we write this, we hope that the reader will recognize the constant 3×10^8 as being a standard value. (The notation 3e08 is the way in which floating point constants are written—the *e* indicates the exponent.) However, if we write this expression in a program as

M * 9e16

then it has become very difficult for anyone reading the program to understand where the number 9×10^{16} came from and what it means. In general it is not a good idea to scatter explicit constants through programs because their meaning is often difficult to deduce from their use.

Many languages allow us to give symbolic names to constants that will appear in the program, just as we give symbolic names to variables to help us remember what they are used for. Fortran is one of these languages. Constants are specified immediately after the variable declarations at the beginning of the program. This is done using a **parameter statement**. It begins with the keyword **parameter** and a list of assignments to variables, enclosed in parentheses. The variables should have been previously declared. If we are writing a program that uses the acceleration due to gravity, then we would include a constant section of the following form:

>   **real** G
>   **parameter** (G = 9.8)

Now we use the symbolic name in expressions (and a good choice of name will make its meaning obvious). It is also a good idea to include a comment at the point where the constant is given its value to explain its purpose, if it's not immediately obvious. A further benefit of constants is that we don't have to worry about mistyping the constant value if it is used in several places. If the value of the constant changes (this doesn't happen to physical constants like G, but does with constants that specify other constraints on real world problems), then we only have to alter its value in one place at the beginning of the program regardless of how many times it is used.

A list of constant assignments can be given, separated by commas.

>   **integer** A, B
>   **parameter** (A = 3, B = 4)

You can mix **parameter** statements in among the declaration statements. We will see later that this is extremely useful. Symbolic names that have been declared as parameters cannot be altered within the program. They retain the values given to them in the **parameter** statement throughout. They can, of course, be used freely in expressions.

To illustrate, we now show a small program that uses the **parameter** statement.

```
*       Program 5.1
        integer A, B
        real X
        parameter (A = 3, B = 4)
        parameter (X = 0.5)
        print *, A * B
        print *, X ** B
        stop
        end
```

As you would expect, the output from this program is

```
12
0.0625000
```

5.6 Operator Precedence

We showed examples in the last two sections of expressions that included more than one operator. In fact, an expression can contain any number of operators. Here are some samples:

```
Num + M * 3
X / 4.5 * 19.2
X + Y / 3.4
```

As soon as we allow more than one operator in an expression, there must be rules about the order in which the operations are going to be performed. For instance, in the expression

```
X / 4.5 * 19.2
```

if the division is done first and then the multiplication, the expression calculates the value of 19.2, times X divided by 4.5. However, if the multiplication is done first, then the expression calculates the value of X over the product of 4.5 and 19.2. These two results are definitely not equal. Similarly, the expression

```
3 + 2 * 5
```

gives the value 25 if the addition is done first but gives 13 if the multiplication is done first.

Every language has rules about the order in which operations are to be performed. These are called **precedence rules**, since they determine which operators are performed before other operators. In Fortran, the arithmetic operations are divided into groups: the operator "**" (exponentiation); the operators "/" (divide), and "*" (multiply); and the operators "+" (add) and "−" (subtract).

Operators in a group have the same precedence and are applied in the left to right order in which they appear in the expression. Operators from different groups have decreasing precedence, so that an operator from the first group will always be applied before any operator from the second. The groups were listed in decreasing order of precedence. These rules mean that the second expression will be evaluated as 13 and not 25 in Fortran.

If an expression has several operators from the same group, they are evaluated left to right. Therefore in

X / 4.5 * 19.2

the division is done before the multiplication (which is perhaps not quite what you might expect from the way it is written).

The exponentiation operator is an anomaly among the Fortran operators. First of all, unlike the other operators, a sequence of exponentiations associates from right to left. The meaning of an expression such as

4 ** 2 ** 3

is obtained by first raising 2 to the power 3, and then raising 4 to the power 8. Thus, this expression has the value 65536. Some care is also needed when the exponentiation operator is mixed with other operators in an expression. For example, the expression

-X ** N

is always negative, because the exponentiation is done first. This can be particularly deceptive in the case where we write

-1 ** 4

which has the value −1, not 1.

Sometimes it is necessary to write an expression where we intend the operations to be performed in some order other than the way the rules would decree. We can always make this happen by surrounding an operator and its operands by parentheses to force that calculation to be done first. For example, if we wished the addition to be done before the multiplication, we could write

(3 + 2) * 5

Operations in parentheses always take precedence over other operations regardless of the precedence rules, so we can always use parentheses to force any order of evaluation we wish.

Simple Calculating

There are pitfalls in translating algebraic formulas to Fortran expressions. Until you get used to doing it, you should always check your expressions carefully. It is also a good idea to include parentheses whenever you are unsure of the order of evaluation even if, strictly speaking, they are unnecessary.

Here are some examples of expressions that are a little tricky. Consider first expressions of the form

8 / 4 / 2

Because of left-to-right evaluation, this expression divides 8 by 4 and then divides the result by 2. The result is therefore 1. This expression is equivalent to

8 / (4 * 2)

Another form of expression that causes trouble is one like this:

8 / 4 * 2

which is evaluated as 8 divided by 4 (giving 2), then multiplied by 2, giving 4. The 2 is part of the numerator rather than the denominator. The most common error to make is to take an algebraic expression of the form *a/bc* and translate it directly as

A / B * C

which corresponds to *ac/b* instead. Another common source of error is caused by the integer division operation.

1 / 3 * 4

evaluates as 0, because the division is an integer division and gives 0, which is then multiplied by 4.

Fortran allows expressions containing both integer and real values. However, it is important to understand the rules that apply when these two types of values are mixed (often called mixed-mode). The essential principle is that when an operation involves an integer operand and a real operand, then the integer operand is converted to the corresponding real value before the operation is carried out. Knowing the precedence rules enables us to determine when this "promotion" of integer values will occur. For example, the expression

2.0 * 3 / 4

has the value 1.5 because the multiplication is done first, giving the value 6.0, which is then divided by 4. On the other hand, the expression

3 / 4 * 2.0

which is mathematically equivalent, has the value 0.0 because the division is done first and is an integer division, with result 0. This is then multiplied by 2.0 giving 0.0. Care must always be taken when mixed-mode expressions are used to ensure that correct values result and that accuracy is not lost.

We finish this section with a program that evaluates some expressions. Before looking at the output, see if you can predict what it will be.

```
*       Program 5.2
        print *, -1 ** 3
        print *, -1 ** 6
        print *, 3 ** 2 ** 3
        print *, (3 ** 2) ** 3
        print *, 3 * 2
        print *, -4 * 5
        print *, 2 / 3
        print *, 3 / 2
        print *, -12 / 5
        print *, -12.0 / 5.0
        print *, -12.0 / 5
        print *, 2 + 3
        print *, 2 + 3.0
        print *, 2 - 3 * 4
        print *, (2 - 3) * 4
        print *, 36 / 4 / 3
        print *, 36 / 4 * 3
        print *, 2.0 * 3 / 4
        stop
        end
```

The output of this program is

```
        -1
        -1
      6561
       729
         6
       -20
         0
         1
        -2
 -2.3999996
 -2.3999996
         5
    5.0000000
       -10
        -4
```

3
27
1.5000000

5.7 The Assignment Statement

Sometimes we want to use the value of an expression in several places in a program and, rather than recalculate it in each place where it is needed, we would like to store its value. This is done using the **assignment statement**. The syntax for an assignment statement is

> *variable* = *expression*

This statement takes the value resulting from computing the expression and stores it in the variable. We read it as "*variable* is assigned *expression*" or "*variable* gets *expression*."

The type of value calculated in the expression must match the type of the variable to which it is being assigned. Thus a character cannot be assigned to a numeric variable. However, it is possible to assign integer expressions to real variables and vice versa. When this occurs, it may not be possible to represent the resulting value as an integer and a condition called **overflow** may occur. This will cause an execution time error on most systems. Even if overflow does not occur, assigning a real value to an integer variable will result in the real value being truncated, that is, all of its decimal places will be discarded. The assignment

> K = 2.3

assigns the value 2 to K (if we assume that K is an integer variable). Similarly the assignment

> K = -3.7

assigns the value -3 to K.

The expression on the right hand side of an assignment statement can contain any variables that have been assigned values, as well as constants and operators.

As an example, consider the following assignment statements. You might try to calculate the values of each of the variables, assuming that I, J, and K contain integers, and X, Y, and Z contain real numbers.

```
I = 3
J = 4
K = I * J
X = 2.5 * K
Y = X * (I / J)
Z = X * (J / I) ** 2
```

The values of the variables would be 3, 4, 12, 30.0, 0.0, and 30.0.

The variable to which the value of the expression is being assigned may also appear in the expression itself. For example, the meaning of the assignment statement

```
X = X + 3
```

is that the current contents of the variable X are added to the constant 3 and the result is placed back in the variable X. This, of course, overwrites the previous contents. The expression is evaluated first and then the assignment is carried out, so although the previous statement may look a little odd, it has a well-defined meaning.

Assignment statements should be laid out so that it is easy to see what is being calculated. Blanks may appear anywhere in the statement if it helps the appearance. In fact, Fortran allows blanks within variable names, although this is *not* recommended. If the assignment statement is so long that a continuation line is needed, the expression should be divided at a natural place.

```
X = (P + Q) / (A + B * C) ** D
&           / (X ** (Y - 3))
```

is much better than

```
X = (P + Q) / (A + B *
&           C) ** D / (X ** (Y - 3))
```

5.8 Inputting Information

We have seen how an output statement can be used to print information produced by programs on an output device. The corresponding statement that allows information to be input to programs is the **read** statement. The input statement looks like this:

> **read** *, *variableList*

When the statement is executed, it attempts to read values for all the variables in the list; the values are expected to appear on an input device, usually the terminal keyboard, in the specified order. The values supplied must be separated by commas or blanks, unless they are character values. Character values must be enclosed in single quotation marks. The implementation is sophisticated enough to skip over blanks when it is trying to find values to read. There is therefore no particular layout required when entering values for programs. However, this can sometimes lead to problems. If a program executes a read statement that is to read three numbers and you only type two numbers on the keyboard, it will wait for the third number to arrive. This can create the appearance of a program that has malfunctioned and can sometimes be difficult to diagnose. For this reason, and because it makes programs generally easier to use, it is a good idea to use an output statement before each read, printing a message explaining what input is expected for programs that are going to interact with a user at a keyboard. This is called *prompting* and the statement that prints the message is called a *prompt*.

A program that wanted the user to input moments of inertia might have statements like these in the program.

```
print *, 'Enter Moment of Inertia'
read  *, Moment
```

The following program illustrates the use of assignment statements and input and output statements. As before, you might like to predict what its output will be before looking at it.

```
*       Program 5.3
        integer J, K, L
        real A, B, C
        J = 2 ** 3
        K = J / 4
        A = 3.0 ** 2
        B = A / 2.0
        C = 2.0 * B
        L = -A / 3
        print *, J, K, L
        print *, A
        print *, B
        print *, C
        print *, ' Enter an integer between 1 and 10'
        read *, J
        K = J + K
        print *, K
        stop
        end
```

The output of this program follows.

```
                    8              2         -3
              9.0000000
              4.5000000
              9.0000000
     Enter an integer between 1 and 10
     5
                    7
```

5.9 Variables and Subprograms

Although subprograms are designed to allow us to describe some part of an algorithm that is a distinct unit, it is often necessary for the subprogram to have access to some of the information in the main program. In particular, the subprogram may need to manipulate the values of the variables in the invoking program.

Consider the following situation. Suppose we write a program to print the square of a number. Although it's not very realistic, let us suppose that we wish to square the number and print it in a separate subprogram. The main part of the program would read the value and then invoke the output subprogram to print it. But the output subprogram has to be told the value of the number that was read in before it can do its job. The following program shows how we do it in Fortran:

```
*      Program 5.4
*      reads a number and prints its square
*
       program Square
       integer Number
       print *, ' Enter Number to be squared'
       read *, Number
       call SqPr(Number)
       stop
       end
*
*      Subroutine to square and print
*
       subroutine SqPr(Num)
       integer Num
       print *, Num * Num
       return
       end
```

The subprogram has some extra information in its header. The extra piece looks like an identifier and the header is followed by a declaration. This means that, when the subprogram is invoked, it will be given an integer value as part of the invocation. The name Num acts as a placeholder for the value that will be used when the subprogram is actually invoked. It is called a **formal parameter** to the subprogram, or simply a *parameter*.

In the main program we write the subprogram invocation in the usual way. However, an actual integer-valued expression is included in parentheses after the name of the subprogram. It is this value (in this case the value stored in the variable Number) that is passed to the subprogram. The names of the variables don't need to be the same because the association between the actual value passed to the subprogram (the *argument*) and the parameter in the subprogram heading is done positionally. The value of the first argument is associated with the first parameter, the second argument is associated with the second parameter (if there is more than one), and so on.

Let us suppose that the previous program is executed. The first statement is an input statement, to which we respond by entering the number 5. This value is stored in the variable Number. The second statement is an invocation of the subprogram and passes the value 5 to the subprogram, where it is known by the name Num. The statement in the subprogram computes the square of the value Num (25) and prints it on the output device. Finally the subprogram returns control to the main program.

We've already suggested that all programs should contain appropriate comments at their start, describing the purpose of the program, who wrote it, when it was written, and anything else that a maintainer will want to know. The beginning of a subprogram is also an appropriate place for comments describing what the subprogram does. These should contain author information if appropriate, as well as explaining the meaning and use of the parameters.

In general, a subprogram may have any number of parameters in its header. For example, here are some equivalent headers for a subprogram with two integer and two real parameters.

> **subroutine** P (I, J, X, Y)
> **integer** I, J
> **real** X, Y
> ...

> **subroutine** P (I, J, X, Y)
> **integer** I
> **integer** J
> **real** X
> **real** Y
> ...

When P is invoked, it expects to be given a list of four values: two integers followed by two reals. Therefore the following header is *not* equivalent because the actual values expected are in a different order.

```
subroutine P (I, X, J, Y)
integer I, J
real X, Y
...
```

Here is a slightly more complex example of a subprogram invocation with values being passed.

```
*       Program 5.5
*       Prints 1 if I = J and 0 otherwise
*
        program Weird
        integer I, J
        I = 3
        J = 4
        call Out(I, J)
        stop
        end
*
*       Print 1 or 0 depending on parameters
*
        subroutine Out(L, M)
        integer L, M
        print *, (L / M) * (M / L)
        return
        end
```

The main program assigns values to integer variables I and J and then passes them to the subprogram. These two values are associated with the names L and M, so that while the subprogram is executing, L contains a copy of the value that was in I, and M contains a copy of the value that was in J.

The output statement in the subprogram evaluates the expression and prints it. If the value of L is smaller than that of M, then L divided by M evaluates to zero and the output is zero. Symmetrically, if the value of M is smaller than L, then M divided by L evaluates to zero and the output is zero. When L equals M, then both of the divisions give the result 1 and the output is 1. So this subprogram prints a 1 if its inputs are equal and 0 otherwise. For the program given, I has the value 3 and J has the value 4. When the subprogram is invoked, these values are passed as arguments. Thus, the value of L in the subroutine is 3 and the value of M is 4. The expression in the subroutine has two parts; the first part evaluates to 0 and the second to 1. Their product is 0.

Subprograms can have their own variables. These are declared at the beginning of the subprogram before the subprogram's executable statements and are called **local variables**. These variables only exist during the time that the subprogram is active. Any subprogram's local variable that has a value assigned to it during one invocation of the procedure will not have a value at the beginning of the next invocation because the variable vanishes when the first invocation completes and returns to the invoking program. (This may not be true in some systems, but since there is no guarantee of values being retained, it is bad practice to program as if they will be.) These rules for the visibility of identifiers are called **scope rules**.

The next program shows a simple example of a local declaration.

```
*       Program 5.6
*       Illustrates local variables
*
        program Local
        integer I
        I = 2
        call P(I)
        print *, I
        stop
        end

        subroutine P(J)
        integer J, K
        K = J * J
        print *, J, K
        J = J + 1
        return
        end
```

The next program shows some of the things that can go wrong when using local declarations.

```
*       Program 5.7
*       Illustrates misuse of local variables
*
        program Local
        integer I, K
        I = 2
        call P(I)
        print *, I, K
        stop
        end
        subroutine P(J)
        integer J, K
        print *, J, K
        K = J
        return
        end
```

This example contains two errors. The output statement inside the subprogram attempts to print the value of the variable K, but the variable has just been created at the beginning of the subprogram and does not yet have an assigned value. The output statement in the main program attempts to print the value of the variable K, but this is also a problem because the value of K is distinct from the one inside the subprogram and has not been assigned a value. In Fortran this will not be detected since it is legal, although very bad practice, to have variables declared implicitly as a result of their appearance in a statement of the program. (See the definition of the **implicit** statement in Appendix C.) The compiler assumes a local definition should be made. In both cases a meaningless value will be printed.

Subprograms are separately compiled. When a compiler is compiling a program consisting of several subprograms, it begins the compilation process from scratch for each new subprogram as it encounters it. No information about the variables of any program unit can ever have any effect on those of some other program unit. This means that, although variables and statement numbers must be unique within each program unit, they need not be between program units.

Programming Example

Problem Statement

Write a program to print the square, cube, and fourth power of a number.

Inputs

None.

Outputs

The value of the square, cube, and fourth power.

Discussion

Although it's not necessary to do it this way, we will use a subprogram to print out each of the values. The subprograms will each be passed the value, compute an appropriate power, and print it out.

Program

```
*       Program 5.8
*       Prints out powers of a constant in the program
*
*        X - value whose powers are calculated
*
        program Power
        integer X
        X = 22
        print *, 'Initial value is', X
        print *, 'Powers are'
        call Square(X)
        call Cube(X)
        call Fourth(X)
        stop
        end

        subroutine Square(Y)
        integer Y
        print *, Y * Y
        return
        end

        subroutine Cube(Y)
        integer Y
        print *, Y * Y * Y
        return
        end

        subroutine Fourth(Z)
        integer Z
        print *, Z ** 4
        return
        end
```

Testing

The output of the program is

```
Initial value is   22
Powers are
           484
         10648
        234256
```

Discussion

As this program executes, it first stores the integer value in the variable X. It then executes two output statements, the first of which prints the value of X, and the second of which acts as a heading for the output that will follow. It then invokes, in turn, the three subroutines that do the actual calculation. Each subroutine consists of a single print statement that evaluates an expression. The value of X is passed as an argument to each, and is known within the subroutine as Y in subroutines Square and Cube, and as Z in subroutine Fourth.

Design, Testing, and Debugging

- Make sure that you understand the exact order in which the various operations will occur in any expression that you write. If you are in doubt, then use parentheses to force the evaluation order you want.
- Always make sure that you understand the order in which conversions will be made when integer and real values are mixed in an expression.
- Watch for integer division and make sure that it doesn't happen in places where you didn't mean it to.
- Always go back and check that you have included declarations for all the variables you have used in your program. Watch for mistypings of variable names, where the same name is spelled differently in different parts of the program.
- Practice your typing skills. Slow typing can waste a lot of your time.
- When you include a division in your programs, always make a mental check that the denominator can never be zero.
- Always check that the parameters declared in each subprogram header match the variables that are used when the subprogram is invoked. They must match in both number and type.
- Always use the **parameter** statement to define constants that will be used throughout the program. This makes it easy to change them if the program is subsequently modified, since the change needs only to be made in one place.

It also prevents you from accidentally typing a different value in different places.
- When considering how best to break up a solution into subprograms, consider the amount of information that has to be moved between the subprogram and its invoking program. A good breakdown will probably not need to pass large amounts of information around when subprograms are invoked.

Style and Presentation

- Choose meaningful names for all of the variables you use in your program. Make sure the name accurately conveys the purpose for which the variable is being used.
- Don't attempt the false economy of using variables for different reasons in different parts of the program. Computer memory is plentiful and your reuse of a name may confuse someone (possibly you). Writing programs is difficult enough already.
- Never write out a value without also printing some text indicating what the value means. Remember that the user doesn't necessarily have the program in front of her and may not know exactly what to expect.

Fortran Statement Summary

Program

A program consists of declarations, followed by parameter statements and executable statements.

> **program** name
> *declarations*
> *parameter statements*
> *statements*
> **stop**
> **end**
> *subprograms*

Declarations

These statements are used to tell the compiler about variables that the program will use.

type variableNames

integer I, J, K
real A, B, C, X
character*80 Line

Parameter Statement

This statement is used to give a symbolic name to a value that will be used in the program. This makes it easy to change the value if necessary and helps prevent typing mistakes in different uses.

parameter *(variable = value)*

parameter (G = 9.8, C = 3.0e08)

Assignment Statement

This statement assigns the value of some expression to a variable. The type of the expression must be the same as the type of the variable, except for the special case of integers and reals.

variable = expression

```
I = J + K - L
A = B * C / D
X = A ** K
```

Read Statement

This statement allows input to be read in by the program.

read *, *variableList*

read *, X, Y, I, J, K

Subroutine with Parameters

Values may be passed to a subroutine. A list of dummy names is provided in the subroutine header and these are matched positionally with arguments used at the point of invocation.

Simple Calculating

```
subroutine name(parameter names)
    declaration of parameters
    statements
    return
end

subroutine Alter(X, Y)
    real X, Y
    X = 3.56 * Y
    return
end
```

Chapter Summary

- Character data is represented inside a computer's memory using an encoding scheme. There are two standard schemes, called ASCII and EBCDIC. On most computers, a character is stored in a single byte, which is usually 8 bits long.
- Floating point values are encoded using a mantissa and an exponent. This means that almost all real numbers have to be approximated by the nearest representable floating point value. The density of representable values in the number line is greatest near zero, and decreases as the magnitudes of the numbers become larger.
- All memory locations to hold data values are allocated by the compiler. The compiler must be told what type of value each location is to contain and what its symbolic name will be. It is referenced using its symbolic name throughout the program.
- A memory location that is referenced symbolically is called a variable. Telling the compiler its type and name is called declaring the variable. This is done using a declaration statement. A different declaration statement is used for each type of variable. The declaration statement begins with the keywords **integer**, **real**, **logical**, and **character** for each type, respectively.
- An expression is a set of operators, variable references, and constants that describes a computation producing a value. An expression always has a type associated with it.
- The arithmetic operators in Fortran are ** (exponentiation), * (multiplication), / (division), + (addition) and − (subtraction).
- Expressions may appear in the list contained within a **print** statement.
- A constant value may be given a symbolic name at the beginning of a program or subprogram using a **parameter** statement. The symbolic name can appear anywhere an expression of its type may appear. The symbolic name cannot appear on the left-hand side of an assignment statement.

- In expressions containing several operators, the order of evaluation is determined by precedence rules. The operators are divided into three classes: **; * and /; and + and −, in decreasing order of precedence. If an expression contains operators from within a single class, then the order of evaluation is left to right. The exception is that a sequence of exponentiations is evaluated from right to left.
- The assignment statement allows the value of an expression to be stored in a variable. The type of the expression should match the type of the variable. Floating point values assigned to integer variables are truncated.
- Input values are read by programs using the **read** statement. The **read** statement continues reading input until it finds enough values to fill its list.
- Input statements that read from the terminal keyboard should be preceded by output statements that inform the user of what information is required. This is called prompting.
- The values of variables can be passed to subprograms. If this is done, the subprogram header must be extended by a list of names by which these values will be known during the subprogram's execution. The values are provided at the point of invocation of the subprogram. The arguments at the point of invocation must match the parameters in the subprogram header in number and type. However, this is *not* checked by most compilers.

Define These Concepts and Terms

ASCII
EBCDIC
Variable
Type
Integer
Floating point
Expression
Parameter statement

Implicit conversion
Operator
Precedence rules
Assignment statement
Overflow
Formal parameter
Local variable
Scope rules

Exercises

1. Look up a table of ASCII representations of characters. What are the ASCII values corresponding to the keys on the top row of your keyboard?
2. Without executing it, what do you expect the value of this expression to be on your machine?

$$123456789 \times 10^{50} - 123456788 \times 10^{50}$$

3. What is the value of this expression

   Big + Small - Big

   if Big contains the value 10^8 and Small contains the value 1.0?
4. Suppose we represent real numbers using a decimal representation with two significant figures. What happens if the value 0.23 is added to itself ten times? What is the result of 10 times 0.23?
5. What is the percentage error on average in the representation of a real value using seven significant figures?
6. What values result from these expressions?
   a. 1 / 3
   b. 3 * 1 / 3
   c. 1 / 3 * 3
   d. 3 + 4 / 4 + 3
   e. (3 + (4 / (4 + 3)))
   f. 5 ** 3 * 2
   g. 5 ** 3 + 2
   h. 5 ** 3 ** 2
   i. 3 + 5 * 2
   j. 17 / 4 + 2
   k. 5 / 2 * 3
   l. 3 * 5 / 2
7. What do the following programs print?

   a.
   program One
   integer I
   I = 2
   call P(I)
   call P(3)
   call P(I+1)
   stop
   end

   subroutine P(X)
   integer X
   print *, X * X
   return
   end

b.
```
program Two
integer I, J
I = 2
J = 3
call P(4)
call P(J)
call P(I)
stop
end

subroutine P (A)
integer A
print *, A * A
return
end
```

8. What is the result of each of the following assignment statements involving the integer variable X?
   a. X = 1 / 3 * 3
   b. X = 1.0 / 3 * 3
   c. X = '1'
9. What does your compiler do with this program?

```
program Bad
integer I
integer I
stop
end
```

This case may seem trivial, but duplicate declarations might occur many lines apart in a large program.

10. Write a program that reads two numbers and computes the value of the first number raised to the power of the second.
11. Give a narrative description of the intermediate values computed and the output produced by the following program, in the same way that we have done for other programs in this chapter.

```
program Watch
integer I, J, K
I = 10
J = 2
K = 3
print *, I, J, K
J = I / J + 1
K = J ** K / 2
I = J - K
print *, I, J, K
stop
end
```

12. Give a narrative description of the intermediate values computed and the output produced by this program, in the same way that we have done for other programs in this chapter.

```
program Watch2
integer I, J, K
I = 10
J = 2
K = 3
print *, I, J, K
call Alpha(I, I)
call Alpha(I, J)
call Alpha(J, K)
stop
end

subroutine Alpha
integer A, B
integer C
C = A / B
print *, A, B, C
return
end
```

13. Write a program to compute the average of five integers that are read in.
14. Modify your program from Exercise 13 to compute the average of 10 numbers.
15. Modify Program 5.8 to work with two values, printing the appropriate powers in two columns on each line.

Control Statements

6.1 Definite Iteration
6.2 Indefinite Iteration
6.3 Selection
6.4 Logical Data
6.5 Numerical Integration
6.6 More General Selection
6.7 Getting Answers from Subprograms

6.1 Definite Iteration

We'll begin by looking at the simplest case of a nonsequential program structure, in which we have some calculation to perform a number of times, but we know in advance how many times it will be repeated. Recall that we referred to this earlier as **definite iteration**. For example, we might have a temperature sensor attached to a piece of machinery that reports temperatures once an hour. If we want to calculate the average temperature over a whole day, then we will have to use exactly twenty-four consecutive readings.

The Fortran statement that causes another statement or statements to be repeated a fixed number of times is the **do** statement. It has the following syntax:

>   **do** *label* *indexVariable* = *start, stop*
>   *statements*
> *label* **continue**

The **index variable** can be any variable. When this structure is executed, the index variable is set to the start value and, if this is no greater than the stop value, the loop body is executed. The index variable is then incremented by 1 and, if it is no greater than the stop value, the body is executed again. This continues until the value of the index variable exceeds the stop value. At this point the next statement in the program is executed.

The *label* is used to associate or match the beginning and the end of the loop. The *label* is a number with up to five digits. In the **continue** statement, it is placed anywhere in columns 1 to 5. The **continue** statement acts as a place-holder to which the *label* can be attached—it has no effect.

As an example consider the following problem: read in ten numbers and print their reciprocals. The program to do this repeats the pair of steps "read in a number, print out its reciprocal" for each of the ten numbers.

```
*       Program 6.1
*       Prints the reciprocals of 10 numbers
*
        program Invert
        integer I
        real Number
        do 100 I = 1, 10
            print *, 'Enter a number'
            read *, Number
            print *, 'Reciprocal is ', 1.0 / Number
100     continue
        stop
        end
```

This program is so short that we don't need to break it up into pieces—everything can be done in the main program.

In the program, the variable Number is being used repeatedly. The first time the loop is executed, the first input value is stored in it. This value is then used to calculate the expression in the output statement. In the next iteration of the loop, the second input value is read and stored in Number and is used for the second output, and so on. (There is a serious problem with this program—when the value of Number is zero, the division is impossible. We postpone discussing how to deal with this until Section 6.3.)

We indent the lines of program code that are repeated to indicate that their execution depends on some condition (in this case the condition that I is between 1 and 10). This is not a language requirement, but a way of making the program more comprehensible to human readers. Indenting a block of statements shows very clearly the structure of the program. There is no rule about how many spaces you should indent, but you should be consistent and use the same number of spaces throughout a program. We now show the output from the program. Follow it through and make sure you understand how it was calculated. The input given to the program was the integers from 1 to 10.

```
Enter a number
1
Reciprocal is              1.0000000
Enter a number
2
Reciprocal is              0.5000000
Enter a number
3
Reciprocal is              0.3333333
Enter a number
4
Reciprocal is              0.2500000
Enter a number
5
Reciprocal is              0.2000000
Enter a number
6
Reciprocal is              0.1666666
Enter a number
7
Reciprocal is              0.1428571
Enter a number
8
Reciprocal is              0.1250000
Enter a number
9
Reciprocal is              0.1111111
```

```
Enter a number
10
Reciprocal is             0.1000000
```

The start value and stop value in the iteration statement don't have to be constants. They can be any expression; specifically, these expressions might be single variables. To see how this works, consider the following extension to the program. Suppose that we want to use the program over and over again to print reciprocals of different sets of numbers. We want to specify, when we run the program, how many numbers we are going to use. The next program allows this.

```
*       Program 6.2
*       Prints the reciprocals of any number of numbers
*
        program Invrt2
        integer I, N
        real Number
        print *, 'Enter the number of reciprocals you want'
        read *, N
        do 100 I = 1, N
            print *, 'Enter a number'
            read *, Number
            print *, 'Reciprocal is ', 1.0 / Number
100     continue
        stop
        end
```

This program has an extra variable, N, that is used to keep track of how many other numbers are going to be read. When the program is run, the first thing that must be entered is the number of other values that are going to be entered. So if we want to print the reciprocals of 20 numbers, we enter 20, followed by the list of numbers. This loop is more general than the one in the first version of the program, but it still carries out a predetermined number of iterations. It is predetermined in the sense that when we come to execute the loop we know how many iterations will occur, even if we don't know when we write the program.

The expressions used for the start and stop values do not have to be integer expressions. They can, in fact, have real values.

If the start value is larger than the stop value, the statement inside the loop is not executed at all and the program simply goes on to the next statement. So, if we had entered −1 as the first input to the preceding program, it would not have produced any output; the repeated statements would never have been executed. This can sometimes be confusing because it's difficult to find a mistake in a program that doesn't seem to do anything.

As a word of warning, earlier versions of Fortran forced a **do** loop to be executed at least once even if that didn't make sense according to the evaluated limits.

The termination condition was effectively tested at the end of the loop. This is one of the major differences between older versions of Fortran and Fortran77.

The index variable takes on successive values in the range from the start value to the stop value as each iteration is executed. These values can be used within the loop body if necessary. For example, the next program prints out the numbers from 1 to 10, one per line.

```
*       Program 6.3
*       Prints the numbers from 1 to 10
*
        program  PrNums
        integer  I
        do  100  I  =  1,  10
            print  *,  I
100     continue
        stop
        end
```

The definite iteration is actually much more general than what we have shown. The examples thus far have used a step of size 1 between successive values of the index variable. The step size can be specified, just as the start and stop values can be, and the appropriate syntax is:

> **do** label indexVariable = start, stop, step
>   statements
> label **continue**

For example, the step size could be negative. The following program prints out the numbers 10, 9, 8, ..., 1 one per line.

```
*       Program 6.4
*       Prints the numbers from 10 to 1
*
        program  PrNum2
        integer  I
        do  100  I  =  10,  1,  -1
            print  *,  I
100     continue
        stop
        end
```

The step size can be an expression giving an arbitrary value. The value can also be integer or real. The following example prints some values between zero and one.

Control Statements   **107**

```
*       Program 6.5
*       Prints some real values between 0.0 and 1.0
*
        program Fract
        real X
        do 100 X = 0.0, 1.0, 0.1
            print *, X
100     continue
        stop
        end
```

As a point of interest, if you run this program you will probably *not* get a list of eleven values 0.0, 0.1, 0.2, ..., 1.0 because of the approximate nature of real number representations.

There are a few other important rules about **do** loops that must be observed.

- The index variable must not be explicitly altered within the body of the loop.
- The lower and upper limits and the step size may be altered within the body of the loop. However, the limits of the loop are determined when the loop begins and are unaffected by any subsequent change in these values.
- You should never assume that the loop index variable will have a value after the loop has completed. There are situations in which it will have a value, and other situations in which its value will vary from compiler to compiler, but you should always assume that it does not have a value.
- The increment cannot be zero (for obvious reasons).

6.2 Indefinite Iteration

In many problems we cannot tell, even as we begin to execute a loop, how many iterations will be required. This is usually because the number of iterations depends in some way either on the data being processed or the results of some computation within the loop. For example, think of finding someone whose birthday is in August. How many people do you have to ask? To handle this case we need a more general way of repeating statements. The Fortran statement that allows this is the **while** statement. It looks like this:

    while (*logicalExpression*) **do**
        statements
    end while

A logical (or Boolean) expression is one whose computed value is either true or false. For example,

3 .lt. 5

is a logical expression whose value is true (because 3 is less than 5). (The .lt. is the Fortran *less than* operator.)

Just like the definite iteration statement, the **indefinite iteration** statement controls the repetition of a list of statements. When the indefinite iteration is executed, the logical expression is first evaluated. If it is false, nothing further happens and the next statement in the program is executed. If it is true, the statement is executed and the logical expression is reevaluated; this continues until the expression is evaluated as false, at which time the loop terminates. Notice that the statements being repeated must change something in the logical expression. If not, then if the expression is true initially, it will be true forever (because it is never changed) and the loop will keep repeating forever.

It's also important to remember that the logical expression is evaluated only *before* each iteration of the loop. Even if one of the parts of the repeated code changes something that would make the expression false, the loop will still complete that iteration before reevaluating the expression and terminating.

In summary, the **while** form of the loop should be used whenever the termination condition cannot be determined when the loop is encountered (because it depends on some calculation in the loop). Definite iteration should be used whenever the number of iterations is known in advance.

Unfortunately, the **while** statement as we have described it is not part of Fortran77. But even though it's not part of the language definition, it is provided by almost all compilers, so you'll probably be able to use it when you write programs on your computer system. The details of the syntax provided may vary a bit from what we've described; you should look this up in the reference manual for your system. If your system doesn't have a **while** statement, then we've shown you how to build your own in Appendix F.

One of the most common ways of building Boolean expressions is to use **comparison operators**. In Fortran these are

.eq.	equal
.ne.	not equal
.lt.	less than
.le.	less than or equal
.gt.	greater than
.ge.	greater than or equal

These can all be used to compare two values and produce the Boolean value true if the comparison is true, and false if it is not. The simplest form of such a comparison is to compare a variable with a constant, like this

$$X \ .lt.\ 35.5$$

which is true if the value of the variable X is strictly less than 35.5. However, both sides of the comparison operator can be expressions, as in this example

$$X\ **\ 2\ .gt.\ Y\ *\ Z\ /\ P\ **\ 2$$

Control Statements

The precedence rules are such that the comparison operators are done last. Thus, the expressions on each side of the comparison operator are evaluated before an attempt is made to compare their values. Again, if there is any possibility of confusion, parentheses should be used to indicate the order of evaluation.

One special case should be mentioned at this point. We have already said that, because of the errors caused by the finite representation of real values, two quantities that ought to be equal may not be represented by the same floating point approximation. Thus, a comparison of two floating point quantities like this

A .eq. B

is likely to be false when we would expect it to be true. When we want to carry out this kind of test for equality, we must be satisfied with a test for approximate equality. It is common to write a test something like this

abs(A - B) .lt. 1.0e-7

This tests to see if the values of A and B are within 10^{-7} of each other. The abs is a function, which we won't formally introduce until chapter 7, but its purpose is to take the absolute value of the difference between A and B. We have to do this because we don't know which of A and B will be smaller, under the accumulation of approximations.

The comparison operators work as you would expect for numeric values, but they can also be used with characters. For example,

'c' .lt. 'd'

is true; this is a statement that says the letter "c" precedes the letter "d" in the character domain. In other words, "c" comes alphabetically before "d".

The ordering of characters is not defined in Fortran as you might expect it would be. Rather, it's defined by the particular machine your programs run on. Different machines have different orderings (depending, for example, on whether they are ASCII or EBCDIC machines). However, it's always true that the lowercase letters are ordered from "a" to "z," that the uppercase letters are ordered from "A" to "Z," and that the digits are ordered from "0" to "9." Thus, it's always possible to write programs that sort words into alphabetical order, for example. Most of the straightforward things (such as sorting) you'll want to do will work fine. But if you want to do esoteric character manipulations, you'll have to get detailed documentation about the character encoding used on your machine, and probably you'll want to consult an expert who has done such things before.

Let us now revise the problem in the previous section that printed the reciprocals of numbers. That program printed the reciprocals of any number of values, but someone had to count them first so that the first input value, telling the program how many numbers to expect, could be entered. We will write a new version that will find the reciprocals of any number of values until the value 0 is entered. The value 0 is a good choice for a flag to signal the end of the input values because it doesn't have a reciprocal and so could never be part of the actual list of values in which we were interested. This next program incorporates these revisions:

```
*       Program 6.6
*       Prints reciprocals using 0 as a sentinel
*       value to terminate the input
*
        program Recips
        real Number
        print *, 'Enter a number (0 terminates)'
        read *, Number
        while (Number .ne. 0) do
            print *, 'Reciprocal is ', 1.0 / Number
            print *, 'Enter a number (0 terminates)'
            read *, Number
        end while
        stop
        end
```

The input and output operations have been shuffled in this new version. The test of the value in Number must be made before entering the loop for the first time. To make sure that there is a value at that time, the first input value is read before the loop begins.

Now, when entering the repeated part, we have already read the first number, so all we have left to do is print its reciprocal. Before testing the Boolean expression again, the next value must be read so it can be checked for equality with zero. Eventually the input value read will be zero and the loop will terminate. Notice that even if the first number read is zero (meaning there is an empty list of numbers), the program works properly (printing no reciprocals) and finishes neatly. This is the first example of a type of loop that we will often use. It is a paradigm for processing a list of data items terminated by a special value called a **sentinel**. We now show the output from a sample run of this program.

```
Enter a number (0 terminates)
1
Reciprocal is              1.0000000
Enter a number (0 terminates)
3
Reciprocal is              0.3333333
Enter a number (0 terminates)
5
Reciprocal is              0.2000000
Enter a number (0 terminates)
0
```

Control Statements

A number of operations that programs need to do can be written as loops. One such operation is to calculate the sum of a set of numbers, another important paradigm. Let's assume that a set of numbers is given and a program is to be written to calculate their sum. Also assume that the list of numbers to be summed has a zero at the end, just as in the reciprocal program. The sum is calculated by keeping a running total of the numbers read at any given time. The next program sums input values.

```
*       Program 6.7
*       Finds the sum of its inputs
*       The input is terminated by a 0 value
*
        program AddUp
        real Total, Number
        print *, 'Enter next number'
        read *, Number
        Total = 0
        while (Number .ne. 0) do
            Total = Total + Number
            print *, 'Enter next number'
            read *, Number
        end while
        print *, 'Sum is ', Total
        stop
        end
```

The structure of this program is very similar to that of the previous one. All that has changed is that a different operation is carried out in the loop and there are some differences at the beginning and end. We have to initialize the variable Total to zero because the running total of no numbers is zero. This means that if the first value read is a zero (so that there were no numbers to be summed), the program prints the correct total. Again, we must read the first input value before the loop so that we can test it before we start the loop for the first time. After the loop has completed, we have read all of the required values; the running total is, in fact, the total of the numbers. This is an extremely important paradigm.

Many loops have the general structure that has been used in the last two examples. It has four parts:

initial part, to get started
while logical expression do
    working part, the repeated step
    get ready for the next iteration

Both the initial part and the part getting ready for the next iteration always affect something that appears in the Boolean expression—the initial part prepares for the first time the Boolean expression is evaluated and the next iteration part prepares for subsequent evaluations. We have constructed a paradigm for looping and have used it to construct a sum paradigm. We will continue to enhance this paradigm in the next few examples.

Another common data operation that uses a loop is the calculation of an average or mean. The mean is obtained by summing a set of numbers and dividing the sum by the number of members of the set. It is often used to get some general result from a large set of data. The program that calculates the sum of a set of numbers should be easily modifiable to do the extra step of calculating the mean. If we use the technique of terminating the input by a zero value, then we will also need to keep track of how many values have been added into the sum for the final division. The next program has been extended in this way.

```
*       Program 6.8
*       Finds the mean of a list of numbers
*
        program Mean
        real Total, Number
        integer Count
        print *, 'Enter number'
        read *, Number
        Total = 0.0
        Count = 0
        while (Number .ne. 0) do
            Total = Total + Number
            Count = Count + 1
            print *, 'Enter number'
            read *, Number
        end while
        print *, 'Mean is ', Total / Count
        stop
        end
```

The program begins by assigning initial values to the Total of the numbers so far, and the Count of the numbers seen. The loop body begins by updating the counters to reflect the most recently read value and then reads in the next number. Thus the first four executable statements are the initial part, the first two statements in the loop body are the working part, and the final two statements in the loop body are the preparation for the next iteration.

This program contains a serious flaw. If the first input value entered is zero (that is, the set of numbers whose mean we want to find is empty), then the Boolean expression in the loop will be false the first time it is encountered. This means that the output statement will try to divide by Count when Count has the value 0. As division by zero is not mathematically well-defined, all computers have the ability to detect attempts to do it. So when this statement tries to execute, a run time error will result, the program will halt, and a message from the operating system will be printed for the user. This is not a good thing for a program to do because in real life situations, the user is not usually the person who wrote the program and will not understand where the problem lies, let alone how to fix it. Programs should always be written so that they don't fail unexpectedly. In the next section we shall see how to improve this program.

It might seem silly, at first, to worry about what happens when there is no input, and that's probably true if a user is sitting at a terminal running the program. But many programs do not run in isolation; they run in groups of programs working together on some task. So it might be quite sensible that the input to our program would come from some other program and there might be situations in which this other program would produce an empty output list. It is easier in the long run to write programs generally than to have to work around the special cases that come up when programs are actually used. You should always think about such boundary cases when you are designing a program.

Another interesting and important application that builds on our program to calculate the mean is to add the capability to calculate the standard deviation. The standard deviation measures how far the values are from the mean. For example, if you are sitting in a sauna with your feet in a bucket of ice, your average temperature might be quite comfortable, but you still might not find it entirely to your liking. That is because, although the average temperature is reasonable, the standard deviation is very large. The standard deviation is usually calculated by taking the squares of the differences between the values in the set and the mean, summing them, dividing by one less than the number of values, and taking the square root, as shown in the following formula:

$$\sqrt{\frac{1}{n-1} \sum_{i=1}^{n} (x_i - \bar{x})^2}$$

Now we have already seen that calculating the mean, $\bar{x}$, requires us to go through the data once to sum them. Calculating the difference of each value from the mean cannot be done until we know the mean, so this would appear to imply that we would have to make two passes through the data. However, if we rearrange the formula a little, we can write it in such a way that we need only look at each value once. We simply expand the squared term and use the definition of $\bar{x}$. Here is the revised formula:

$$\sqrt{\frac{1}{n-1} \left(\sum_{i=1}^{n} x_i^2 - \frac{1}{n} \left(\sum_{i=1}^{n} x_i \right)^2 \right)}$$

Now we can write a program that calculates both the mean and the standard deviation of a set of data using only one pass through the data.

```
*       Program 6.9
*       Finds the standard deviation of a list of numbers
*
        program StdDev
        real Std, Mean, SumX, SumXSq, X
        integer Count
        SumX = 0.0
        SumXSq = 0.0
        Count = 0
        print *, 'Enter number (0 terminates)'
        read *, X
        while (X .ne. 0) do
            SumX = SumX + X
            SumXSq = SumXSq + X * X
            Count = Count + 1
            print *, 'Enter number'
            read *, X
        end while
        Mean = SumX / Count
        Std = sqrt((SumXSq - (SumX*SumX) / Count) / (Count-1))
        print *, 'Mean is ', Mean, ' Standard Dev is', Std
        stop
        end
```

Two things are calculated from the set of input values: their sum and the sum of their squares. These terms are then used in the formula to calculate the standard deviation. This program adds one more variable to the previous version. This new variable SumXSq is treated in exactly the same way as SumX.

The sqrt in the third last executable line is another function, which we will discuss more in the next chapter. In this program it calculates the square root of the expression that follows it.

This program is becoming large. We first of all show what its output looks like for a representative set of inputs.

```
Enter number (0 terminates)
17.0
Enter number
36.3
Enter number
14.7
Enter number
10.3
```

Control Statements

Enter number
15.2
Enter number
0.0
Mean is   18.6999969   Standard Dev is   10.1422434

Notice the value obtained for the mean. If you work the mean out yourself you will see that it is actually 18.7, but the finite representation of the values causes the computed value to be slightly different.

Sometimes we design and write a program and it appears to work correctly, or perhaps it is correct most of the time. One useful way to check that it is doing what you think it is doing is to determine the values of variables that are internal to the program. We can obtain these values by inserting extra print statements into the program, solely for purposes of debugging and testing. We can insert some extra print statements into Program 6.9. These extra output statements would allow us to check the values of the sums during each iteration of the loop. The output might now look like this

Enter number (0 terminates)
  SumX   17.0000000 SumXSq   289.0000000
Enter number
  SumX   53.3000031 SumXSq   1606.6901855
Enter number
  SumX   68.0000000 SumXSq   1822.7800293
Enter number
  SumX   78.2999878 SumXSq   1928.8698730
Enter number
  SumX   93.4999847 SumXSq   2159.9096680
Enter number
  Mean is   18.6999969 Standard Dev is   10.1422434

The input has been omitted in this sample output display. From now on we will not show the input typed by the user. It is good programming practice to print out the values entered, and most of our examples do this when it is appropriate. Further, if the output is going to a file rather than a screen, the input and output values are not interleaved as we have shown them thus far.

When we use an iteration construct with a predetermined number of iterations—a definite iteration—we can always be sure that the loop will eventually terminate. With the more general form of iteration—indefinite iteration—we no longer have this confidence. Therefore we always have to explicitly check that such loops will eventually terminate.

The loops in the programs we have written in this section have all depended on some sentinel value of the input (zero) to act as a flag to signal the end of the repetition. There are two things that we have to consider in deciding whether each of these loops will terminate. The first is that we must ensure that, somewhere in the body of the loop, at least one of the variables that appears in the controlling

Boolean expression has its value changed. If this doesn't happen, then the loop will never terminate. This is called an *infinite loop* and is generally easy to detect. However, if you write a program with a loop that does not produce any output but just does some lengthy calculation, it may be hard to judge when the loop should have finished. So it may be difficult to decide if your program contains an infinite loop. The solution to this kind of problem is to get into the habit of always checking, as you finish each loop, that the Boolean expression actually gets altered in a way that will eventually terminate the loop. In the programs in this section, and in other cases where we're checking in the **while** statement for a sentinel value in the input, we ensure that every path through the loop will cause a new value to be read.

The second reason why the loops in the examples in this section might never terminate is that they never receive a zero input. This, of course, is not under the programmer's control. There are several things to remember though: if the program doesn't appear to terminate, then you can immediately suspect that the correct sentinel is never being entered. Also, your program will be easier to use if you always make clear to the user (in the documentation and by an appropriate printed message) the kind of input required for termination.

Because a loop may contain any statement, it may in particular contain another loop. Such interior loops are called *nested loops*. Nested loops are important parts of programs because they are typically executed many times. In fact, a single nested loop body may sometimes constitute 80 percent of the statements executed by a program. Consider the simplest form of nested loop:

```
        Program 6.10
        program Nested
        integer I, J
        do 100 I = 1, 10000
          do 200 J = 1, 10000
            print *, I, J
200       continue
100     continue
        stop
        end
```

The output statement is executed 100,000,000 times! For each value of I, the interior loop executes 10,000 times. Since I takes on 10,000 values, the total number of executions of the output statement is the product of the number of iterations of each loop. You can see that even very small parts of programs can be very important when it comes to considering how long a particular program will take to execute.

Here is a more useful example of a program with a nested loop. This program reads in a list of pairs of numbers X and N and computes X^N for each pair. It must be the case, of course, that N is nonnegative. The program contains two loops. The outer loop reads in each pair of numbers until it encounters an X value of 0. The inner loop multiplies the value of X by itself N times to give the required

Control Statements 117

power. (Since Fortran has an explicit exponentiation operator, this is not the best way to do this calculation.)

```
*       Program 6.11
*       Raises numbers to powers. Input is
*       terminated by a zero
*
        program Powers
        real X, XPower
        integer I, N
        print *, 'Enter number and power'
        read *, X, N
        while (X .ne. 0.0) do
            XPower = 1.0
            do 100 I = 1, N
                XPower = XPower * X
100         continue
            print *, 'Result is ', XPower
            print *, 'Enter number and power'
            read *, X, N
        end while
        stop
        end
```

The last line of input to this program must have an X value of 0 because that is what is used to terminate the loop. However, some value for the variable N must also be provided on this last line or else the input statement will wait for it; the input statement expects to find X and N values in pairs.

Notice that if N is zero, the answer is still correct. The iteration statement will not execute since N will be less than 1 and so the answer will be 1.0.

6.3 Selection

In the preceding sections we have seen how to repeat parts of a program. The other important ability we need so we can write useful programs is the ability to carry out **selection**, that is to be able to select different things to do under different circumstances.

The Fortran statement that allows us to select from two alternate courses of action is the **if** statement. It has the following syntax:

```
if (logicalExpression) then
    statements1
else
    statements2
endif
```

If the logical expression evaluates to true, then statements1 are executed; if the logical expression is false, then statements2 are executed. Precisely one of the two sets of statements can be executed. By choosing an appropriate logical expression, we can use the **if** statement to select between two sets of operations depending on the data that the program is handling. The first part is often called the *then clause*, and the second part is called the *else clause*.

There is also a short form of the **if** statement for the common case where something needs to be done under certain conditions, but nothing is to be done if those conditions don't occur. In this case, we can write an **if** statement that doesn't have an **else** clause. It looks like this:

> **if** (*logicalExpression*) **then**
>   *statements*
> **endif**

To see how these statements work, let's rewrite the program that sums numbers so that it only includes positive ones in the sum. Negative numbers are ignored and a zero value indicates the end of the input. The new program looks like this:

```
*       Program 6.12
*       Sums the positive numbers in a list
*
        program SumPos
        integer Sum, Number
        Sum = 0
        print *, 'Enter number'
        read *, Number
        while (Number .ne. 0) do
           if (Number .gt. 0) then
              Sum = Sum + Number
           endif
           print *, 'Enter number'
           read *, Number
        end while
        print *, 'Sum is ', Sum
        stop
        end
```

The input statement inside the loop is not part of the **if** statement. We use indentation to indicate which parts of the program are conditional on the Boolean expression, exactly as we did for the iteration statements. There are several other conditional statements in Fortran. These are described in Appendix C.

Now we can write the improved version of the program that calculated the mean in the previous section. This version includes a test to see if the number of values summed was zero. If so, a warning message is printed and the mean is not calculated.

Control Statements   **119**

```
*       Program 6.13
*       Finds the mean of a list of numbers and
*       includes a check for a zero length list
*
        program Mean
        real Total, Number
        integer Count
        print *, 'Enter number'
        read *, Number
        Total = 0
        Count = 0
        while (Number .ne. 0) do
            Total = Total + Number
            Count = Count + 1
            print *, 'Enter number'
            read *, Number
        end while
        if (Count .ne. 0) then
            print *, 'Mean is ', Total / Count
        else
            print *, 'No input values'
        endif
        stop
        end
```

Now that we can repeat parts of programs and select different parts conditionally, we can write (at least in theory) a program to do anything that is computable. There are a number of particularly useful applications that use loops and conditionals and we will now look at some of these. They should all become part of your set of paradigms.

The first problem is that of finding the largest element in a set of numbers. We'll assume that a zero value will be used to signal the end of the input values, so the loop part of the program will be much the same as the examples that we have done. Now let's consider the business of finding the largest number among the input values. Clearly, we can't decide which number is largest until we have looked at all of the numbers (because the last one entered could be the largest). Does this mean that we have to keep all the numbers until they have all been entered? The answer is that we don't need to—it is sufficient to keep the largest value that we have seen so far in the input list. As we read each new value, we compare it to the largest value we have seen. If it is larger, then it becomes the new "largest value so far." When we have looked at all the values, the largest so far is in fact the largest in the list.

```
*       Program 6.14
*       Finds the maximum of a list of numbers
*
        program Largst
        integer Large, Number
        print *, 'Enter number'
        read *, Number
        Large = Number
        while (Number .ne. 0) do
            if (Number .gt. Large) then
                Large = Number
            endif
            print *, 'Enter number'
            read *, Number
        end while
        print *, 'Largest was ', Large
        stop
        end
```

We have to give the variable Large some initial value so that the comparison in the loop makes sense the first time it is executed. We can't simply choose some value such as zero because, for instance, if all the values entered were negative, then the largest value wouldn't even be one of the values entered. This forces a redundant comparison on the first iteration of the loop because Number and Large hold the same value at that point.

Because the statements within either the *then* or *else* clause can be any statement, one of them might be another **if** statement. This allows us to deal with situations in which there are more than two choices. For example, suppose that we have three different things that we wish to do, depending on whether the value of a variable Size is positive, negative, or zero. We could write an **if** statement containing an **if** statement to do one of the three possibilities (we call this a *nested if statement*).

```
        if (Size .gt. 0) then
            positive case
        else
            if (Size .lt. 0) then
                negative case
            else
                zero case
            endif
        endif
```

There are a number of different but equivalent ways in which we could have written this statement. The next statement shows one of them.

```
if (Size .ge. 0) then
    if (Size .gt. 0) then
        positive case
    else
        zero case
    endif
else
    negative case
endif
```

The only things that make a difference in the way we express something like this are whether one is easier to understand than the other and whether one requires less work than the other. The second version is probably a little harder to understand than the first. If we know something about how likely each of the possibilities is, then we can use that information to determine the best way to write the statement. For instance, in the first version we only have to perform one comparison to check whether the value of Size is positive, but it takes two comparisons to distinguish whether it is zero or negative. Thus, if almost all of the values were negative, this particular way of writing the statement would be inefficient, whereas writing it as

```
if (Size .lt. 0) then
    negative case
else
    if (Size .gt. 0) then
        positive case
    else
        zero case
    endif
endif
```

would be much better.

Although we have been talking about nested **if** statements, it is the case that some of our examples are essentially sequential decisions. Fortran provides another way of writing these structures, using the **elseif** construct. The next statement is an example.

```
if (X .lt. 0) then
    print *, 'negative'
elseif (X .eq. 0) then
    print *, 'zero'
else
    print *, 'positive'
endif
```

There can be any number of **elseif** clauses before the final **else** clause. (For another way of doing this, see the arithmetic **if** statement in Appendix C.)

6.4 Logical Data

Some of the control statements presented in this chapter lead to a natural consideration of computing in another domain. Just as we can compute with numeric and character data, so we can compute with logical data. This notion may be less familiar to you, but we hope you have encountered it before (in secondary school) or can pick it up from the description we give here. Careful use of logical data can be a significant enhancement to programming and can allow you to fully exploit the power of selection and repetition constructs.

Logical data represent the truth value of statements. For example, the statement "Canada is to the north of the United States" is true, and the statement "pigs have wings" is false. In a more computational setting,

3 .lt. 5

is true, and

'c' .eq. 'C'

is false.

There are facilities in Fortran to allow us to compute with logical data. Specifically, we can declare variables using the type **logical**, as follows

    logical L, M

There are logical constants, written as .true. and .false., which can be assigned to logical variables, as follows.

    L = .false.
    M = .true.

The comparison operators take numeric or character operands and produce logical results. The logical constants, together with the logical results produced by the comparison operators, can be manipulated using the logical operators. The operator .not. inverts the truth value of the operand to which it is applied. The operators .and. and .or. produce the logical conjunction and disjunction, respectively, of their two operands. They have the usual meaning from mathematical logic, where "A and B" is true precisely when both A and B are true, and "A or B" is false precisely when both A and B are false.

We can thus write statements such as the following.

L = 0 .lt. X .and. X .le. 10
M = .not. L .or. Y .eq. 0

With these facilities we can write interesting control statements in our programs. Consider the problem of monitoring output values from an instrument measuring the voltage on a line. The instrument takes a measurement every 50 milliseconds and writes the measured value and the time in a file. A program is to read the file, stopping at the end (indicated by a voltage and time of zero), when it detects a voltage above a specified threshold, or when it encounters an entire second (20 samples) of positive voltage. Here is a program to carry out this task.

```
*       Program 6.15
*       This program reads until it encounters either a voltage
*       above a threshold or a second of positive voltage
*
        program Sensor
        integer Volts, Time, NumPos, Thresh
        logical Done
        NumPos = 0
        read *, Thresh
        Done = .false.

        while (.not. Done) do
            read *, Volts, Time
            Done = (Volts .eq. 0) .and. (Time .eq. 0)
            Done = Done .or. Volts .gt. Thresh
            if (.not. Done) then
                if (Volts .gt. 0) then
                    NumPos = NumPos + 1
                    Done = NumPos .eq. 20
                else
                    NumPos = 0
                endif
            endif
        end while

        print *, Time, Volts, Thresh, NumPos
        stop
        end
```

The program makes use of a logical variable, Done, to determine when to stop reading values. It would be possible, but tricky and unattractive, to reconstruct the program without this variable.

Even though we will not make frequent use of them, it is good to have logical data values and variables in your programming toolkit to use when appropriate. You can even read and write logical values but we find it unattractive and not particularly useful, so we will not use those facilities in this book. We need to know the precedence of the **logical** operators so that we can predict how **logical** expressions will be evaluated. In Fortran the comparison operators have the highest precedence, followed by the operator .not., then .and. and then .or.. Thus the parentheses in the line

    Done = (Volts.eq.0) .and. (Time.eq.0)

are actually redundant.

6.5 Numerical Integration

By way of illustrating some of the concepts we have discussed, we turn to examples of programs that are useful to engineers. Some of these examples are illustrative and are simplified beyond reasonable limits to illustrate techniques in programming. Others are actually useful in the form in which we present them. As a first application we examine numerical integration.

There are many techniques from calculus that can be applied to find definite integrals of mathematical functions. However, many functions that arise in engineering do not lend themselves to integration using these techniques. Some are simply too complicated, while others are not sufficiently continuous. It is still important to be able to find the definite integral of such functions, and there is a branch of numerical analysis devoted to this problem. We will look at some simple, but useful, methods for calculating definite integrals.

Suppose that we have some function $f(x)$ as shown in Figure 6.1, and we want to find its integral between two points a and b. One way to approximate the area

Figure 6.1
Function to Integrate

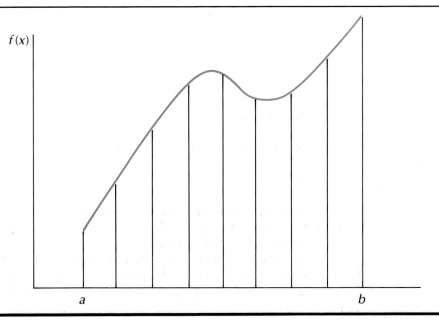

is to divide the interval between *a* and *b* into subintervals. If we draw lines from each of these points vertically to intersect the function, then we have a set of strips. The sum of the areas of the strips is the area under the curve between *a* and *b*.

Now the exact area of the strips is hard to calculate. But we can approximate the area of each one by a rectangle whose width is the size of the subinterval and whose height is the value of the function at the left endpoint of each subinterval. This is shown in Figure 6.2.

Of course, if the function is increasing, then the area of each rectangle is slightly smaller than the true area of the strip, and if the function is decreasing, then the area of the rectangle is a little larger than that of the strip. But we can hope that if the strips are sufficiently small, then the error in each strip will be very small, and if the function oscillates a bit, then some of the errors may cancel each other out.

We now know enough about Fortran to write a program that will do this kind of calculation. It's definitely a useful program to write because we will want to have a large number of rectangles to get good accuracy and it quickly becomes tedious to calculate by hand. This program is really a variation of the summation program except that we are summing areas instead of simple values. Suppose we want to calculate the area under the quadratic

$$f(x) = x^2$$

The next program performs that calculation.

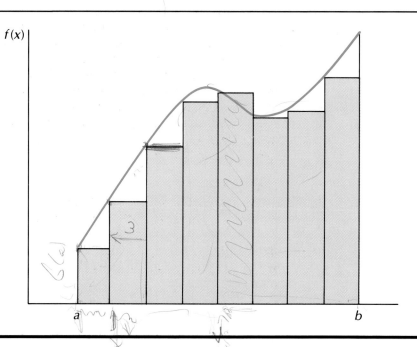

Figure 6.2 Rectangles for Integration

```
*       Program 6.16
*       Finds approximations to integrals by computing
*       the area as a sum of rectangles
*       approximating the function
*

        program Size
        real A, B, Width, Height
        integer N, I
        real Area, X
        print *, 'Enter integral limits and number of intervals'
        read *, A, B, N
        Area = 0.0
        Width = (B - A) / N
        do 100 I = 0, N - 1
            X = A + I * Width
            Height = X * X
            Area = Area + Height * Width
    100 continue
        print *, 'Integration limits were ', A, B
        print *, 'Number of intervals was ', N
        print *, 'Area approximation is ', Area
        stop
        end
```

The number of rectangles to be used should be an input to the program for two reasons: if the function is steep, then the number of rectangles needed to get a given accuracy will be larger and the user may know enough about the function to exploit this. Also, the only way for a user to have confidence in the accuracy of the answer produced by a program of this type is to run it with different numbers of intervals and get answers that are close together. For example, if we want to calculate the integral of the function between 0 and 1 and we get an answer of 0.3332 using 500 intervals and an answer of 0.3333 using 1000 intervals, then we know that our answer must be reasonably accurate. We could, of course, modify the program to carry out the computation with two different numbers of rectangles and compare the results.

The next program is a slightly different version.

```
*       Program 6.17
*       Finds approximations to integrals by
*       approximating the area by sums of rectangles
*

        program Size
        real A, B, Width, Height, Area, X
        integer N, I
```

Control Statements

```
      print *, 'Enter limits of integral and number of intervals'
      read *, A, B, N
      Area = 0.0
      Width = (B - A) / N
      X = A
      do 100 I = 0, N - 1
         Height = X * X
         Area = Area + Height * Width
         X = X + Width
100   continue
      print *, 'Limits of integration were ', A, B
      print *, 'Number of intervals was ', N
      print *, 'Approximate area is ', Area
      stop
      end
```

The only difference is in the way that X is calculated. In the first version of the program, X was calculated by adding a multiple of the width to A. In this second version, X is calculated by adding the width to the last value of X. The two methods are equivalent mathematically but, because of the limited accuracy of the representation of real numbers, they are not equivalent computationally. To see how this works, consider what happens if we take a small quantity like 0.33 and use a representation that allows us to keep two significant figures. If we add 0.33 to itself, maintaining only two significant figures, then we get the sequence 0.33, 0.66, 0.99, 1.3, 1.6, 1.9, 2.2, 2.5, 2.8, and 3.1 (ten terms). However, if we multiply 0.33 by 10 (which should give the same result mathematically), then we get 3.3, which we can in fact represent. The multiplication method is more accurate because the small errors caused by the inaccuracy of the representation do not accumulate. This kind of accumulating error is called **round-off error**.

This example shows how a programmer has to be concerned with matters that are outside the province of mathematics and in the realm of engineering. Round-off errors are a real world problem. It is always wise to consider the effects of round-off errors both when designing a program and when using a program written by someone else.

If we want to alter our integration program so that it finds the integral of some other function, then we need to alter the line computing Height to calculate the value of some other function. For example, if we want to find the integral of a particular cubic, we could substitute a line of the form:

```
      Height = X ** 3 + 2 * X + 5
```

In the next chapter we will look at other ways of doing numerical integration and more general ways of allowing different functions to be integrated by the same program.

6.6 More General Selection

When we have to do many different things depending on the value of some expression, we could express our program using many nested **if** statements. However, if the possibilities are really all at the same level, then this is misleading to anyone reading the program and makes the actual program code very difficult to understand.

We've seen one way to avoid this by using **elseif** clauses, which keep the various alternatives at the same "level." We could also encode a more general selection paradigm using the **go to** statement, but we advise against that as a general solution, and we will not do it in this book. (The **go to** statement is discussed in Appendix C.) Some languages have another construct for these cases. Unfortunately, Fortran does not; we have to make do with the nested **if** structures.

6.7 Getting Answers from Subprograms

We saw in the last chapter how we can pass values into a subprogram for the subprogram to use. We can reverse this and get values from a subprogram back to the program that invoked it.

Here is an example of a program that reads in a number, uses a subprogram to calculate its square, gets the square back from the subprogram and prints it out.

```
*       Program 6.18
*       Illustrates how to pass a value to a subprogram
*       and get a value back from a subprogram
*
        program Square
        integer Number, NumSq
        print *, 'Enter a number'
        read *, Number
        call Sqrlt(Number, NumSq)
        print *, NumSq
        stop
        end

        subroutine Sqrlt(J, JSq)
        integer J, JSq
        JSq = J * J
        return
        end
```

The subprogram has two different variables declared in its header. The first one, J, is a parameter just like those we have seen before. The second parameter, JSq, is used to get the result back. When the calling program invokes the subprogram, the values of the variables used as arguments (Number and NumSq) are effectively copied to both the variables J and JSq. This results in J having the value that was read into Number and JSq having the value of NumSq (currently undefined). In addition, when the subprogram has completed and before it returns to the invoking environment, the current values of all the parameters are effectively copied *back* to the corresponding arguments. Since the subprogram has given JSq a value, this value is copied back to become the value of NumSq. Hence the **print** statement in Square is sensible.

Another way to think of this is to imagine a wall between the invoking program and the subprogram. There are holes cut in the wall, one for each parameter. The holes are labeled by one name on the subprogram side (the name within the subprogram) and by another name on the invoking side (the argument name). Each hole is of a particular shape, depending on the type of the variable to which it belongs.

When the subprogram is invoked, the invoking program pushes a value through each of the holes. The subprogram then carries out its computation and pushes values back through all of the holes to the invoking program. Parameters which are passed back are called *variable* parameters. In Fortran all the things we've seen thus far are passed as variable parameters. The existence of variable parameters means that we can divide programs up into functional units that can be given some information, will process it, and return some answers to the invoking program. We now know how to divide up programs to reflect the structure of the solution that we discussed in Chapter 3.

Because the value of a variable parameter is copied back into the argument, if the argument is an expression, a temporary variable is created by the compiler to store its value. This variable is discarded when the subprogram terminates. If the corresponding argument for a parameter that is changed is not a variable, but rather is an expression, then the value should not be changed.

```
*       Program 6.19
*       Illustrates what happens when a constant
*       is passed to a subprogram and modified
*
*       WARNING: This program will behave differently on
*       different machines
*
        program Change
        print *, '0'
        call Modify ('0')
        print *, '0'
        stop
        end
```

```
subroutine Modify (C)
character*1 C
C = '1'
return
end
```

The output from this should be

0
0

Some older Fortran compilers actually would change the value of the constant argument, which seems an obviously wrong approach. Your compiler might not do so. Some compilers will cause this program to fail with a fatal error at execution time.

Programming Example

Problem Statement

Write a program to calculate the values of the function

$$\frac{x^4 + x^3 + x^2 + x + 1}{x^2 - 3x + 2}$$

for N evenly spaced values between 0 and 4 inclusive.

Inputs

The value N of the number of points at which to evaluate the function.

Outputs

The values of x and the function at x for each of the N values required.

Discussion

We need a loop that will allow us to evaluate the function at each point. The difference between the successive x values can be calculated by dividing the length of the interval (0 to 4) by the number of intervals N minus 1 (because the number of intervals is one less than the number of points).

Some care is needed because the denominator will be zero at two points that lie between 0 and 4. We need to ensure that we do not carry out the division of the numerator by the denominator if the denominator has the value zero. We therefore calculate the numerator and denominator separately and test for this before the division.

Program

```
*       Program 6.20
*       Evaluates a function at evenly spaced
*       points in the interval [0, 4]
*       It checks for a zero denominator in the function
*
*       A, B - limits of evaluation
*       X - present x values
*       Numer - function's numerator
*       Denom - function's denominator
*       Step - step size between successive x values
*
        program EvalFn
        real A, B
        real X, Numer, Denom, Step
        integer N
        parameter (A = 0.0, B = 4.0)
        print *, 'Enter number of sample points'
        read *, N
        print *, N, ' samples'
        print *, '          X value              Function value'
        Step = (B - A) / (N - 1)
        X = A
        while (X.le.B) do
           Numer = 1 + X + X * X + X ** 3 + X ** 4
           Denom = X * X - 3 * X + 2
           if (Denom.ne.0.0) then
              print *, X, '   ', Numer/Denom
           else
              print *,'    Value is Undefined. Zero Denominator'
           endif
           X = X + Step
        end while
        stop
        end
```

Testing

The output of this program is

```
Enter number of sample points
      5 samples
      X value                    Function value
      0.0000000                       0.5000000
      Value is Undefined. Zero Denominator
      Value is Undefined. Zero Denominator
      3.0000000                      60.5000000
      4.0000000                      56.8333282
```

You may not get exactly these values with this arrangement because that will depend on which compiler and computer you use. This will be the case for all of the floating point output that we show throughout the book.

Discussion

Notice the use of the **parameter** statement. The way in which we have written the expression to evaluate the numerator is clumsy. Come up with a nicer way to write this line by rearranging the mathematical expression. It would also be more reliable to use a proper test for checking whether the denominator is equal to zero. Strictly speaking, the function value is only undefined when the denominator is exactly zero. However, if it is almost zero, then the function value will be extremely large.

Design, Testing, and Debugging

- Always make a mental check when using a **while** statement to ensure that the loop will terminate. If there are circumstances under which the loop may not terminate, guard against them using an **if** or warn the user of the circumstances under which the program will not work correctly.
- Always prompt for input and check that the input values provided are reasonable. For example, if your program will only work if the user enters a number between 1 and 100, check that the number actually is between 1 and 100.
- Think about the order in which the comparisons are made whenever you use nested **if** statements.
- Use **logical** expressions when they are appropriate. Don't be afraid of them.
- Be aware of which values within a subprogram can be changed and which cannot. Never surprise a user of a subprogram by altering values that should not be altered.

Style and Presentation

- Always indent the statement(s) within a control structure to show that they are conditional. Pick a standard indentation spacing and always use it. It should be possible to look at the left margin of your program and discern most of its logical structure. Statements that are within nested structures are indented by a multiple indentation step.
- Keep the labels of **continue** statements such that they increase through the program. This makes it easier for someone reading it to quickly find the matching **continue** for a **do** statement.
- Plan your comparisons so that they make sense in the context of the problem being solved. Test $x \leq 10$ rather than $10 \geq x$.

Fortran Statement Summary

Do Statement

This statement allows definite iteration. The *step* is optional—if omitted it defaults to 1.

```
       do label id=start, stop, step
           statements
label  continue
```

```
       do 100 I = 1, N, 2
           read *, A, B
           print *, A, B
100    continue
```

While Statement

This statement allows indefinite iteration. The loop body is executed as long as the *logicalExpression* is true.

```
       while (logicalExpression) do
           statements
       end while
```

```
       while (I.lt.50) do
           print *, I * I
       end while  I = I + 1
```

Chapter 6

Logical Expression

A logical expression is one whose value is either true or false. It is usually constructed of values and comparison operators.

> *variable   comparisonOperator   variable*

where

> *comparisonOperator* is one of
> .eq.
> .ne.
> .lt.
> .le.
> .gt.
> .ge.

If Statement

This statement allows selection of different statements to be executed. The **else** and **elseif** clauses are optional.

> **if** (*logicalExpression*) **then**
>   statements
> **elseif** (*logicalExpression*) **then**
>   statements
> **else**
>   statements
> **endif**
>
> **if** (X.lt.0) **then**
>   **print** *, 'Negative'
> **elseif** (X.eq.0) **then**
>   **print** *, 'Zero'
> **else**
>   **print** *, 'Positive'
> **endif**

Logical Declaration

This statement declares variables of type **logical**

> **logical** *variableList*
>
> **logical** Done, OK

Control Statements

Subroutine Header with Parameters

Subroutines can have parameters passed to them and can pass values back. When this is done, the parameters are listed following the subroutine name. As the **return** statement is executed, values for each of the parameters are copied back to the invoking environment.

> **subroutine** *id* (*parameterList*)
> *declarations*
> *statements*
> **return**
> **end**
>
> **subroutine** Prntit (X, Y)
> real X, Y
> X = Y * Y * Y
> return
> end

Chapter Summary

- Definite iteration is implemented using the **do** loop. The **do** loop allows an iteration to occur for a number of times that must be specified when the loop is first entered.
- The loop index and the variables that are used to determine the loop limits may not be altered within the body of the loop. The value of the loop index may be defined after the loop terminates, but programs should never assume that it will be.
- If the lower limit of a **do** loop exceeds the upper limit when the loop is first encountered, then the body of the loop is not executed.
- A **logical** expression is one whose value is either true or false. Logical expressions are usually generated by comparing the values of two expressions using a comparison operator.
- Indefinite iteration is implemented using the **while** loop. The **while** loop does not exist in Standard Fortran77. When the loop is first encountered, the expression in the **while** statement is evaluated. If it is false, the whole body of the loop is skipped. If it is true, the body of the loop is executed and the expression in the **while** statement is re-evaluated. Some value in the logical expression must be altered within the body of the loop or it will never terminate.
- All indefinite iterations contain four parts: an initialization part, a loop condition, a repeated part, and a preparation for the next iteration.
- Two floating values should never be compared using the .eq. operator because of the problems of approximate representation. Two values that ought to be

- mathematically equal, but that have been derived using different computations, may not actually be represented by the same floating point number. Tests for equality are done by determining if two values differ only by some specified tolerance. It is important to remember that the absolute value of the difference must be taken.
- Sequences of input are often conveniently terminated by a special value which can be distinguished from the legitimate input. This value is called a sentinel, and is used to signal the end of the input sequence.
- Selection of different statements to execute, based on some condition, is done using the **if** statement. It contains two subordinate parts, exactly one of which is executed depending on the value of a logical expression.
- Fortran77 allows us to compute with the **logical** values .true. and .false.. You can use **logical** variables as **logical** expressions; **logical** expressions can also be built up from **logical** operators and comparisons. The values of **logical** expressions can be assigned to **logical** variables.
- Round-off error is that error caused in numerical calculations by the cumulative effects of the approximation of real values by floating point ones. Algorithms can often be written carefully to reduce the effects of round-off on the accuracy of their results.
- Values can be passed back from subprograms using the same mechanism used to pass values to subprograms. In fact, almost all values passed to subprograms are also passed back.
- Integrals can be estimated using straightforward numerical techniques.

Define These Concepts and Terms

Definite iteration   Sentinel
Index variable   Selection
Indefinite iteration   Round-off error
Comparison operators

Exercises

1. How many times is the output statement executed in each of the following programs?

a.
```
      program P
      integer I, J
      do 100 I = 1, 6
         J = I
         while (J/2 .gt. 1) do
            print *, I, J
            J = J - 1
         end while
100   continue
      stop
      end
```

b.
```
      program Q
      integer I, J
      do 100 I = 20, 1, -1
         if ((I/2)*2 .eq. I) then
            print *, I
         endif
100   continue
      stop
      end
```

2. By modifying program AddUp, write a program to find twice the sum of a list of numbers read in.
3. Write a program to find the second-largest number in a list. You may assume there is a zero value at the end of the list.
4. What is the value of the variable Sum after the loop in the following program?

```
      program S
      integer I, Sum
      Sum = 0
      do 100 I = 1, 10
         Sum = Sum + I * I - 1
100   continue
      print *, Sum
      stop
      end
```

5. The Fibonnaci sequence is a sequence beginning

   1, 1, 2, 3, 5, 8, 13, ...

   in which each element, after the first two, is the sum of the previous two elements. Write a program that will read a number *n* and calculate the *n*th

Fibonnaci number. You may have to watch for overflow (that is, computing a number too big to be represented) because the Fibonnaci numbers become large very quickly.

6. Write a program to find the mean of a list of positive real numbers, omitting zero values from the calculation of the mean. You may assume that the list is terminated by a negative value.

7. Consider the problem of calculating powers of a value using only multiplications. Suppose we are given two numbers X and N and we want to calculate X^N. Doing this in the straightforward way requires doing $N - 1$ multiplications, which can be very expensive.

However, for values of N which are powers of 2, we can do better. For example, suppose we want to calculate X^8. We can do this by calculating X^2 (one multiplication), X^4 (one more multiplication, multiplying X^2 by itself) and finally multiplying X^4 by itself (1 more multiplication). Thus we can calculate X^8 in a total of three multiplications instead of 7.

In general, if N is a power of 2, then we can calculate X^N in $\log_2 N$ multiplications. For values of N which are not powers of 2, we can still use combinations to calculate the value we want. For example, we can calculate X^7 as $N^4 \times N^2 \times N$ in a total of four multiplications.

Write a function that, given X and N, will calculate X^N in the smallest possible number of multiplications.

8. Write a program to demonstrate that

    .not. (A .and. B)

    is equivalent to

    .not. A .or. .not. B

    where A and B are of type **logical**, by evaluating all possible combinations.

9. Rewrite Programs 6.1 through 6.5 to use indefinite iteration.

10. Show a general form for rewriting a definite iteration as an indefinite one. Be careful of the difference between a positive and negative step size.

11. Write a program to determine if a number is a perfect square. Check all numbers up to half its size to see if any factor squared gives the number. How can you easily improve this algorithm?

12. Write a program to implement the greatest common divisor algorithm of Exercise 13 of Chapter 1.

13. Write a program to read an integer and print the indicated number of copies of this pattern.

```
*
 *
  *
   *
    *
   *
  *
 *
*
```

14. Write a program to read a single digit and then print the digit 10 lines high. Use a subprogram for each different digit.
15. Modify Program 6.15 so that it reads two threshold values, a low value and a high value, and checks for either a value that is too high or too low, or a second of positive voltage.

Functions

7.1 Motivation for Functions
7.2 Constructing a Function
7.3 Invoking a Function
7.4 Intrinsic Functions
7.5 Using Functions
7.6 Tracing Programs

7.1 Motivation for Functions

We have shown how to use Fortran to break a problem solution into pieces; to implement sequencing, selection, repetition, and abstraction; and to do simple input and output. In this chapter we show how to develop subprograms that produce values rather than carry out sequences of operations.

We saw in the last chapter that there are situations where we want to manipulate not just pieces of data but also functions. Fortran provides a language mechanism that allows us to work with mathematical functions. For instance, if we have a function

$$f(x) = x^2$$

we often think of it as describing a curve in the plane. We can also think of it in a more active way, as a "black box" that, whenever it is given a value, produces a new value as output. For example, if we give the quadratic function above 2, we get 4, and if we give it 7, then we get back 49. This active picture of a function fits naturally with a picture of an executing program. We have already seen how a subprogram can be given a value and execute its own statements to compute a result that is then passed back to the invoking program.

Fortran provides a special kind of subprogram, called a **function**, that does this kind of calculation in a particularly useful way. Whenever it is invoked and given data values, it computes a new value, the result of the function.

7.2 Constructing a Function

A function looks very similar to a subroutine both in the way it is laid out and where it is placed in a program. A function begins with a header like this:

>   **integer function** F(X)
>   **integer** X

The **function header** is identified by a keyword identifying the type of the function (in this case **integer**), followed by the keyword **function**, followed by the name of the function (in this case F). This is followed by a parenthesized list of the parameters of the function. These are declared with exactly the same format and follow the same rules as the parameters for other subprograms.

A function always computes a single value as its result. The type of this value is indicated in the header line.

The body of the function is like that of a normal subprogram, and in particular it contains the declaration section at the beginning. This is followed by the executable statements making up the function. The last executable statement should be **return**. An **end** comes after the executable statements. The only other difference between a normal subprogram and a function is that, somewhere inside the function body, a value must be assigned to the name of the function exactly as if it were a variable. It is legal to assign a value to the function name more than once, in which case the most recent value is the one returned to the invoking environment. We will occasionally use this in later programs.

This next Fortran function implements the mathematical squaring function.

```
integer function F(X)
integer X
F = X * X
return
end
```

This function has no internal (local) variables; X is a parameter. It has only a single assignment statement that takes the value of X and squares it, assigning the resulting value to the name of the function.

Although it is legal, it is considered poor style for a value other than the principal result to be returned from a function.

7.3 Invoking a Function

We have already seen how to invoke a normal subprogram by writing the word **call** followed by the subprogram name, together with its actual arguments. Invoking a function is a little different. Because a function always produces a value of a particular type, a function invocation behaves like an expression of that type. It wouldn't make sense just to write the function invocation like a normal subroutine invocation because that doesn't specify what to do with the value the function computes. Instead, a function invocation appears anywhere that an expression can appear. A function name cannot appear in an expression within its own body, even if the function name has previously been given a value. Thus, a function name may appear on the left-hand side of an assignment statement inside its body, but may not appear on the right-hand side.

In the following program, we illustrate how to invoke the function we wrote earlier. In this case, the function name appears as an expression in an output list.

```
*       Program 7.1
*       Shows how to invoke functions
*
        program UseFun
        integer I, F
        do 100 I = 1, 10
            print *, F(I)
100     continue
        stop
        end
```

```
*
*       Calculates the square of its input
*

        integer function F(X)
        integer X
        F = X * X
        return
        end
```

The main part of this program uses a simple loop, iterating ten times. The variable I takes on the successive values 1, 2, ..., 10. When the output statement is to be executed, the function F has to be invoked to get the value to be printed. At this point, the output statement waits while the function is invoked, with the current value of the variable I being passed to the function as the parameter X. The function executes, calculating the square of this value. This computed value then appears in the expression in the main program as if it were the value of a variable. This program prints out the values 1, 4, 9, 16, ..., 100, each on a separate line.

A function invocation is caused by the need for its value in an expression that is being computed. When it is invoked, the calculation of the expression is temporarily halted while the statements in the function are executed. Whatever value was assigned to the function name inside the function is returned to the invoking program, where the calculation of the expression can then continue.

The next program invokes the function slightly differently.

```
*       Program 7.2
*       Illustrates function use
*

        program UseFun
        integer I, ISq, F
        do 100 I = 1, 10
           ISq = F(I)
           print *, ISq
  100   continue
        stop
        end

        integer function F(X)
        integer X
        F = X * X
        return
        end
```

In this version, the function is invoked when the expression on the right-hand side of the assignment statement is to be calculated. This causes the function to execute and the value it produces to be assigned to the variable ISq. The value of this variable is then printed in the output statement. This version is preferable when the value of the function is to be used several times, since the function need only be invoked once.

7.4 Intrinsic Functions

Because there are many calculations that are done frequently by many programs, all Fortran systems come with a library of built-in or **intrinsic functions**. If you invoke these functions in your program and you don't write a function of your own with the same name, then the compiler understands that you want to invoke one of these built-in functions and includes it with your program. These functions are thus a library of useful pieces of code, and you can use them in programs without having to write them each time.

Some of these functions provide standard numerical tools. For example, most language systems provide functions that calculate the trigonometric functions (sin, cos, arctan) and the exponential functions (exp and log). (Appendix D shows the intrinsic functions available in Fortran.) These are sufficient to build many other functions. For example, all of the trigonometric functions can be built using sin, cos, and arctan. Specifically, the tan function can be built as

```
real function Tan(X)
real X
Tan = sin(X) / cos(X)
return
end
```

Because of the way the sin and cos functions are calculated, this particular method of calculating the tan function is not a good one to use in practice.

We now describe some of the most useful intrinsic functions that Fortran provides. We have already seen some of them. The abs function returns the absolute value of its argument. The sqrt function calculates the square root of its argument.

Several functions are provided to allow explicit conversions between different types. For example, the assignment statement

```
A = I / J
```

will carry out an integer division and assign the result, converted to the corresponding floating point value to A. We assume that A is declared to be real and I and J are declared integer. If we want to force the division to be a real division then we can use the function real to force the conversion of either I or J to type real, like this

```
A = real(I) / J
```

An analogous function allows real values to be converted to integer values by rounding to the nearest integer, rather than truncating. For example, after this expression

$$K = 32.5 / 1.5$$

K has the value 21, whereas after this assignment

$$K = int(32.5 / 1.5)$$

K has the value 22.

Another useful function is the mod function. It takes two arguments and returns the remainder on dividing the first argument by the second.

Fortran77 intrinsic functions are *generic*, which means that the compiler can infer the type of the function from the type of the arguments. For example, we can write abs(10) and abs(10.0) and the compiler will infer that the first invocation refers to the function that takes an integer and returns an integer, while the second refers to the function that takes a real value and returns a real value. This is, of course, similar to ordinary algebra, where you expect the result of the absolute value operator to be of the same type as its argument, and similar to the generic use of primitive numeric operators like "+" in Fortran, which we expect to produce an integer result for integer operands and a real result for real operands. Earlier versions of Fortran did not have generic intrinsic functions. If the programmer wanted to take the absolute value of a real operand, the function abs was used, but for an integer operand, the function iabs was used. You may encounter programs written in earlier versions of Fortran that use these older versions of the intrinsic functions. In fact, Fortran77 allows you to use them, but we will not do so in this book.

Most systems will have subprograms other than the intrinsic functions available for you to use. These are stored in *program libraries* and are simply subprograms someone else has written and made available to you. We cannot tell you what subprograms will be available since it will differ from one system to the next. You should consult your system documentation to determine what is provided on your system.

7.5 Using Functions

In this section we will develop a more interesting example using functions. The roots of the quadratic equation

$$ax^2 + bx + c$$

are given by the equation

$$x = \frac{-b \pm \sqrt{b^2 - 4ac}}{2a}$$

We will write a program that will calculate the roots of an equation, given the values of the coefficients *a*, *b*, and *c*. Of course, there may not be any real roots and our program will have to deal with this contingency.

Let us begin by writing a function that calculates the discriminant

$$b^2 - 4ac$$

given the values of *a*, *b*, and *c*. Such a function looks like this:

> **real function** Disc(A, B, C)
> **real** A, B, C
> Disc = B * B - 4 * A * C
> **return**
> **end**

Now we need two functions that calculate the larger and smaller roots, respectively, assuming that the discriminant is nonnegative.

> **real function** Pos(A, B, Discrm)
> **real** A, B, Discrm
> Pos = (-B + sqrt(Discrm)) / (2 * A)
> **return**
> **end**

> **real function** Neg(A, B, Discrm)
> **real** A, B, Discrm
> Neg = (-B - sqrt(Discrm)) / (2 * A)
> **return**
> **end**

The sqrt function is an intrinsic function that calculates the square root of its argument. Now we can put together the whole program.

> *   Program 7.3
> *   Finds and prints the roots of a quadratic
> *
>
> **program** Quad
> **real** A, B, C, D, Disc, Pos, Neg
> **print** *, 'Enter coefficients (descending order)'
> **read** *, A, B, C
> D = Disc(A, B, C)

Functions

147

```
        if (D .ge. 0.0) then
            print *, ' First root is ', Pos (A, B, D)
            print *, ' Second root is ', Neg (A, B, D)
        else
            print *, ' No real roots'
        endif
        stop
        end
*
*       Calculates discriminant
*
        real function Disc(A, B, C)
        real A, B, C
        Disc = B * B - 4 * A * C
        return
        end
*
*       Calculates larger root
*
        real function Pos(A, B, Discrm)
        real A, B, Discrm
        Pos = (-B + sqrt(Discrm)) / (2 * A)
        return
        end
*
*       Calculates smaller root
*
        real function Neg(A, B, Discrm)
        real A, B, Discrm
        Neg = (-B - sqrt(Discrm)) / (2 * A)
        return
        end
```

Notice how we have used a function invocation inside an output statement. This is legal because an expression can appear there and a function invocation can appear wherever an expression can.

7.6 Tracing Programs

Now that we have seen enough of Fortran to write more complicated programs, it's important to have a disciplined approach to understanding what happens when a program executes. You'll want to do this when you encounter programs you have not seen before, or when you try to fix a program of your own that is not working properly.

Ideally we would like to have an appreciation of what a program does in the general case. This would allow us to generate the kinds of formal statements about its behavior that we mentioned in Chapter 1. But a common way to come to such a general understanding is by looking at some specific cases. That is what program testing tries to do. We examine specific cases to increase our understanding about the program's general behavior.

One way to understand a program, then, is to consider what it does in specific cases. Using a paper and pencil, we can mimic the behavior of the machine, step by step, following through the program and watching the values of variables as the program is executed. This is called **tracing** a program's execution. Our general approach will be to keep a table of values of program variables. We'll organize this table with a section for the main program, which is always present, and a section for each active subprogram, which is created when the subprogram is invoked and removed when the subprogram terminates. We'll illustrate this, using the quadratic formula program in the previous section.

When the program begins to execute, none of the variables A, B, C, or D is defined. So our table for the main program looks like this:

Main:
   A: undefined
   B: undefined
   C: undefined
   D: undefined

The first statement executed prints a message and therefore does not affect the values of any of the variables. The next statement is an input statement and thus defines the values of A, B, and C. Let us suppose that the user entered the values 1, −3, and −4. Our new table is

Main:
   A: 1
   B: -3
   C: -4
   D: undefined

The next statement is an assignment to D. Before the assignment can be made, the function Disc must be evaluated. We expand our table to include the variables of the function Disc.

Main: Disc:
   A: 1   A: 1
   B: -3   B: -3
   C: -4   C: -4
   D: undefined

The execution of the function Disc causes the evaluation of an expression that assigns the value 25 to Disc. Thus we have

Main: Disc: 25
  A: 1   A: 1
  B: -3   B: -3
  C: -4   C: -4
  D: undefined

Now the value returned by the function is assigned to D so that we have

Main:
  A: 1
  B: -3
  C: -4
  D: 25

Since this value is greater than or equal to zero, we execute the then clause of the **if** statement. We start to print an output line, at which point we need to evaluate the function Pos. When Pos is invoked, our table becomes

Main: Pos:
  A: 1   A: 1
  B: -3   B: -3
  C: -4   Discrim: 25
  D: 25

Evaluating the expression in Pos gives a result of 4, which is passed back to the main program and printed. We now print the second line, which necessitates invoking the function Neg. Our table becomes

Main: Neg:
  A: 1   A: 1
  B: -3   B: -3
  C: -4   Discrim: 25
  D: 25

The expression in Neg is evaluated, giving a result of −1, and this value is passed back to the main program and printed.

This simple approach to following what a program is doing can be one of the most powerful debugging tools. It is especially effective when the program seems to be doing something that you did not expect. Rather than try and puzzle it out, a simple trace will usually shed some light on what is happening. A word of warning: it's crucial that you do exactly what the computer would do when you trace through your program. You must understand what each statement does and be careful not to use common sense.

Programming Example

Problem Statement

Write a program to tabulate the ranges achieved by a shell fired with constant muzzle velocity as the initial elevation (that is, angle at which it is fired) changes.

Inputs

Muzzle velocity and range of angles of elevation (in radians).

Outputs

A table of elevations and ranges at 0.05 radian steps.

Discussion

We will use a function that will calculate the range, given an elevation and a muzzle velocity. The equations of motion of a projectile in a gravitational field can be used to show that the range of a projectile is given by

$$Range = \frac{2 v^2 \sin \theta \cos \theta}{g}$$

Program

```
*       Program 7.4
*       Calculates the ranges reached by a shell
*       fired at different angles of elevation
*
*       A, B - limits and angles
*       Angle - current angle of elevation
*       Vel - muzzle velocity
*
        program Shell
        real A, B, Angle, Vel
        real Range
        print *, 'Enter start and stop angles and muzzle velocity'
        read *, A, B, Vel
        print *, A, B, Vel
        print *
        print *, '                    Angle           Range'
```

Functions

```
         Angle = A
         while (Angle.le.B) do
             print *, Angle, Range(Vel, Angle)
             Angle = Angle + 0.05
         end while
         stop
         end
*
*        Computes the range given the elevation
*        and muzzle velocity
*
*         Muzzle - muzzle velocity
*         Elev - angle of elevation
*         G - accelaration due to gravity
*
         real function Range (Muzzle, Elev)
         real Muzzle, Elev
         parameter(G = 9.8)
         Range = 2 * Muzzle * Muzzle * sin(Elev) * cos(Elev) / G
         return
         end
```

Testing

Enter start and stop angles and muzzle velocity
    0.5000000   1.3999996   30000.0000000

Angle	Range
0.5000000	7.7277904E+07
0.5500000	8.1845520E+07
0.5999999	8.5595360E+07
0.6499999	8.8490000E+07
0.6999998	9.0500432E+07
0.7499998	9.1606608E+07
0.7999997	9.1797536E+07
0.8499997	9.1071232E+07
0.8999996	8.9434976E+07
0.9499996	8.6905088E+07
0.9999995	8.3506896E+07
1.0499992	7.9274368E+07
1.0999985	7.4249792E+07
1.1499977	6.8483392E+07
1.1999969	6.2032704E+07

```
     1.2499962              5.4962256E+07
     1.2999954              4.7342640E+07
     1.3499947              3.9250016E+07
     1.3999939              3.0765232E+07
```

Discussion

Notice that the finite representation causes the printed numbers to differ slightly from their expected mathematical values. As expected, the maximum range is achieved when the angle of elevation is at $\pi/4$ radians.

Design, Testing, and Debugging

- Check that the type of a function in the invoking program matches the type declared in the function's header.
- Always check that the types of parameters in the function heading match those of the invocation for all functions you use.
- Remember that there is an overhead associated with invoking a function. If the value of the function is needed in several places, it is more efficient to invoke it once and store the value it produces in a variable that can subsequently be used again.

Style and Presentation

- Function names are just as important as subprogram names. They should also be chosen to reflect their meaning.
- Never use the name of an intrinsic function as the name of a function you define yourself. Although it is legal and will work, it can be very misleading to someone reading your program.

Fortran Statement Summary

Function Header

A function header defines a function and its type and also indicates the parameters that the function takes.

```
type function id (parameters)
declarations
statements
return
end

real function Divide (X, Y)
real X, Y
Divide = X / Y
return
end
```

Chapter Summary

- A function is a form of subprogram that produces a special, distinguished value. A function is invoked by writing its name in an expression. The requirement for its value causes it to receive parameters, execute its statements, and return.
- Because it produces a value, each function has a type. Within its executable statements, the function name must be assigned a value. It may be assigned a value multiple times. A function's name may not appear on the right-hand side of an assignment statement within its own body.
- Intrinsic functions are provided as part of the language system. They consist of functions that are commonly used. The compiler determines which intrinsic functions a program invokes and arranges for them to be accessible from a library.
- Tracing the execution of a program is a powerful way to find subtle errors. The values of each variable are recorded as each statement of the program is nominally executed.

Define These Concepts and Terms

Function header   Tracing
Intrinsic function

Exercises

1. Write functions Plus and Minus that each have two integer arguments and return the sum and difference of their arguments, respectively.
2. Using the intrinsic functions, write functions to calculate
   a. $e^{-x^2/2}$
   b. $\sin^2 x$
3. Look up a list of the intrinsic functions your compiler provides. Compare it with the list in Appendix D.
4. Write programs to verify each of the following trigonometric identities by evaluating both sides and checking to see if they are equal (remember that they may not be exactly equal because of the finite accuracy of the representation). Use a range of values between 0 and 2π.
   a. $\sin^2 x + \cos^2 x = 1$
   b. $\sin 2\theta = 2 \sin \theta \cos \theta$
5. Write a **logical** function that determines whether or not a given integer is prime.
6. Rewrite the numerical integration program in Chapter 6 as a function. Use it to find the area between

   $$f(x) = x^2 + 3$$

   and

   $$g(x) = x^2$$

   in the interval [0, 3].
7. Write a program to simulate a simple calculator. Input is of the form

   \<integer\> \<integer\> \<operator\>

   where \<operator\> is one of +, −, * or /. Use functions as appropriate.
8. For what parameter values does this factorial function return the appropriate result on your machine? Consider both mathematics and machine representation limits.

   integer function Fact(N)
   integer N
   integer P, I
   P = 1

```
        do 100 I = 2, N
            P = P * I
100     continue
        Fact = P
        return
        end
```

9. Write a function with two integer parameters that returns a **logical** value indicating whether the first parameter is the second parameter raised to some integer power. Thus the result for the pair of values 32, 2 is *true* but for the pair of values 160, 10 the result is *false*.
10. For which integer parameters is P true?

```
logical function P(X)
integer X
P = X / 2 * 2 .eq. X
return
end
```

11. Modify the programming example to find the elevation giving the maximum range. Add checks to prevent elevations below zero and greater than $\pi/2$.
12. Modify the programming example into a game in which the user enters an elevation and the program responds with the distance under or over a target a fixed distance away.
13. Modify Program 7.3 by removing the variable D.
14. Trace Program 6.20 when the input is the value 5.
15. What happens when you run the following program using your compiler?

```
program Weird
print *, F(2)
stop
end

integer function F(X)
integer X
print *, X
F = X
return
end
```

Engineering Problem 1

Electron Emission

Many materials emit electrons when they are bombarded by photons. This is called the photoelectric effect. As photons strike a material, electrons are emitted with a kinetic energy that depends on the frequency of the incident radiation, that is on the energy of the photons. It was this effect that started Albert Einstein thinking about what later became quantum mechanics. In fact, he published a paper on the photoelectric effect in 1905. The surprising observation about this effect is that the energy of the electrons depends on the frequency of the incident radiation rather than the number of photons. This started Einstein thinking about energy as something that was quantized, a key idea in the development of quantum mechanics.

The energy requirements of the photoelectric process are described by the photoelectric equation

$$e\phi + 0.5\, mu^2 = hf$$

where e is the charge on an electron (1.602×10^{-19} coulombs), ϕ is called the work function in electron-volts and depends on the material, m is the mass of an electron (9.11×10^{-31} kg), u is the initial velocity of electrons, h is Planck's constant (6.63×10^{-34} joule-seconds), and f is the frequency of the incident photons. The first term on the left-hand side is the energy cost of emitting electrons. This cost depends on the material being illuminated. The second term is the kinetic energy of the electrons that are produced. The right-hand side describes the energy supplied by the photons. Remember that the product of the frequency and wavelength of electromagnetic radiation equals the speed of light (3×10^8 ms^{-1}).

Programming Problems

1. Write a program to read in work functions for various materials and calculate the longest wavelength for which they will emit electrons. At this frequency the kinetic energy of the emitted electrons is zero.
2. For each material, calculate the kinetic energy of the emitted electrons if the material is illuminated with light of wavelength 6000 angstroms (an angstrom is 10^{-10} meters). Use the following work functions:
   - Copper 4.1 eV
   - Cesium 1.71 eV
   - Tungsten 4.54 eV
   - Oxide-coated nickel 1.0 eV

Engineering Problem 2

Heat Transfer through Windows

One of the major points of heat transfer in buildings is through the glass panes of windows. In this problem we investigate this form of heat transfer.

There are three ways in which heat can be lost or gained by an object: conduction, convection, and radiation. We will discount radiation as a major source of heat transfer, leaving conduction and convection to be considered.

It is clear that not all of the heat transfer through a window is by conduction. Consider a window of thickness 3 mm, where the temperature inside the building is $+20°C$ and the temperature outside is $+35°C$. If all of the heat flow occurred through conduction, then we would expect the outer surface of the glass to be at $+35°C$ and the inner surface at $+20°C$. This is not the case — the inner surface will be warmer than the room and the outer surface will be cooler than the outside air, but not to those extremes. The remaining heat loss occurs through convection at both the inner and outer surfaces of the glass.

The amount of heat transfer by conduction by a solid can be calculated using the heat transfer equation

$$H = \frac{k A (t_2 - t_1)}{L}$$

where H is the amount of heat transferred, k is a heat transfer constant which depends on the material, A is the area, L is the thickness of the block of material, and t_2 and t_1 are the temperatures at the surfaces of the solid. For glass, the conduction constant k has the value 0.8372 joules/sec m °C.

Convection can be described by the equation

$$H = h A (t_2 - t_1)$$

where h is a convection constant and t_1 and t_2 are the temperatures of the plate and the air. In general, h depends on the orientation of the material since convection depends on movement of air. For a vertical plate of glass, the value of h is

$$1.775(t_2 - t_1)^{1/4} \text{ joules/sec m}^2 \text{ °C}$$

To calculate the heat transferred per unit area of glass, we can assume that the temperatures of the inside and outside of the glass are equal and such that the temperature difference between the inside air and the inside of the glass is the same as the difference between the outside air and the outside of the glass. Using the preceding values, this would mean assuming that the inside and outside of the glass are at 27.5°C. Using the convection equation, the convection coefficient is

$$h = 2.9374$$

and hence the heat transferred per unit area is

$$\frac{H}{A} = 22.031 \text{ joules / m}^2$$

Now the glass is not really at a uniform temperature. If it were, then no heat would flow across it. We calculate the change in temperature across the glass that would produce a rate of heat flow equivalent to that required by the convection. For this we use the conduction equation. This gives

$$\Delta t = 0.0789 °C$$

where Δt is the temperature difference between the surfaces of the glass. This heat difference is enough to cause the required heat flow through the glass.

Now we can use this to alter our assumption about the initial temperatures. We now use the approximation that the temperature of the outside surface is 27.5395°C and the temperature of the inner surface 27.4705°C.

This implies that the temperature difference between the air and the surface of the window is smaller than it was. We can repeat the original calculation to get a new convection coefficient and then new approximations for the glass surface temperatures. We continue this repetition until it converges to an accurate solution. Remember that Δt is the temperature difference across the glass. Therefore the temperatures of the glass surfaces are always symmetric about the starting temperature of 27.5°.

Programming Problems

1. Write a program that will read internal and external temperatures and produce a good approximation to the heat flow across the glass. Assume that the thickness of the glass is also a program input.
2. Modify the program to calculate the temperature at points through the glass. Sketch, by hand, a graph showing it.

Mathematics for Engineering

8.1 More Numerical Integration
8.2 Simpson's Rule for Integration
8.3 Evaluating a Series
8.4 Solving Differential Equations
8.5 Passing Functions as Arguments
8.6 Root Finding

8.1 More Numerical Integration

We saw at the end of Chapter 6 how we could approximate definite integrals by summing the areas of geometric objects. We used rectangles as the objects whose areas approximated the area under the function. This technique for integration is called the **rectangle rule**.

Now that we know how to use functions, we can make our rectangle integration more general and easier to use. Instead of calculating the height of the function using an expression inside the summation loop, we will invoke a Fortran function. Here is the revised version of the program:

```
*       Program 8.1
*       Calculates approximations to integrals using rectangles
*
        program Size
        real A, B, Width, Height, Area, X, F
        integer N, I
        print *, 'Enter limits of integration and number of intervals'
        read *, A, B, N
        Area = 0.0
        Width = (B - A) / N
        do 100 I = 0, N-1
            X = A + I * Width
            Height = F(X)
            Area = Area + Height * Width
100     continue
        print *, 'Limits of integration ', A, B
        print *, 'Number of intervals ', N
        print *, 'Approximate area is', Area
        stop
        end

        real function F(X)
        real X
        F = X * X
        return
        end
```

Remember that, according to the rules about variable scope, the two variables named X are separate variables. However, in this case, the value of X in the main program is passed to the function, where the corresponding formal parameter is also called X.

The rectangle integration program uses the height of the function at the left-hand end of each subinterval to determine the height of the rectangle. This leads to an error in approximating the area of the corresponding strip unless the function is absolutely flat in the subinterval. There is no particular reason to suppose that using the left-hand end of the subinterval is better than using the value of the function at the right-hand end. We would have obtained a slightly different answer as the approximation to the integral for most functions, but there's no way in general to tell which would be more accurate.

Here is a version of the same program that uses the right-hand end of the subinterval to calculate the function height. You can see that it is similar to the first version.

```
*       Program 8.2
*       Integrates using rectangles
*
        program Size2
        real A, B, Width, Height, Area, X, F
        integer N, I
        print *, 'Enter limits of integration and number of intervals'
        read *, A, B, N
        Area = 0.
        Width = (B - A) / N
        do 100 I = 1, N
            X = A + I * Width
            Height = F(X)
            Area = Area + Height * Width
100     continue
        print *, 'Limits of integration ',A, B
        print *, 'Number of intervals ', N
        print *, 'Approximate area is ', Area
        stop
        end

        real function F(X)
        real X
        F = X * X
        return
        end
```

Now we might expect that we would get a more accurate approximation to the definite integral if we took the average of the answers produced by these two versions, because when one rectangle overestimates the area, the corresponding rectangle in the other version will tend to underestimate it. This turns out to be true, but we don't actually need to do both calculations to get the same result. Consider the rectangles in Figure 8.1. The area given by averaging the areas of the two rectangles is exactly the same as the area of the trapezoid with its upper edge connecting the function values at the left and right ends of the subinterval. So, rather than calculating the area of the rectangles at all, it is enough to calculate the area of trapezoids based on each subinterval. It is not surprising that the approximation given by the trapezoids is better than either of the rectangle methods because the top edge of the trapezoid follows the function more closely.

We are really using the function made up of straight lines in each of the subintervals to approximate the function we are interested in. This method of numerical integration is called the **trapezoid rule**. We want to calculate the sum of the areas of the trapezoids between the points a and b. The summation is

$$\sum_{i=0}^{n-1} \frac{f(a + iw) + f(a + (i + 1)w)}{2} w$$

The width of the strips is a constant so that we can take it outside the summation and multiply it in at the end. This means a considerable computational saving because we may have a thousand strips. Doing the multiplication after the summation means that we do one multiplication instead of many. In addition, by recognizing that each intermediate point is used twice and manipulating the summation, we can write

$$w \left[\frac{f(a) + f(b)}{2} + \sum_{i=1}^{n-1} f(a + iw) \right]$$

Functions can themselves call functions and, in this example, we will find the definite integral of a complicated trigonometric function.

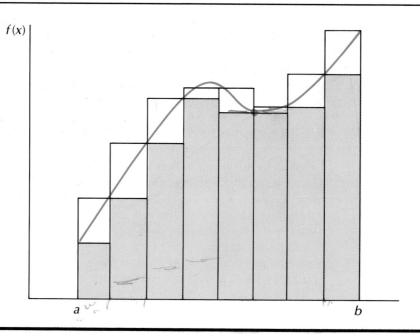

Figure 8.1
Integration with Rectangles

```
*       Program 8.3
*       Integrates using trapezoids
*
        program Trap
        real A, B, Width, Area, X, F
        integer N, I
        print *, 'Enter limits of integration and number of intervals'
        read *, A, B, N
        Width = (B - A) / N
        Area = (F(A) + F(B)) / 2.0
        do 100 I = 1, N-1
            X = A + I * Width
            Area = Area + F(X)
100     continue
        Area = Area * Width
        print *, 'Limits of integration ', A, B
        print *, 'Number of intervals ', N
        print *, 'Approximate area is ', Area
        stop
        end

        real function F(X)
        F = sin(X) * cos(X) / exp(X)
        return
        end
```

It is interesting to consider what happens as we increase the number of intervals used in this program. If we use only a small number of intervals, then we are not counting large areas in the summation, because the straight lines at the tops of the trapezoids do not follow the function very closely. As we increase the number of intervals, the approximation gets more and more accurate. However, perhaps surprisingly, if we continue to increase the number of intervals, the approximation gets worse, rather than better. The reason, of course, is that round-off error becomes more and more significant. There comes a point where increasing the number of intervals adds more error to the summation than it gains by more accurate approximation to the function. Using programs such as these requires finding this point. That is one reason why the user selects the number of intervals, rather than having the program decide. For this particular program we get the following results for the integral between 0 and 1.

10 subintervals	0.1960662
90 subintervals	0.1971525
500 subintervals	0.1971613
750 subintervals	0.1971613
1000 subintervals	0.1971611
1500 subintervals	0.1971451
2000 subintervals	0.1971229

You can see that, until about 500 subintervals, adding extra subintervals improves the accuracy. However, after we use about 1000 subintervals, we start to see round-off errors creeping in and the approximation gets steadily worse as we continue to increase the number of subintervals used.

8.2 Simpson's Rule for Integration

The last two methods approximated the shape of the curve being integrated by straight line segments. **Simpson's rule** is the name given to the equivalent technique using quadratic curves (parabolas) as the approximating curve in each pair of adjacent subintervals.

Let us begin by considering a parabola

$$ax^2 + bx + c$$

in a small interval $[-w, w]$ around the origin. Since this is a quadratic polynomial, it is easy to integrate and we can calculate the definite integral

$$\int_{-w}^{w} (ax^2 + bx + c)dx = \left. \frac{ax^3}{3} + \frac{bx^2}{2} + cx \right|_{-w}^{w}$$

$$= \frac{2}{3}aw^3 + 2cw$$

Now we can do some simple algebra on this result as follows:

$$\frac{2}{3}aw^3 + 2cw$$
$$= \frac{w}{3}(2aw^2 + 6c)$$
$$= \frac{w}{3}\left[(aw^2 - bw + c) + 4c + (aw^2 + bw + c)\right]$$

which is simply

$$\frac{w}{3}[f(-w) + 4f(0) + f(w)]$$

Thus, the area under a parabola in an interval such as this is given by the last expression. It is clear that the area does not change if we slide the parabola along parallel to the x-axis, so we can treat this as a general expression for the area. That is, for any parabola, if we take an interval of size $2w$ (as the interval $[-w, w]$ is), then we can calculate the area under the parabola using the value of the function at the left endpoint, four times the value of the function at the midpoint, and the value of the function at the right endpoint.

Suppose we want to calculate the definite integral of the function $f(x)$ in the interval $[a, b]$. Then we divide the interval up into an *even* number of subintervals, each of size w. These subintervals give rise to strips, the sum of whose area is the area under the curve. Now we take the strips, two by two, starting from the left. The area under the function in the first two strips can be approximated by a parabola passing through the three points $f(a)$, $f(a + w)$, and $f(a + 2w)$. The area under this parabola is given by the previous expression and is

$$\frac{w}{3}[f(a) + 4f(a + w) + f(a + 2w)]$$

Now we take the next two strips in exactly the same way. The area under these two strips is

$$\frac{w}{3}[f(a + 2w) + 4f(a + 3w) + f(a + 4w)]$$

The total area under the function in the interval $[a, b]$ is given by the summation

$$\frac{w}{3}\left\{f(a) + f(b) + 4\sum_{i=1}^{n/2} f(a + (2i - 1)w) + 2\sum_{i=1}^{n/2-1} f(a + 2iw)\right\}$$

Points at the boundary between two adjacent pairs of strips get their function values counted twice in the sum. Points in the middle of a pair of strips have their function values counted in the sum four times. The endpoints of the interval are only added into the sum once.

We can now write a program that uses this summation to calculate the area under a function. We need to calculate the two sums involved and then use them to compute the preceding expression.

```
*       Program 8.4
*       Integrates using Simpson's rule
*
        program Simp
        real  A,B,Width,EvenS,OddS,Area,EvenX,OddX,F
        integer N, I
        print *, 'Enter limits of integration and number'
        print *, 'of intervals (must be even)'
        read *, A, B, N
        EvenS = 0.0
        OddS  = 0.0
        Width = (B - A) / N
```

```
      do 100 I = 1, N/2
         EvenX = A + 2 * I * Width
         OddX = A + (2 * I - 1) * Width
         EvenS = EvenS + F(EvenX)
         OddS = OddS + F(OddX)
100   continue
*     correct by subtracting extra even term
      EvenS = EvenS - F(EvenX)
      Area = Width * (F(A) + F(B) + 2*EvenS + 4*OddS) / 3.0
      print *, 'Limits of integration ', A, B
      print *, 'Number of intervals ', N
      print *, 'Approximate area is ', Area
      stop
      end

      real function F(X)
      real X
      F = sin(X) * cos(X) / exp(X)
      return
      end
```

The variable EvenX takes on all the even step values, while OddX takes on all the odd values.

8.3 Evaluating a Series

Many functions can be expanded using an infinite series. Under very general conditions, a function $f(x)$ that has derivatives of all orders can be expressed in the form

$$f(x) = \sum_{i=0}^{\infty} \frac{x^i f^{(i)}(0)}{i!}$$

If we know the value of the function and its derivatives at 0, then this summation gives us an infinite series that converges to the value of the function. For example, suppose that $f(x)$ is sin x. The derivatives of sin x are cos x, $-$ sin x, $-$ cos x, sin x, cos x , $-$ sin x, and so on, infinitely. Now the value of sin(0) is 0, the value of cos(0) is 1, and the value of $-$ cos(0) is -1. Substituting in the preceding summation, we get that

$$\sin x = \sum_{i=0}^{\infty} \frac{-1^i x^{2i+1}}{(2i + 1)!}$$

This expansion is called the MacLaurin series of the function.

Many of the common trigonometric functions can be expanded in this way. Expansions of this kind are interesting because they are computational in nature. We can use the summations to calculate the approximate value of the functions at various points.

Now of course the preceding expansion is an infinite one and we cannot compute infinite summations (they take too long). However, we are dealing with an environment in which all real values are approximated. If we only sum a finite part of the sequence, we can get an approximation to the function. If the terms that are not included in the sum are sufficiently small that they could not be represented anyway, then our finite summation will give as good an answer as an infinite summation would have produced. For example, if we are limited to a two decimal digit mantissa such as 0.34, then adding in a term such as 0.00056 does not change the first term, even though the second term can be represented using a two-digit mantissa (0.56×10^{-3}).

The MacLaurin series for the exponential function is

$$e^x = \sum_{i=0}^{\infty} \frac{x^i}{i!}$$

The next program calculates it.

```
*       Program 8.5
*       Calculates exponentials by evaluating
*       the MacLaurin series expansion
*
        program Expon
        real X, Sum, Term
        integer I
        Sum = 1.0
        Term = 1
        print *, 'Enter X value'
        read *, X
        do 100 I = 1, 100
            Term = Term * X / I
            Sum = Sum + Term
100     continue
        print *, 'Value of exp at ', X, ' is ', Sum
        stop
        end
```

Now the *i*th and *i* + 1st terms of the preceding series are

$$\frac{x^i}{i!} \qquad \frac{x^{i+1}}{(i+1)!}$$

Mathematics for Engineering

and there is an easy transformation that allows the $i + 1$st term to be calculated when you know the ith. It is this transformation that allows the algorithm to be reasonably efficient. The program uses this, keeping the value of the most recent term and using it to calculate the next term to be added in. We get better accuracy, as well as a performance advantage, by doing the calculation this way. Both the numerator and denominator of the terms are relatively large and will tend to cause overflow. However, none of the variables in the program ever contains a large value since the size of a term is a ratio of two large quantities and is of moderate size. This technique of using the values of previous terms should be used for all series evaluations.

The preceding program simply calculated the first 100 terms of the series using a definite iteration and then stopped. In general, however, we would like to use a more intelligent mechanism for deciding when we have added enough terms to the series to get a reasonable approximation. One way is to examine the value of each new term and stop when the added term is small enough. Many computers use a representation that allows about seven decimal digits to be represented in the mantissa. Thus, stopping when the new term to be added in is smaller than 10^{-8} is a sensible terminating condition.

It is also wise to use some fixed iteration counter as a supplementary terminating mechanism in case the series does not converge quickly. For example, the expansion for the natural logarithm function is given by

$$\log(1 + x) = \sum_{i=1}^{\infty} \frac{-1^{i+1} x^i}{i}$$

The terms of this series get small very slowly compared to those of the previous exponential function, where the denominator is a factorial. In fact, this series is almost useless for actual calculation of the logarithm because it converges so slowly. When several thousand terms have been added into the sum, round-off errors have accumulated to such an extent that very few accurate significant figures remain.

The next program is a revised version of the exponential series summation that uses a magnitude-of-the-term test and an indefinite iteration to decide when to terminate.

```
*       Program 8.6
*       Evaluates the exponential function using the
*       MacLaurin series expansion. Terminates
*       when the term being added is small
*

        program Expon2
        real X, Sum, Term
        integer I
```

```
            Sum = 1.0
            print *, 'Enter X value'
            read *, X
            Term = X
            I = 1
            while (abs(Term).gt.1.0e-9 .and. I.le.100) do
                I = I + 1
                Sum = Sum + Term
                Term = Term * X / I
            end while
            print *, 'Value of exp at ', X, ' is ', Sum, ' in ', I, 'terms'
            stop
            end
```

We have had to rearrange the program a little so that the first part of the condition in the while loop could be tested when the while statement is first encountered. The Boolean expression is in two parts, joined by the word *and*. (Notice that periods precede and follow the word *and*, as is the case with other Boolean operators and constants.) The while loop will only be executed if the absolute value of the term is larger than 10^{-9} and if there haven't already been 100 iterations. The function invocation *abs* uses the intrinsic function that returns the absolute value of its argument.

The output statement at the end of the program writes out the value obtained by summing all of the terms, but it also prints out the number of terms that were used in the summation. Therefore, if the series should happen not to converge, an approximation will be printed, but the output will show that the full 100 terms were used and so the answer should be regarded with suspicion. This kind of series evaluation is one way in which the intrinsic mathematical functions can be implemented.

8.4 Solving Differential Equations

We now move on to consider differential equations, a mathematical concept used in modeling many engineering systems. Suppose we have a first-order, ordinary differential equation such as

$$\frac{dy}{dx} = f(x, y)$$

Then we solve this equation analytically by separating the two variables and integrating. We get a solution of the form

$$y = g(x, c)$$

where c is some constant of integration. Thus, a differential equation of this kind really describes not one single function, but a whole family of functions, one corresponding to each value of c.

For example, if we have the differential equation

$$\frac{dy}{dx} = xy$$

then we can separate variables and integrate to get

$$\log y = \frac{x^2}{2} + c$$

or

$$y = e^{x^2/2} e^c$$

The particular value of c determines which member of this family of curves we are concerned with.

Often the differential equation is complex enough that an analytical solution is not possible. Rather than solve it using analytic techniques, it may be sufficient in some situations to determine the values of the function $g(x, c)$ at some set of points. Usually we know some boundary condition that allows us to determine which one of the family of functions is of interest.

One useful approach is to use **Euler's method**. Suppose that we are given some differential equation and want to find the value of the function passing through

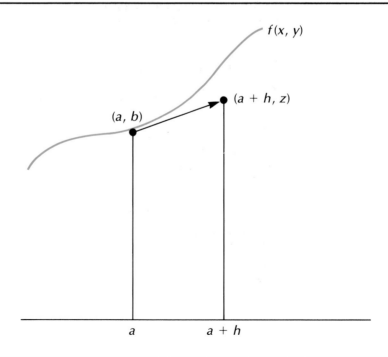

Figure 8.2
Differential Equation

the point $x = a, y = b$. The slope of the function at the point (a, b) can be determined using the differential equation, which states that

$$\frac{dy}{dx} = f(a,b)$$

This gives us the slope of a tangent from the point (a,b). Thus we can construct the line from (a, b) to a point $(a + h, z)$, for some small h. The value of z can be determined by calculating the intercept of the tangent with the line $x = a + h$ as in Figure 8.2. If the value of h is sufficiently small, then the new point will lie close to the function we are tabulating. We can then repeat the step using the new point as the new start.

If the function g is increasing at this point, then the value of z will be too small and our new point will be slightly below the curve of the function we are trying to calculate. Thus, when we iterate, we will be starting from a point that is not on the function and using a tangent to another function in the same family. The second point we calculate will tend to be even more in error than the first one. The errors are cumulative, so that after a large number of steps we may be a long way from the function we are trying to follow. Two things may help: using very small intervals to minimize the difference between the tangent and the function in the interval; and approximating the function only for a small distance away from the point (a,b).

The next program implements the algorithm we have described.

```
*       Program 8.7
*       Evaluates a differential equation
*       using Euler's method
*
        program ODE
        real X, Y, Deriv, H, DyDx
        integer I, Steps
        print *, 'Enter initial point, step size and # of steps'
        read *, X, Y, H, Steps
        print *, ' Table of X, Y values'
        print *, X, Y
        do 100 I=1, Steps
            Deriv = DyDx(X, Y)
            X = X + H
            Y = Y + Deriv * H
            print *, X, Y
100     continue
        stop
        end
```

```
real function DyDx(X, Y)
real X, Y
DyDx = X * Y
return
end
```

The program produces the following output, starting from a point (1,1) and taking ten steps of size 0.1.

```
Enter initial point, step size and # of steps
  Table of X, Y values
            1.0000000              1.0000000
            1.0999994              1.0999994
            1.1999989              1.2209988
            1.2999983              1.3675184
            1.3999977              1.5452948
            1.4999971              1.7616348
            1.5999966              2.0258789
            1.6999960              2.3500185
            1.7999954              2.7495203
            1.8999949              3.2444324
            1.9999943              3.8608723
```

We are following the function that passes through the point (1,1). From the preceding equation, when the value of x has increased from 1 to 2, the value of y increases from 1 to about 4. This is reflected in the values.

An improvement to this algorithm can be made by using more information about the derivative of the function. Suppose we start, as before, from the point (a,b) and follow a tangent to the function to the point $(a + h, z)$. Now we take the average of the derivative of the function at the point (a,b) and $(a + h, z)$. We use this new gradient to generate a third point $(a + h, t)$, where t is the intercept of the tangent from (a,b) with the line $x = a + h$. We use this third value as the new approximation to the function and repeat the procedure. This is illustrated in Figure 8.3.

This is called a **predictor-corrector** method. The first new point we generate is the predictor point. We use some information about the derivative at this point to get a better approximation for the derivative in the interval and use this to generate a new approximation. This is the corrector step. The next example program uses the predictor-corrector method on the same function as the previous program.

```
*       Program 8.8
*       Evaluates a differential equation
*       using the predictor-corrector method
*
        program ODE2
        real X, Y, Deriv, H, DyDx, TX, TY
        integer I, Steps
        print *, 'Enter initial point, step size and # of steps'
        read *, X, Y, H, Steps
        print *, 'Table of X, Y values'
        print *, X, Y
        do 100 I=1, Steps
*           Predictor step
            Deriv = DyDx(X, Y)
            TX = X + H
            TY = Y + Deriv * H
*           Corrector step
            Deriv = (Deriv + DyDx(TX, TY))/2.0
            Y = Y + Deriv * H
            X = X + H
            print *, X, Y
100     continue
        stop
        end
```

Figure 8.3
Predictor-Corrector Method

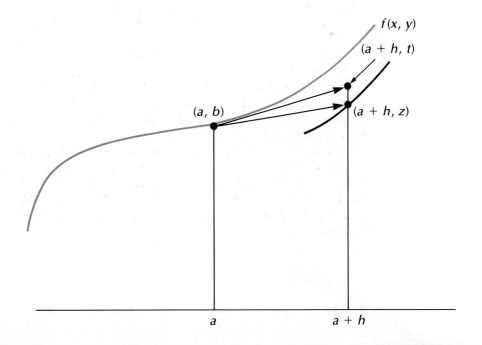

```
real function DyDx(X, Y)
real X, Y
DyDx = X * Y
return
end
```

We next show the output of this program on the same input data values as were used with the previous version.

```
Enter initial point, step size and # of steps
Table of X, Y values
            1.0000000           1.0000000
            1.0999994           1.1104994
            1.1999989           1.2455359
            1.2999983           1.4109421
            1.3999977           1.6142578
            1.4999971           1.8652735
            1.5999966           2.1767731
            1.6999960           2.5655432
            1.7999954           3.0537643
            1.8999949           3.6709280
            1.9999943           4.4565039
```

You can see that the values obtained are noticeably different from the previous version, even over this small range. The predictor-corrector version is more accurate because it tends to follow the function more closely.

8.5 Passing Functions as Arguments

It is possible in Fortran to pass a function to a subprogram as if it were an ordinary argument. This makes it possible to write very general subprograms that can be passed information not only about data to be used but functions to be used as well. For example, we could write an integration function that could be passed both the limits of the integral and the function to be integrated.

When a function is going to be used as a parameter, it appears in the subprogram heading as a dummy name in exactly the same way that the other parameter names do. It will be replaced by the name of an actual function at the time that the function is invoked. The name of the function must also appear in an **external** statement in the invoking subprogram. This allows the compiler to distinguish a function name from a variable. The form of the **external** statement is

   external *functionNameList*

If the function that is to be passed as a parameter to a subprogram is an intrinsic function, then the **intrinsic** statement must be used to declare that the function name is being used for this purpose. The form of this statement is the same as that of the **external** statement except that it begins with the word **intrinsic** instead of **external**.

As an example we'll write a function that will use Simpson's rule to calculate the definite integral of a function in a given range.

```
*       Program 8.9
*       Illustrates how a function can be passed as a parameter.
*       A function to integrate using Simpson's rule is used
*
        program Test
        real A, B, G, Simp                      F/G)        fog
        external G
        integer N
        print *, 'Enter limits of integration and number of intervals'
        read *, A, B, N
        print *, A, B, N, Simp(A, B, N, G)
        stop
        end

*
*       Integrates a function that it is passed
*
        real function Simp(A, B, N, F)
        real A, B, Width, EvenS, OddS, Area, EvenX, OddX, F
        integer N, I
        EvenS = 0.
        OddS = 0.
        Width = (B - A) / N
        do 100 I = 0, N/2
            EvenX = A + 2 * I * Width
            OddX = A + (2 * I + 1) * Width
            EvenS = EvenS + F(EvenX)
            OddS = OddS + F(OddX)
100     continue
        EvenS = EvenS - F(EvenX)
        Simp = Width * (F(A)+F(B) + 2*EvenS + 4*OddS) / 3.0
        return
        end

        real function G(X)
        real X
        G = X * X
        return
        end
```

This language feature allows the development of helpful packages of functions that can be used in many different contexts.

If we want to calculate the integral of an intrinsic function, then we can use the same function as before but we need to invoke it differently. The next example shows a main program that calculates the integral of the sine function, using the intrinsic function.

```
*       Illustrates how a function can be passed as a parameter.
*       A function to integrate using Simpson's rule is used
*
        program Test2
        real A, B, Simp
        intrinsic sin
        integer N
        print *, 'Enter limits of integration and number of intervals'
        read *, A, B, N
        print *, A, B, N, Simp(A, B, N, sin)
        stop
        end
```

8.6 Root Finding

Another useful mathematical technique is root finding. Given a function $f(x)$, a zero of the function is a value of x at which the function's value is zero. The zero of a function corresponds to the **root of an equation** of the form $f(x) = 0$, and we will use the term *root* throughout this section. Such points are important in themselves because the places where a function is zero often have some physical meaning. In this section we will investigate algorithms for finding the roots of arbitrary functions. Once we can find the root of a function, we can solve the function for any value. For example, suppose we want to find the value of x such that $f(x) = a$. Then this value of x is a root of the function $f(x) - a$.

We will describe two major algorithms. Which one is used depends on how much we know about the function whose root we are trying to find.

The first method is called the **bisection method**. The only constraint the bisection method imposes on the function is that it must be continuous. Suppose we have an interval $[a, b]$ and we know that $f(a)$ is negative and $f(b)$ is positive. If the function is continuous we may conclude that between a and b the function crosses the x-axis. This point is the root for which we are searching.

Now suppose that we look at the function value at the point midway between a and b. If the function value is negative at this new point, then we know that the root must lie between the new point and b. On the other hand, if the function value is positive at the new point, then we conclude that the x-axis has already been crossed and the root lies between a and the new point.

Now we have a new interval (either $[a, newpoint]$ or $[newpoint, b]$ depending on the sign of the function) that has the same property as the original interval. The function value is negative at one end and positive at the other and so it must contain a root. We now consider the value of the function at the midpoint of the smaller interval. We can repeat exactly the same reasoning and conclude that the root must lie in one of the two halves of the smaller interval.

We repeat this sequence of steps, each time narrowing the possible interval in which the root must lie. Eventually the size of the interval will become so small that we can treat it as effectively zero and say that the root lies at the point that results. This is shown in Figure 8.4.

We must consider what happens if there is more than one root in the interval we start with. It is easy to see that if there are two roots in the interval $[a, b]$, then it cannot happen that the function value at one end of the interval is negative while the function value at the other end is positive. The function crosses the x-axis twice in the interval and so must end up on the same side that it started on. If there are three roots in the interval, then it will be the case that the function will have opposite sign at the ends of the interval. When we consider the two subintervals created by the midpoint, then either all three roots will be in one half and none in the other, or two will be in one half and one in the other. If we follow the algorithm described earlier, we will always choose the subinterval with either one or three roots for the next step. Arguing repeatedly, it is clear that eventually we will find one of the roots.

In general, if there are an odd number of roots in the interval, then the bisection algorithm will find one of the roots. If there are an even number of roots in the interval, then the initial conditions are not satisfied and we cannot use the algorithm. Of course, we must be careful about functions that just touch the axis and may have multiple roots at that point. We will not worry explicitly about this, but a program to go in a library would have to.

Figure 8.4
Root Finding by Bisection

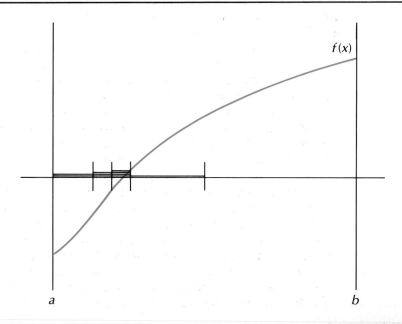

We have not said anything about how an appropriate initial interval can be found. In general, we must know something about the shape of the function to be able to find such an interval. For example, we can sample the function values at a sequence of points until we find a pair of subsequent points where the function changes sign. This should, of course, be done with care since it is possible to miss a double sign change completely. We could include this option in the program itself, giving it an initial x value and allowing it to step along the x-axis in fixed-size steps until the function changes sign.

Let us now write the algorithm as a function that is passed the name of a function, two endpoints of an interval in which a root must lie, and a tolerance on the answer indicating when the root is considered to be found.

```
*       Program 8.10
*       Finds a root of an equation
*       using the bisection method
*
        program Root
        real A, B, Tol, G, Bisec
        external G
        print *, 'Enter interval and required tolerance'
        read *, A, B, Tol
        print *, 'Limits of interval ', A, B
        print *, 'Root is ', Bisec(A, B, Tol, G),' to tolerance ', Tol
        stop
        end
*
*       Does the bisection
*
        real function Bisec(A, B, T, F)
        real A, B, T, F, Mid
        Mid = (A + B) / 2
        while (abs(B - A) .ge. T) do
           if (F(Mid) .gt. 0.0) then
              B = Mid
           else
              A = Mid
           endif
           Mid = (A + B) / 2
        end while
        Bisec = Mid
        return
        end
```

```
real function G(X)
real X
G = (X - 1) * (X + 4)
return
end
```

In each iteration, either the value of the midpoint is placed into variable A or variable B. Since the midpoint must lie between the values in A and B, the distance between A and B must decrease with each iteration. Since the loop stops when this distance is sufficiently small, it must be that the loop will eventually terminate.

There are several special situations to consider. What happens if the root is in fact at the midpoint of the initial interval? The preceding function will select the right subinterval as the one containing the root and go on to the next iteration. The next thing to ask is: what happens if the root is at the endpoint of the interval? The function still works in this case. However, it might be worth modifying the function to check these special cases and stop the iteration if they occur. This could save some execution time, particularly if the initial interval is very large.

There is still a serious problem with the function as currently written. The function assumes that the endpoints will be given to it in such a way that $f(a)$ is negative and $f(b)$ positive. However, it might be the other way around—$f(a)$ might be positive and $f(b)$ negative. The next version of the program is a better one.

```
*      Program 8.11
*      Another bisection root finding program
*

       program Root
       real A, B, Tol, G, Bisec
       external G
       print *, 'Enter interval and required tolerance'
       read *, A, B, Tol
       print *, 'Limits of interval ', A, B
       print *, 'Root is ', Bisec(A, B, Tol, G),' to tolerance', Tol
       stop
       end
*
*      This function does not assume that the endpoints of the
*      interval are in order
*
       real function Bisec(A, B, T, F)
       real A, B, T, F, Mid
```

```
        Mid = (A + B) / 2
        while (abs(B - A) .ge. T) do
            if (F(A) * F(Mid) .le. 0.0) then
                B = Mid
            else
                A = Mid
            endif
            Mid = (A + B) / 2
        end while
        Bisec = Mid
        return
        end

        real function G(X)
        real X
        G = (X - 1) * (X - 4)
        return
        end
```

The Boolean expression in the **if** statement in Bisec now tests whether the values of the function at the midpoint and A are of opposite sign (if they have the same sign, positive or negative, then their product will always be positive). Now it doesn't matter in which order the values for A and B are given.

We can calculate in advance how many iterations will be required to find the root to any given accuracy. For instance, suppose that the interval from A to B has length 1. After one iteration we have reduced the possible size of the interval in which the root must lie to 0.5. After two iterations we have reduced the interval to 0.25, after three iterations to 0.125 and so on. The size of the search interval is halved after each iteration. In general, if we start with an interval of size y, then after n iterations we have reduced the possible interval to $y/2^n$. Since $1/2^{10}$ is about 1/1000, it takes ten iterations to reduce the interval to one thousandth its previous size. Therefore, if we use a tolerance of 10^{-6}, it will take about twenty iterations to get an answer. Notice that this does not depend on the shape of the function at all. The same number of iterations gives a root to the same accuracy for all functions.

This suggests another possible termination condition that takes into account the behavior of the function. We could examine the value of the function at the midpoint on each iteration and terminate when the value is sufficiently small. This is intuitively satisfying because the termination condition is related to the calculation we are actually doing (the root is the point at which the function's value is zero). It also means that if the function is very flat, then the algorithm will take fewer iterations, whereas if the function is very steep, it will take more. The change to the program is very small.

```
*       Program 8.12
*       Uses bisection and a vertical tolerance
*
        program Root
        real A, B, Tol, G, Bisec
        external G
        print *, 'Enter interval and required vertical tolerance'
        read *, A, B, Tol
        print *, 'Limits of interval ', A, B
        print *, 'Root is ', Bisec(A, B, Tol, G),' to tolerance', Tol
        stop
        end

        real function Bisec(A, B, T, F)
        real A, B, T, F, Mid
        Mid = (A + B) / 2
        while (abs(F(Mid)) .ge. T) do
            if (F(A) * F(Mid) .le. 0.0) then
                B = Mid
            else
                A = Mid
            endif
            Mid = (A + B) / 2
        end while
        Bisec = Mid
        return
        end

        real function G(X)
        real X
        G = (X - 1) * (X - 4)
        return
        end
```

Notice that the tolerance is now a tolerance in the y direction, whereas in the previous version it was a tolerance in the x direction.

A variation of the bisection method is *walking bisection*. In this method, we start with a point and move along the x-axis in fixed-size steps, until the sign of the function at one of the sample points is the opposite of the sign of the previous point. Then we reverse directions and go backwards using steps of, say, half the previous size until we see another sign change in the values of the function. This triggers a further reverse in direction and a further halving of the step size. Eventually, the step size becomes so small that we stop instead of reversing.

The bisection method and its variants are reliable and guaranteed to terminate for continuous functions. However, a faster algorithm exists for functions that are both continuous and differentiable. It is called the **Newton-Raphson method** for root finding. The idea behind the Newton-Raphson method is that if we are at point x, then the distance from x to the root can be guessed by considering the slope of the function at x. If the function is steep, then the root may be quite close. If the slope is small, then the root is probably some distance away.

The method generates a sequence of points that (it is hoped) converge to the root. Starting from the point x_1, we calculate the tangent to the function at x_1 and then find its intercept with the x-axis. This becomes the point x_2 and is often closer to the root than x_1. We then repeat the calculation, this time starting from x_2 and get a new point x_3. This continues until the value of the function at the point x_i is sufficiently close to zero for us to conclude that x_i is a good approximation to the root. This is shown in Figure 8.5.

Given a point x_1, the equation of the tangent to the function at x_1 is

$$y - f(x_1) = f'(x_1)(x - x_1)$$

and the x-intercept of this line is

$$x_2 = x_1 - \frac{f(x_1)}{f'(x_1)}$$

This is the recurrence that we use to get x_{i+1} from x_i. In general

$$x_{i+1} = x_i - \frac{f(x_i)}{f'(x_i)}$$

Because we need to calculate the derivative, the function f must be differentiable.

It may happen that the denominator $f'(x_i)$ will be zero and that the division required in the recurrence is impossible. The derivative $f'(x_i)$ is zero exactly when

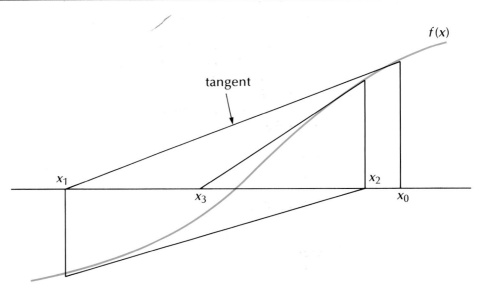

Figure 8.5
Root Finding by Newton-Raphson Method

x_i is a turning point of the function. This corresponds to a tangent that is parallel to the x-axis and whose x-intercept is therefore infinite. Clearly we must be able to handle this situation, since some otherwise well-behaved functions have many turning points (for example, $\sin(x^2)$). The algorithm is modified to check for a zero value for the derivative before doing the division and altering the value of x if one is found. In fact, it is a problem even if the value of the derivative is quite small, because the tangent is then very flat and the next x-intercept will be a very long way from the current value of x. The next program finds a root using the Newton-Raphson method.

```
*       Program 8.13
*       Finds roots using the Newton-Raphson method. Requires
*       the function to be differentiable
*
        program Root
        external G, GPrime
        real A, Tol1, Tol2, G, GPrime, Newton
        print *, 'Enter initial estimate, vertical tolerance'
        print *, 'and tolerance in closeness to turning points'
        read *, A, Tol1, Tol2
        print *, A,Tol1,Tol2
        print *, 'Root is', Newton(A,Tol1,Tol2,G,GPrime)
        stop
        end

*
*       This function uses the recurrence to calculate the root
*
        real function Newton(A, T1, T2, F, FPrime)
        real A, T1, T2, F, FPrime, X
        X = A
        while (abs(F(X)) .ge. T1) do
            while (abs(FPrime(X)) .le. T2) do
                X = X + 1.0
            end while
            X = X - F(X) / FPrime(X)
        end while
        Newton = X
        return
        end

        real function G(X)
        real X
        G = (X - 1) * (X - 4)
        return
        end
```

```
real function GPrime(X)
real X
GPrime = 2 * X - 5
return
end
```

If the value of the derivative is too small (no greater than Tol2), then the value of X is incremented by 1. There is no particular sequence of X values to follow to find the root, so it doesn't cause any numerical problems to change X in a fairly arbitrary way. This program evaluates both the function and its derivative at the same point twice for each value of X. You should try to rewrite it so that each function is never evaluated more than once at the same point.

There are several engineering applications of these concepts. One very common situation in real world problems is that the state of some physical systems is the result of two different forces, say $f(t)$ and $g(t)$. We can find the actual state of the system by looking for the points at which these two functions are equal. We are therefore interested in the roots of the function $f(t) - g(t)$.

Another important application is in finding the maxima or minima of functions. These turning points correspond to the roots of the derivative of the original function.

Programming Example

Problem Statement

Write a subprogram to calculate integrals using Simpson's rule. It should calculate an approximation to the integral using a given number N of intervals and also using $2N$ intervals.

Inputs

A function to be integrated, the limits of integration, and the number of intervals to be used.

Outputs

Two estimates of the value of the integral.

Discussion

Because it may be expensive to recalculate the function, we will write this subprogram so that the function is not calculated more than once at any point. We do this by observing that, if the number of intervals is doubled, then all of the function values that were calculated at first can also be used in the second evaluation. The only new points that need to be evaluated are those interval boundaries that fall in the middle of the old intervals.

Program

```
*       Program 8.14
*       Uses Simpson's rule to integrate.
*       Uses N and 2N intervals and prints both
*       results for comparison
*
*       A, B - limits of integration
*       N - number of intervals used initially
*       Area1 - approx to integral with N intervals
*       Area2 - approx to integral with 2N intervals
*
        program Integ
        real A, B, G
        real Area1, Area2
        integer N
        external G
        print *, 'Enter limits of integral and # of intervals'
        read *, A, B, N
        print *, 'Limits of Integration', A, B
        print *, 'Number of Intervals', N
        call Simp(A, B, N, G, Area1, Area2)
        print *, 'First Approximation to integral = ', Area1
        print *, 'Second Approximation to integral = ', Area2
        stop
        end
```

```
*
*       Calculates an approx to the area using Simpson's rule
*
*       A, B - limits of integration
*       N - number of intervals (even)
*       Width - width of each interval
*       EvenS, OddS - sum of function values at even and odd
*               numbered points
*       EvenX, OddX - even and odd numbered points
*       Area1, Area2 - approx areas with N and 2N intervals
*
        subroutine Simp(A, B, N, F, Area1, Area2)
        real A, B, Width, EvenS, OddS
        real Area1, Area2, EvenX, OddX, F
        integer N, I
        EvenS = 0.0
        OddS = 0.0
        Width = (B - A) / N
        do 100 I = 1, N/2
            EvenX = A + 2 * I * Width
            OddX = A + (2 * I - 1) * Width
            EvenS = EvenS + F(EvenX)
            OddS = OddS + F(OddX)
  100   continue
*       make correction
        EvenS = EvenS - F(EvenX)
        Area1 = Width * (F(A) + F(B) + 2*EvenS + 4*OddS) / 3.0
        EvenS = EvenS + Odds
        OddS = 0.0
        Width = Width / 2
        do 200 I = 0, N-1
            OddX = A + (2 * I + 1) * Width
            OddS = OddS + F(OddX)
  200   continue
        Area2 = Width * (F(A) + F(B) + 2*EvenS + 4*OddS) / 3.0
        return
        end

        real function G(X)
        real X
        G = sin(X) * cos(X) / exp(X)
        return
        end
```

Testing

Here is the output on the function shown in the preceding program.

```
Enter limits of integral and # of intervals
Limits of Integration        0.0000000      2.0000000
Number of Intervals            500
First Approximation to integral  =     0.2279248
Second Approximation to integral =     0.2279233
```

Design, Testing, and Debugging

- Numerical integration should never be done by using one set of strips and printing an answer. Instead, the user should be permitted to redo the calculation using several different numbers of strips to check that the value obtained for the integral is reasonable.
- Numerical programs should always provide the user with as much information as possible about the calculations carried out. This helps to increase the user's confidence in the program because failures, caused by round-off error or program bugs, can often be detected from this other information.

Style and Presentation

- Floating point values are often hard for users to absorb, especially if there are many such values printed at once. Extra thought should be given to the output of numerical programs.

Chapter Summary

- Numerical integration can be done by summing the areas of rectangles or trapezoids. A better approximation can usually be achieved by approximating the function being integrated by a set of parabolas and using Simpson's Rule.
- Many useful functions, including most trigonometric ones, can be calculated by summing a series. As these series are infinite, we approximate by summing a large number of terms. For most series, a new term can be calculated efficiently from a previous term.
- Differential equations can be approximated using either Euler's method or an improved version called the predictor-corrector method. Both these methods

use local information about the slope of the function in question to try and track its path.
- Function names can be passed as arguments to subprograms. This allows very general subprograms to be written, since they can operate on values and functions that are not determined until run-time. Functions that are to be passed as arguments must be designated in an **external** statement if they are user-defined, or in an **intrinsic** statement if they are intrinsic functions.
- The roots of equations can be found numerically using the bisection method or the Newton-Raphson method. The bisection method is slower to find a root in general, but requires only that the function be continuous. The Newton-Raphson method requires that the function also be differentiable.

Define These Concepts and Terms

Rectangle rule
Trapezoid rule
Simpson's rule
Euler's method

Predictor-corrector method
Root of an equation
Bisection
Newton-Raphson method

Exercises

1. Describe the class of functions for which each of the three numerical integration methods (rectangle rule, trapezoid rule, Simpson's rule) will give exact answers, within the limits of the accuracy of your computer.
2. Calculate the MacLaurin series expansion for the function cos x. Write a program that calculates the cosine function using this expansion.
3. Try both the trapezoid and Simpson's rule example programs on some actual functions.
4. Write programs to evaluate some of the series mentioned in this chapter. Use the values of the intrinsic functions corresponding to each series to determine how many terms of the series are required to get the same accuracy as the intrinsic functions.
5. Find the values of e^{10} and e^{-10} accurate to seven decimal places, assuming you can get this accuracy on your system. How many terms of the series are needed to calculate each one? Is there a difference?
6. The square roots of any number a are the roots of the function $x^2 - a$. The Newton-Raphson method can be used to find the root of this function since it is continuous and differentiable. Write a function that will calculate the square root of any real number.
7. The sine and cosine functions have the same shape but are shifted relative to each other. Suppose that we had to write a set of intrinsic functions for sine,

cosine, and tangent. We would only need to actually write a sine function since we could calculate the cosine function using the series expansion of the sine function. For example, we can calculate the cosine of x by calculating the sine of $x + \pi/2$. Furthermore, the sine function has the same basic shape in each of these four intervals: $[0, \pi/2), [\pi/2, \pi), [\pi, 3\pi/2), [3\pi/2, 2\pi)$. Therefore, if we can calculate the sine in the first of these intervals, we can always calculate the sine of any other value by choosing the point in $[0, \pi/2)$ at which the function has the same value. In fact, we can do even better. The trigonometric identity $\sin 2\theta = 2 \sin \theta \cos \theta$ can be used to calculate any sines that fall outside the interval $[0, \pi/4)$. Thus we can concentrate all of our programming efforts on computing sines in this interval. We will then be able to calculate all of the other trigonometric functions using this one series. Fortunately, in this interval the value of x is small and the series will converge quickly. Write a set of functions that use these optimizations to calculate sine, cosine, and tangent. Remember that $\tan x = \sin x / \cos x$. (In practice, calculating tan this way isn't a particularly good idea.)

8. Calculate the value of $\sin x$ three ways: using the intrinsic function, using the series expansion described in the chapter, and using one of the integration programs and the intrinsic function $\cos x$. Check the values at 0 and $\pi/2$. Explain any differences you observe.

9. Write a walking bisection program to find a root of any fourth-degree polynomial. The user should not have to input the starting point or step size.

10. Compute the integral of $\sin (100x)$ over the range 0 to π using first 10 steps, then 100, then 1000 steps. Explain the results you obtain.

11. Consider the family of functions

$$f(x) = x^3 - 6x^2 + 12x - a$$

where a is an integer in the range 0 through 8. All nine functions have a root between -1 and 5. Write a program to find these roots and print a table of the values of a and the corresponding root. Explain what is happening as the constant varies in this range.

12. Modify the bisection method for root finding to find the point of intersection of two curves defined by functions F(X) and G(X).

13. Modify Program 8.14 to accept a tolerance rather than a number of steps, and to iterate, doubling the number of intervals at each step, until two successive approximations differ by less than the given tolerance. Build in some other stopping criterion as well (such as a maximum number of iterations) in case the specified tolerance cannot be achieved in a reasonable amount of time.

14. Modify Programs 8.12 and 8.13 so that they count the number of steps taken to find a root.

15. Experiment with the program produced in Exercise 14, using the set of polynomials of degree 1 through 5 with all coefficients having the value unity. Make a table of the number of steps required for each degree for each method. Use several starting points and take the average number of steps.

Engineering Problem 3

Escape Velocity of a Space Vehicle

An object, such as a spacecraft, on the surface of any body has to leave the surface with a certain velocity (the escape velocity) if it is to escape from the body's gravitational well. In this problem, we investigate how the escape velocity depends on the body's mass and determine the Earth's escape velocity. We will assume that the bodies in question are airless to remove the complication of the atmospheric resistance.

The force acting on a spacecraft near the Earth is given by

$$F = \frac{mgR^2}{(R+h)^2}$$

where h is the distance that the spacecraft is above the planet's surface. If we take the positive direction to be upward, then the equation of motion of the spacecraft is

$$F = m\frac{dv}{dt} = -\frac{mgR^2}{(R+h)^2}$$

We now have a differential equation giving the rate of change of velocity as a function of time. However, the right-hand side of

$$\frac{dv}{dt} = -\frac{gR^2}{(R+h)^2}$$

does not contain any explicit references to time (although h is a function of t). We convert this differential equation into one that describes the change of v with h. From elementary calculus we know that

$$\frac{dv}{dt} = \frac{dv}{dh} \cdot \frac{dh}{dt} = \frac{dv}{dh}v$$

Therefore

$$\frac{dv}{dh} = -\frac{gR^2}{v(R+h)^2} \quad (1)$$

We can use this differential equation to investigate the motion of a spacecraft that takes off from a planetary surface with a specified initial velocity. (In actual fact, most spacecraft do not actually leave the ground with a determined velocity, but accelerate during the initial minutes of lift-off. However, we will ignore this complication.) This gives an initial condition, namely that the initial velocity of the spacecraft at time $t = 0$ is $v = v_i$. Euler's method can be used to solve the differential equation at different heights and so determine the velocity of the spacecraft. If the initial velocity is small, the spacecraft will move upward for some time, but will eventually start to fall back toward the planet's surface. If the initial velocity is at least as large as the escape velocity, then the spacecraft will continue to move away from the planet forever.

Of course, in the world of physics, two bodies exert a gravitational force on each other at any distance, so that there isn't really an escape velocity. We can assume that a spacecraft that has reached a height of two Earth radii and still has positive velocity has escaped the Earth's gravitational well.

Programming Problems

1. Write a function beginning

    real function DvDh (V, H)
    real V, H

    that will calculate the value of differential equation (1). Use the following data for the Earth: radius = 6378260 meters, $g = 9.8$ m/s^2.
2. Write a subroutine that is passed an initial value for v, a step size, a number of steps, a starting height, and a function describing a differential equation. It then uses Euler's method to determine the values of v at heights determined by the step size. Your subroutine should print the velocity, the height, and the force on the spacecraft after each step. Remember that you should stop as soon as the velocity becomes zero.
3. Write a main program that uses the previous subprograms and can be used to detect the escape velocity of the Earth.
4. Sketch graphs by hand illustrating the velocity of a spacecraft as a function of height for initial velocities below the escape velocity, the escape velocity itself, and velocities above the escape velocity.
5. Experiment with the effect of the step size chosen on the accuracy of the answers produced by the program. What can you conclude?

You might want to look up the radii and masses of some other planets and moons. You can use the equation

$$M = gR^2$$

to calculate the value of g for each body and then calculate the escape velocity of each one using your program.

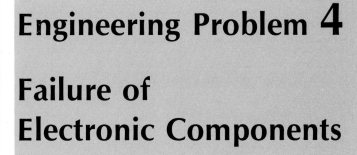

Engineering Problem 4

Failure of Electronic Components

If we construct any circuit with a large number of components, we must consider what will happen when one or more of the components fail. This is particularly important in computer circuits where large numbers of components are used.

There are many potential causes that might make a part of such a circuit fail. However, if the circuit is sufficiently large, we can model the failure rate using a *Poisson distribution*. This distribution describes what we think of as random failures. If the average number of failures in a year is m, then this distribution describes what we intuitively expect, that the number of failures in a year is unlikely to be either very much smaller than m or very much larger. More formally, the Poisson distribution states that the probability of x random failures in a specified time period is given by

$$P = \frac{m^x e^{-m}}{x!}$$

where m is the mean number of failures in the same time period.

For example, if we know that a failure anywhere in the circuit occurs, on average, once per year, then the probability of having three failures in a year is

$$P[\,3\ failures\,] = \frac{1e^{-1}}{3!}$$

or 0.0613132 (about 6%).

Circuits that have to perform under conditions where repair is difficult or impossible are often designed so that they can continue operating even in the presence of several failures. Such circuits are called *fault tolerant*. The Poisson distribution is very useful in determining how reliable such circuits are.

We can use the Poisson distribution to determine the likelihood of a small number of failures. The probability of at most f failures is given by

$$P[\text{at most } f \text{ failures}] = \sum_{x=0}^{x=f} \frac{m^x e^{-m}}{x!}$$

and the probability of more than f failures is given by

$$P[\text{more than } f \text{ failures}] = \sum_{x=f+1}^{x=\infty} \frac{m^x e^{-m}}{x!}$$

Programming Problems

1. Sketch, by hand, the shape of the Poisson distribution for some different values of the mean number of failures m. If you have access to a spreadsheet program, you may find that it provides an easy way to calculate the function values. If not, you can write a simple program to calculate them.

2. Write a program to calculate both of the preceding sums. Use values for the mean number of failures per year of 1, 10, and 50. Use the function exp to calculate the exponential in the summations. Because the second sum is infinite, you will have to calculate a finite but large number of terms. Remember that your machine has a limited range of real values that it can represent. You can do this problem without underflow or overflow, but only with some care. Check that the probabilities of there being no more than f failures and more than f failures add up to 1.

Engineering Problem 5

Time between Failures

We saw in the previous problem that the failure rates of circuit components can be described using a Poisson distribution. In this problem we look at the effect of the failure rate on the expected time between failures. This time is called the *mean time between failures* (MTBF) and is a very important measure of reliability. It can be used to determine the appropriate intervals between service or maintenance and, in situations where repair is not possible (for example, deep-space vehicles), can be used to determine the circuit lifetime.

For a system in which failures have a Poisson distribution, the probability that the interval between failures will be t is

$$P[t] = Me^{-Mt}$$

where M is the mean failure *rate*. This is called the inter-failure density function.

The expected time between failures is obtained by calculating the integral

$$E[t] = \int_{t=0}^{t=\infty} tMe^{-Mt}\,dt$$

We would like to be able to calculate this integral for circuits. Simpson's rule can be used to get an approximation to the integral, except that the integral we require is an indefinite one. To get around this, we calculate the integral over the interval $[0, y]$, where y is a sufficiently large number.

Programming Problems

1. Sketch, by hand, the shape of the inter-failure probability density function for different failure rates.
2. Write a program that reads a mean failure rate and calculates the expected time between failures using Simpson's rule. The upper limit for the integral

should also be input to the program. The subinterval size should be constant regardless of the size of the interval used for the integration.
3. Plot, by hand, the results of your program for a range of upper limits on the integral. Is there a point beyond which there is no improvement in the answer? Is there a point beyond which the answer gets worse, not better?
4. Plot, by hand, the expected time between failures as a function of the mean failure rate. Does it agree with the shape of the function you calculated in Problem 1?

Engineering Problem 6

Finding a Set of Sample Points

In sampling the effect of large-scale underground phenomena, it is common to plant detectors along some sampling line. In this problem we investigate ways of choosing points at which to place sampling instruments along such a line, given that we have a finite number of sampling instruments.

Suppose we have been given a line of length L and a special point l_r on that line. We wish to select a total of N sample points along the line, with the property that the selected points become denser (closer together) as we get close to l_r. The general idea is illustrated in Figure P6.1.

Suppose that we consider the function $n(l)$ representing the number of points that occur between one end of the line and the point l. Then we can describe the property of "becoming denser" by examining the differential equation dn/dl. When this function is constant, we come across new points at a constant rate as we move along the line, so the points are evenly spaced. When it increases, we encounter new points at a greater rate and so the points are closer together. We want this function to be flat when l is far from l_r and to become steep when it gets closer to l_r.

To make sure that we end up with N sample points, we need to also ensure that the integral

$$\int_0^L \frac{dn}{dl} \, dl = N$$

holds. This is done by using a scaling constant in front of the differential equation and calculating the value of the constant that will satisfy the integral.

Now to actually find the locations of suitable points, we must find values such that

$$\int_0^l \frac{dn}{dl} \, dl = 1, 2, \ldots, N$$

Finding a Set of Sample Points 199

Programming Problems

1. Construct an appropriate differential equation for a sampling line that is 10 km in length. The maximum distance between sample points should be 1 km, decreasing to a minimum of about 0.1 km near l_r.
2. Write a program to find the constant that will allow you to use this function to calculate 35 sample points.
3. By calculating appropriate integrals, find 35 sample points when l_r is 5 km from the end of the line (in the middle) and again at 2.5 km from one end.

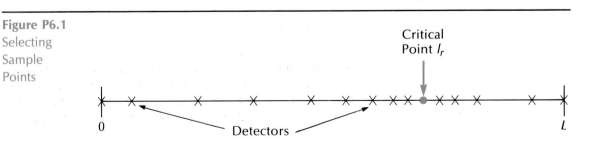

Figure P6.1
Selecting Sample Points

Engineering Problem 7

Calculating the Force of Impact

We know from Newton's Second Law that

$$F = m \frac{dv}{dt}$$

Rearranging this slightly we see that

$$F\,dt = m\,dv$$

and integrating both sides we get

$$\int F\,dt = \int m\,dv$$

The left-hand side of this integral is called the impulse of the force F over the time that it acts. The right-hand side measures the change of momentum produced by the action of the force over a period of time.

If we graph the force as a function of time in a real collision (as opposed to a collision between ideal bodies that takes place in zero time), it starts at zero, rises to a maximum, and then drops to zero again. The equation above shows that the area under such a curve equals the change in momentum during the collision.

Suppose that we consider the collision of a ball with a wall. The ball has mass m, approaches the wall with velocity v_1, and leaves the wall with velocity v_2 (of opposite sign). Using this information we can calculate the impulse of the force exerted on the ball.

$$Impulse = m(v_1 - v_2)$$

If we want to calculate the *maximum* force exerted on the ball, then we must know the shape of the force versus time curve and the duration of the collision. No matter what the duration of the collision, the area under the force-time curve must be constant because the change of momentum is constant. If the ball is relatively hard, then the collision duration will be small and hence the maximum force will be large. If the ball is yielding, then the duration of the collision will be larger and the maximum force will be less.

We can approximate the shape of the force versus time function by

$$ae^{-x^2/b^2}$$

with the origin at the time of maximum force and b a parameter that measures the sharpness of the curve. This function does not reduce to zero at times before and after the collision but can be effectively treated as zero when its value is smaller than 10^{-3}.

Programming Problems

1. Calculate the impulse of the force exerted on the ball if its mass is 0.2 kg and its initial and final velocities are 15 m/s and -10 m/s.
2. Find the maximum force exerted on the ball if the duration of the collision is 3 msec, 2 msec, and 1 msec, respectively.

Input, Output, and Formatting

9.1 Simple Output
9.2 Formatting of Output
    Format Codes
    Carriage Control
    Printing Multiple Lines
    Repeating Format Codes
    Reusing Format Codes
    The Implied Do Loop
9.3 Simple Input
9.4 Structure in the Input Stream
9.5 Files and the File System
9.6 Plotting

Almost all programs written for computers perform some interaction with their environment. You might conceive of programs that do not (perhaps a program to simply consume a given amount of time), but they are few and mostly uninteresting. The most common means of passing values to and from programs is via input and output statements. You can think of input statements as being assignments of values coming from "outside" to variables inside the program and you can think of output statements as being assignments of the values of expressions coming from the program to variables (files or output streams) that are "outside" the program.

9.1 Simple Output

We have already seen the simplest form of the output statement in Fortran, namely the **print** statement. A list of expressions, separated by commas, can be specified for output on the screen, to a file, or to a printer, depending on the particular defaults of your computer system. An expression in the list can be a variable, a constant (numeric or character), or a combination of these and appropriate operators. Here is an example.

    print *,'There are ',I,' values averaging ',Sum/I

Each **print** statement begins a new line of output.

The **print** statement is actually a special way to specify the most frequently used case of a more general statement. The more general statement is **write**. Its form is

    write (*,*) expressionList

So the more general form equivalent to the example is

    write(*,*) 'There are ',I,' values averaging ',Sum/I

9.2 Formatting of Output

The details of how to display the expressions in the output statements we have written thus far were all decided by the computer system. Strings are printed in a field of exactly the required width, and there is also a predetermined width for numeric values. Integer values appear as you would expect, but in a field of this predetermined width, so that they often seem oddly spaced. Floating point numbers come out in scientific notation in a field of the predetermined width. This notation can be difficult to read, but it can flexibly represent numbers of a wide range of magnitudes. It is also similar in structure to the internal representation of the values in the machine, so it's a good choice as a default.

Although these defaults allow a program to communicate, they do not allow it to do so as attractively and effectively as is usually required. Accordingly, there is a mechanism in Fortran for overriding the defaults and specifying formatting requirements explicitly. This is done by associating format codes with the **write** statement; these codes specify how each expression is to be laid out in the output line. In the **print** statement we have always used an asterisk as the first item after the keyword. In fact, this asterisk is an indication to the compiler that the default format codes are to be used. The defaults can be overridden in two ways. The first is that the asterisk can be replaced by a string containing the format codes to be used.

> **print** '*(formatCodeList)*', *expressionList*

The second is that a **format** statement can be written containing the codes. A **format** statement has this form:

> *statementNumber* **format** *(formatCodeList)*

The statement number is used as a label for the line containing the **format** statement, and is substituted in the **print** statement in place of the asterisk. The number on the **format** statement must appear in columns 1 to 5 on the line. For simple cases the first approach (putting the codes into the **print** statement) is attractive. However, sometimes format codes contain quotation marks and these have to be doubled (just as for a quotation mark inside a string constant); when this happens the statement can be messy and difficult to write and maintain.

It's also possible to use either of these two approaches with the **write** statement. The general form of the **write** statement that we showed previously contained two asterisks. The first asterisk can be ignored for now, but the second one plays the same role that the single asterisk does in the **print** statement. That is, the second asterisk in the **write** statement is an indication to the compiler to use the default formatting codes. These can be overridden either by putting the list of codes into the **write** statement itself, or by using a separate **format** statement and putting its number in place of the asterisk.

Statement numbers must be unique in a program unit (main program, subroutine, or function). Thus if you use a **format** statement, the number you use for it cannot appear on any other statement in that program unit. Several output statements can refer to the same **format** statement number.

The placement of **format** statements in a program is a matter of taste. If a **format** statement is used only once, it's probably a good idea to keep it with the input or output statement that uses it. If the **format** statement is used several times, you may want to move it to the end of the subroutine or program. Some programmers prefer to put all **format** statements together at the end of a program, and you may encounter that style if you modify someone else's code.

Format Codes

Now we consider the **format codes** that can be used. The list of codes comprises individual format codes separated by commas (with an exception we will point out). A format code can specify spacing on a line or it can specify how to transfer an expression to the line.

To transfer an integer use

　　　I*width*

This indicates the transfer of an integer expression into a field of the specified width on the output line. If the field is larger than the space required to print the integer value, then the value is placed at the right-hand end of the allowed space (right justified) and blanks are used to fill the beginning of the field. If, when the program is executed, the width is not sufficient to print the value, the field is filled with asterisks to indicate this. These two properties are true for all of the format codes: extra space is padded with blanks, and insufficient space results in asterisks. This program

```
*       Program 9.1
        program Show
        print '(I4, I3)', 1, 1000
        stop
        end
```

produces the following output:

　　　1***

since the first integer is put into a field of width 4, and the second will not fit in a field of width 3, so asterisks are printed instead. This property can be very annoying when a program has done a substantial amount of calculation and then is unable to display its results. The only way to avoid this irritation is to always think about what the largest reasonable data value can be and allow a generous amount of room for it.

To transfer a string use

　　　A*width*

This indicates the transfer of a string expression into a field of the specified width on the output line. When the program is executed, if the width is not sufficient to print the value, the value is truncated on the right and the specified number of characters is printed. If the width is more than is required, the value is printed at the right of the field and sufficient blanks are printed on the left to fill the field. The numeral is actually optional in this code, since if it is omitted the computer system will allow exactly as many character positions as are required to print the expression. Here is an example.

```
*       Program 9.2
        program Show
        print '(A4, A4, A4, A, A1)',
    *      'Help','me','quickly','please','!'
        stop
        end
```

This example produces the following output:

Help   mequicplease!

The first string is transferred exactly into the field; the second is short, so it's padded on the left with blanks; the third is truncated on the right; the fourth uses exactly the number of characters required; and the last character shows where the generated characters end on the page.

A literal string appearing in a **format** statement indicates that the string value itself is to be transferred into a field of precisely the required width. For example,

```
*       Program 9.3
        program Show
        print 100
100     format ('Hi there.')
        stop
        end
```

results in the printing of a single line

Hi there.

To transfer a floating point value use

F*width*.*fractionWidth*

This indicates the transfer of a floating point value as a fixed point representation. The value is to be printed as a decimal fraction in a field of the specified width, with the specified number of digits after the decimal point. The width must be enough to accommodate the fractional digits, the decimal point, the integer part of the expression value, and a minus sign if the number is negative. The decimal part of the value is *rounded* to the specified number of decimal places before it is printed. When the output statement is executed, if the value cannot be represented in the specified field, the field is filled with asterisks to indicate this. The next example illustrates this code.

```
*       Program 9.4
        program Show
        print 100, 1.5, 1e10, 123456.78, -.1
100     format (F6.2, F6.2, F10.2, F3.1)
        stop
        end
```

It generates the following output:

```
1.50******  123456.75-.1
```

The first item goes into a field of width 6 with two digits after the decimal point; the second item cannot fit into a field of width 6, so the field is filled with asterisks; the third item fits into its field, but we see that the representation (at least on the machine on which the example was executed) could not store all the significant digits; the last item fits into its field, using one character for the minus sign.

To transfer a very large or very small floating point value use

*E*width.*fractionWidth*

This indicates transfer of a floating point value in scientific notation. The width must be at least seven greater than the fractionWidth since it must provide room for a leading plus or minus sign, a zero, a decimal point, the fraction digits, an exponent indicator (the letter "E"), a sign for the exponent, and two digits of exponent. If the value will not fit in the field, the field is filled with asterisks. If the field width is more than the minimum required, the value is placed to the right of the field. Here is an example:

```
*       Program 9.5
        program Show
        print 100, 1.5, 1e10, 123456.78, -.1
100     format (E12.5, '!', E14.5, '!', E14.10, '!', E8.1)
        stop
        end
```

The output produced is

```
0.15000E+01!   0.10000E+11!**************!-0.1E+00
```

The third value transferred will not fit in the field specified, since there is not room for all the characters required in addition to the fractional digits, so the field is filled with asterisks.

The most commonly used spacing format code is

*width*X

which is used to insert blanks into the output. A width must be specified, even when only one blank is required.

*   Program 9.6
    program Show
    print 100
100 **format** ('Help', 1X, 'me', 2X, 'please!')
    stop
    end

The output produced from this is:

Help me please!

It is not legal to leave out the "1" in "1X".

The codes presented so far are reasonably straightforward, and they are sufficient for printing most of the things you'll need. But there are several more codes. You can think of them as more convenient mechanisms for doing common things that can only be done with some trickery using the format codes that have been presented thus far.

First we'll present some additional codes for transferring data items. There are several variants of a tab code.

    T*column*

is used to tab to a specific column on the output line. Movement can be either right or left. Thus

*   Program 9.7
    program Show
    print 100, 12345, 67890, 3
100 **format** (T10, I5, T20, I5, T5, I1)
    stop
    end

transfers three integers.

    3   12345   67890

The first goes into a field of width 5 at column 10, the second into a field of width 5 at column 20, and the third into a field of width 1 at column 5.

The tab specification can be relative instead of absolute.

    TR*positions*
    TL*positions*

These cause movement of the specified number of positions to the right or left, respectively. For example,

```
*       Program 9.8
        program Show
        print 100, 1234, 5678, 9101
    100 format (T16, I4, TL8, I4, TR8, I4)
        stop
        end
```

produces this output:

```
            56781234      9101
```

An integer goes in a field of width 4 printed at column 16; then there is a shift of 8 columns to the left followed by an integer printed in a field of width 4; then there is a shift of 8 columns to the right followed by a third integer printed in a field of width 4.

There is a code that gives more flexibility with respect to numeric representations.

$Gwidth.fractionWidth$

is a general numeric format, allowing a value to be printed without an exponent if it can fit in the specified field in that form (like an F code) or with an exponent if not (like an E code). The width value specifies the number of spaces to be used. The rightmost four spaces are reserved for an exponent (the letter "E", a plus or minus sign, and two digits); they are left blank if a fraction is printed and the exponent is not needed (in other words, if the fixed point representation can be used). If there are d spaces reserved for the fractional part, then a value will be printed using the F format code if it has magnitude between 0.1 and 10^d. Thus the d field plays a slightly different role in the G format code than it does in the E and F codes. (G format can also be used for other types such as integer.) As an example, this output statement

```
*       Program 9.9
        program Show
        print 100, 1.2345, 12.345, 123.45, 1234.5, 12345
    100 format (G10.3, G10.3, G10.3, G10.3, G10.3)
        stop
        end
```

produces in the following output:

```
    1.23      12.3      123.      0.123E+04      12345
```

When the values transferred fit in the fixed point representation, that is what is used; but when the values won't fit that way, they are presented in the floating point form.

Carriage Control

There's a significant fact about output that we have been ignoring up to this point, having to do with how lines are placed on the output page of a printer. Fortran was designed when output devices were simpler than the wide variety of devices that we now have for printing. The "model" of printing understood by Fortran includes a printer feeding a continuous form of paper past a print mechanism. The paper is perforated into pages. When producing an output line, the programmer had to tell the printer how to move the paper relative to the print mechanism. This was done by sending a single character telling the printer how to control its carriage — how much to move up or down on the page — followed by the actual characters to be printed. Thus the first character generated by each output statement in a Fortran program is consumed for **carriage control** when the output is going to a printer. When the output is going to a file, the character actually appears in the generated output — it is not consumed anywhere in the computer system.

There are several different carriage control characters with predetermined meanings. If the carriage control character is a blank, the meaning is to space one line. Thus, printing a series of lines with initial blank characters results in single-spaced output, which is what one might expect as a reasonable default. Here are a few carriage control characters that are interesting and useful; your system might support others but they tend to be less common and less useful.

- The blank (" ") means space one line before printing, and thus allows you to produce single-spaced output.
- The character zero ("0") means space two lines before printing, and thus allows you to produce double-spaced output.
- The character one ("1") means skip to the first line of a new page.
- The plus sign ("+") means do not space before printing, and thus allows for overprinting of the previously printed line.

Each output line directed to a printer will have its first character consumed as carriage control. This can obviously have surprising results. The safest approach is to always explicitly provide carriage control when printing lines.

The next example shows some output using carriage control.

```
*       Program 9.10
        program Show
        print 100, 'One'
100     format (1X, A)
        print 200, 'Two'
200     format (1X, A)
        print 300, 'Three'
300     format ('0', A)
        print 400, 'Four'
400     format ('+', T10, A)
        stop
        end
```

When the output from this program is directed to a file, the carriage control characters are kept on the lines that are produced. Here's what the file contents look like.

```
 One
 Two
0Three
+        Four
```

But when the lines are directed to a printer that understands the carriage control conventions that Fortran uses, this is what is printed:

```
One
Two

Three    Four
```

The leading character of each line has been stripped off and "interpreted" in the manner we specified. All the examples we've shown previously have actually been cheats to some extent; we've just looked at the contents of each line produced as if it had gone into a file, rather than considering how the first character of each line would be interpreted as carriage control if the lines were printed.

Printing Multiple Lines

So far everything we have shown has had a single output line created from a single output statement. It is often convenient to let a single output statement generate more than one output line, and Fortran allows this with the "/" format code. The interpretation of the slash is "start a new record." Unlike other format codes, the slash does not have to be separated from adjacent codes with a comma.

The next example shows some uses of the slash format code.

```
*         Program 9.11
          program Show
          print 100, 123456789, 1234, 5678
100       format (1X, I10, /, 1X, I5, I5)
          print 200, 123456789, 1234, 5678
200       format (1X, I10/1X, I5, I5)
          print 300, 123456789, 1234, 5678
300       format (1X, I10, ///, 1X, I5, I5)
          stop
          end
```

The first output statement controls the transfer of three integers into two output lines. The carriage control for the first line is a blank (generated by the X code) and the line contains a single integer in a field of width 10. The slash indicates that the first output line is ended, and then a second line is built with a blank for carriage control (generated by the second X code), followed by two integers each in a field of width 5. The second statement is equivalent to the first; it has simply omitted the commas around the slash. These statements used one slash to generate zero blank lines (it simply ended one partially filled line, and the next line began to be filled). The third statement causes two blank lines to be produced, since it begins three new records — the two blank ones and then the one that actually receives the last two output values. Here is what the output would look like if it were stored in a file.

```
123456789
1234  5678
123456789
1234  5678
123456789


1234  5678
```

Repeating Format Codes

All the codes we have shown, except for the slash and T (tab) codes, can be preceded by an integer constant indicating a repetition factor. Repetition factors can be applied not only to a single code, but also to any bracketed group of codes. These repetition factors are often called **group counts**. The next program illustrates the use of group counts, both for single format codes and for bracketed groups of codes.

*	Program 9.12
	program Show
	print 100, 123456789, 1234, 5678
100	**format** (1X, I10, /, 1X, 2I5)
	print 200, 1, 2, 3, 4, 5, 6, 7
200	**format** (1X, I10, 3(2(/), 1X, 2I5))
	stop
	end

produce the following output:

```
 123456789
 1234 5678
         1

    2    3

    4    5

    6    7
```

The first output statement is equivalent to those in the previous example program. The second statement has two levels of repetition. The two slashes actually generate one empty line, and the larger bracketed group is repeated three times to handle the amount of output specified.

Reusing Format Codes

It may happen that a format statement does not contain as many transfer codes as there are values specified to be transferred. In this case some or all of the **format** statement will be reused. Each time part of the statement is reused it is as if a new format statement were being used in the sense that a new line of output is begun. The rule for deciding what to reuse is simple to state, but can be confusing in use. It is this: scan right to left, and reuse the codes inside the first paired set of parentheses or its multiple. Here are a few examples.

*	Program 9.13
	program Repeat
	print '(1X, I3)', 1, 2, 3, 4
	print '(1X, (I3))', 1, 2, 3, 4
	print '(1X, (I3), 1X)', 1, 2, 3, 4
	print '(1X, (I3, I2))', 1, 2, 3, 4
	stop
	end

Here is the output produced from the program.

The Implied Do Loop

The final feature we want to present in this section is the **implied do loop**. This is a Fortran input and output statement option that is essentially a shorthand notation for a list of data items. Instead of writing a single data item in an output list, we can write something of the form

(itemList, variable = start, stop, step)

where the step specification is optional, as it is in the regular **do** construct. Since an implied do loop replaces an item in the list, you might expect that an implied do loop could be placed inside another implied do loop, and this is indeed the case. Here is an example:

```
*       Program 9.14
        program Repeat
        print '(15I3)', 1, (2, I=1, 3)
        print '(15I3)', 1, (2, 3, I=1, 3)
        print '(15I3)', 1, (2, I=1, 2), 3, J=1, 3)
        stop
        end
```

This program produces three lines of output:

```
1  2  2  2
1  2  3  2  3  2  3
1  2  2  3  2  2  3  2  2  3
```

Input, Output, and Formatting 215

The first line includes the value 1 one time, followed by the output generated by the implied do loop, which is the value 2 appearing three times. The second line includes the value 1 one time, followed by the output generated by the implied do loop, which is the sequence of values 2, 3 appearing three times. The third line shows a nested implied do loop output.

It is also possible to use implied do loops in input statements, but this doesn't really have any practical use unless it is combined with arrays, which we introduce in Chapter 10.

9.3 Simple Input

In Chapter 5 we introduced the simple input statement, which has the form

    read *, variableList*

This allows us to read values from an input stream. Numeric values are separated by commas or blanks.

The asterisk in the simple input statement stands for a default format control, just as it did in the simple output statement. We can write explicit format control using either a list of format codes or a **format** statement number in place of the asterisk. For example, these two **read** statements have the same effect.

    read '(10X, I5, 5X, I5)', A, B
    read 100, A, B
100 **format** (10X, I5, 5X, I5)

In each case the input is assumed to be ten characters that are to be ignored, followed by an integer in a field of width 5, followed by another five characters to be ignored, followed by another integer in a field of width 5.

Most of the format codes have an obvious interpretation when used for input. One complication arises when the data item being read does not exactly match the assumption in the format code. The rules to use in these cases are really quite simple. First, a field consisting entirely of blanks is read as a zero; this means that if the five-character field expected to contain an integer in the example actually contained "   ", it would be read as 0. Second, a decimal point or an exponent explicitly given in a numeric field will override the specification given by the format code; this means that if the code F6.2 were used to read the string "123456", the result would be 1234.56, but if the input were "1.2345", the result would be 1.2345.

The form of the input statement that we have been using is actually a simplified form. The more general form is

    read (*, *) *variableList*

Here the second asterisk corresponds to the default format specification and can be overridden by replacing it either with a list of codes or the statement number of a **format** statement. We will discuss the relevance of the first asterisk later in the chapter. Formatted input is not used very often when the input comes from the standard input because it's tedious for humans to have to place input in particular columns. However, it can be useful when the program's input is itself the output of some other program. It permits only certain parts of the input to be read and can make such interactions easier.

9.4 Structure in the Input Stream

Programs often get their input from files. In fact, one example of a file is the standard input stream, which usually comes from the keyboard. There can be many such files used in a program. In addition to the contents of the file, there is an explicit end-of-file indicator that can be detected by your program. This is done in Fortran as an option of the **read** statement. We can actually write an additional optional item in the statement in the following manner.

```
*         Program 9.15
          program EndFil
          integer A
          while (.true.) do
             read (*, *, end=100) A
             print *, A
          end while
100       continue
          stop
          end
```

Here the **end**= specification in the **read** statement specifies the statement number of a statement to which control passes when the end-of-file condition is detected. This condition is detected when the program attempts to read past the end of the data to get the next (nonexistent) record. If this program were applied to an input file containing these two lines

1 2
3 4

the output produced would be

1
3

and the program execution would terminate normally.

This explicit end-of-file indicator is very useful. It allows us to write simple programs for some common tasks. For example, here is a program to read a list of samples comprising a voltage reading and a time and print all those with negative readings.

```
*          Program 9.16
*          Illustrates handling end of file
*
           program EndFil
           integer Volts, Time
           while (.true.) do
               read (*, *, end=100) Volts, Time
               if (Volts .lt. 0) then
                   print *, ' At', Time, ' reading was', Volts
               endif
           end while
      100  continue
           stop
           end
```

9.5 Files and the File System

All the **input-output** we have described thus far has been input from what we have called the **standard input** and output to what we have called the **standard output**. While many useful programs can be written assuming a single input stream and a single output stream, there are other things you'll want to do where this is not enough. For example, if you had a file containing information about the steam turbines installed in a power plant and received daily information about maintenance activity on the turbines, you might want to **merge** the daily information with the existing file and at the same time produce a summary report on the activity so far this month. This would require three files: the pre-existing file on the turbines, the file (possibly the standard input) of daily information that is to be merged in, and the file (possibly the standard output) containing the report. Fortran has features that allow us to define and manipulate many files. We'll introduce these features in this section.

First, let's consider the form of input statements. We have already said earlier in the chapter that the form we have been using throughout the book is a simplified one and that there is a more general one available. This is

   read (*, *format*) *variableList*

This causes input to be taken from the standard input. This is implied by the asterisk in the statement. In fact, we can make the indication of the standard input explicit by replacing the asterisk with the number 5. We can do this because Fortran allows an integer representing a file at this point. This particular value has historically been associated with the standard input.

> **read** (5, *format*) *variableList*

The implicit or explicit indication of the standard input can be replaced by an explicit indication of another input file. The integers indicating files are sometimes called **unit descriptors**.

The output statement can be generalized in an analogous manner, replacing the implicit indication of the standard output (the explicit file designator is the integer 6) with some other file designator. Thus,

> **write**(*, *formatSpecification*) *expressionList*

is equivalent to

> **write**(6, *formatSpecification*) *expressionList*

Consider the following situation. We have an electronic device with two voltage measurement devices attached to it. Each device periodically produces output comprising the time and the measured voltage. Each device has its own file for these data. We want to write a program that merges the two files, producing a single file of data with the values interleaved as appropriate to produce a single time line. It may happen that the two measurement devices both produce output for a time that is the same, as far as the resolution of their internal timers is concerned. In that case we will assume that either reading is acceptable and keep one and delete the other.

Merging data streams of various kinds is a common paradigm in computing. This particular example is simple enough to state, but it requires a bit of care to handle the cases where one of the files has run out of data items. The next program does the merging. You should read it carefully to be sure you understand it.

```
*       Program 9.17
*       Merges two files
*       Each file is a list of (time, voltage) pairs
*
        program Join
        call Merge(1, 2, 3)
        stop
        end

        subroutine Merge(In1, In2, Out)
        integer In1, In2, Out
        integer Time1, Volts1
        integer Time2, Volts2
        read (In1, *, end=200) Time1, Volts1
        read (In2, *, end=300) Time2, Volts2
```

```
      while (.true.) do
         if (Time1 .lt. Time2) then
            write (Out, *) Time1, Volts1
            read (In1, *, end=100) Time1, Volts1
         else
            write (Out, *) Time2, Volts2
            if (Time1 .eq. Time2) then
               read (In1, *, end=200) Time1, Volts1
            end if
            read (In2, *, end=300) Time2, Volts2
         endif
      end while
*     copy the remaining items
  100 continue
      write (Out, *) Time2, Volts2
  200 continue
      while (.true.) do
         read (In2, *, end=400) Time2, Volts2
         write (Out, *) Time2, Volts2
      end while
  300 continue
      while (.true.) do
         write (Out, *) Time1, Volts1
         read (In1, *, end=400) Time1, Volts1
      end while
  400 continue
      return
      end
```

The explicit file designators In1, In2, and out matching the file numbers 1, 2, and 3 must be associated with appropriate files using the techniques required by your operating system. You'll have to consult your system's documentation to see how to do this, since it varies considerably from system to system. (In Appendix E we show how to do it for a few common systems.) The input and output statements used are much like those we've used before, and as long as we are careful in the use of file designators, there should be no problem with programs like this that use several files.

Since each of the while statements consumes input from a file that can be assumed to be finite, and since there is an **end=** option on the input statement that causes a jump out of the iteration, we can be confident that this program will terminate.

If we use the following input files:

In1:

1 10
2 15
4 20
6 25
8 30
10 35

In2:

1 5
3 6
4 20
5 7
10 5

the following output file will be produced:

1	5
2	15
3	6
4	20
5	7
6	25
8	30
10	5

Let's consider a variant of this problem. Suppose now we're interested in looking at only those times when both measurement devices took readings at the same time, and the readings were different. This might be useful if we suspected one of the devices was not functioning correctly. The basic framework of the previous program can be used, but there are some changes to be made to avoid producing the merged output stream except in these special cases.

```
*       Program 9.18
*       Prints the differences between 2 files
*       Each file is a list of (time, voltage) pairs
*
        program Join
        call Diff(1, 2, 3)
        stop
        end
```

Input, Output, and Formatting

```
      subroutine Diff(In1, In2, Out)
      integer In1, In2, Out
      integer Time1, Volts1
      integer Time2, Volts2
      read (In1, *, end=100) Time1, Volts1
      read (In2, *, end=100) Time2, Volts2
      while (.true.) do
         if (Time1 .lt. Time2) then
            read (In1, *, end=100) Time1, Volts1
         else
            if (Time1 .eq. Time2) then
               if (Volts1 .ne. Volts2) then
                  write (Out, *) Time1, Volts1, Volts2
               endif
               read (In1, *, end=100) Time1, Volts1
            endif
            read (In2, *, end=100) Time2, Volts2
         endif
      end while
  100 continue
      return
      end
```

If we run it with the same input files as before, it will produce this output:

```
     1         10         5
    10         35         5
```

We said earlier that to associate unit numbers with external files you would use operating system facilities. Fortran provides language facilities that are a second way of dealing with making this association. Each method has its own advantages and disadvantages—using the operating system means learning a new way to do it on every new system. There are two statements in Fortran that are useful in programs of this type. The first is the **open** statement. Its form is

   open (*fileDesignator*, **file**=*fileName*)

Here *fileName* refers to a file known to the operating system.

Suppose we had the two input files from the previous example in files named "ONE FILE" and "TWO FILE", respectively, on our machine. We could replace the main program of the previous example with the following one.

```
*       Program 9.19
*       Merges two files
*
        program Join
        open (1, file='ONE FILE')
        open (2, file='TWO FILE')
        open (3, file='THREE FILE')
        call Merge(1, 2, 3)
        stop
        end
```

There is also a **close** statement in Fortran. It is used to dissociate an internal designator from an external designator.

close (*fileDesignator*)

Suppose we wanted to read some data from a file and then read additional data of the same kind from another file. If the data streams were highly structured, it might make sense to use the same pieces of program to do both. This could be easily done by opening one file and reading the data it contains, then closing it, opening another, and reading the data that one contains.

Here is a simple example to illustrate this concept. This example doesn't do anything interesting apart from providing the illustration. If we have two files, "ALPHA FILE" and "BETA FILE" containing the integers 157 and 637, respectively, this program

```
*       Program 9.20
*       Illustrates opening and closing files
*
        program OpenIt
        integer I, J
        open (1, file='alpha file')
        read (1, *) I
        close (1)
        open (1, file='beta file')
        read (1, *) J
        print *, I, J
        stop
        end
```

would produce this single line of output.

           157   637

The names that are used for the files must conform to the rules of the file system on your computer.

Input, Output, and Formatting

In addition to the unit descriptors 5 (for the standard input) and 6 (for the standard output), some programs will use 7 as another output stream. Years ago this was assumed to be an output stream directed to a card punch, just as the standard input was assumed to be a card reader and the standard output was assumed to be a line printer. Some older programs that you encounter might contain references to unit 7.

9.6 Plotting

We now have enough language mechanisms at our disposal to write plotting routines. Graphical output of computational results is often important in engineering problems; sometimes it's easier to see what is happening when data are displayed graphically rather than as tables of numbers; at other times it's virtually impossible for a human to make sense of the data without a graphical display, as in the case of a stress analysis being done on the wing of an aircraft.

As a first step, let's write a simple program to step along the x-axis and print out the value of a specified function at some points. We'll assume that the program is to be given a function to use, a minimum x value, the increment to be used between successive values, and the number of steps to be taken along the x-axis. The next program accomplishes this. We have used $\sin x$ as the function.

```
*       Program 9.21
*       Illustrates the printing of function values in a table
*
        program ShowMe
        external FOfX
        real FOfX
        call Show(FOfX, 0.0, 0.4, 15)
        stop
        end

        real function FOfX(X)
        real X
        FOfX = sin(X)
        return
        end
*
*       This subroutine handles the actual printing
*
        subroutine Show(F, Start, Step, Number)
        real F, Min, Step
        integer Number
        real Here, FHere
        integer I
```

```
      do 200 I = 0, Number
         Here = Start + I * Step
         FHere = F(Here)
         print '(2(1X, G12.5))', Here, FHere
200   continue
      return
      end
```

This program is very easy to understand. Unfortunately, its output is not very attractive; even for this simple bit of mathematics, the list of numbers is not particularly effective at conveying a sense of what is happening. The output looks like this.

```
0.00000E+00    0.00000E+00
0.40000        0.38942
0.80000        0.71736
1.2000         0.93204
1.6000         0.99957
2.0000         0.90930
2.4000         0.67546
2.8000         0.33499
3.2000         -0.58374E-01
3.6000         -0.44252
4.0000         -0.75680
4.4000         -0.95160
4.8000         -0.99616
5.2000         -0.88346
5.6000         -0.63127
6.0000         -0.27942
```

It would be a marked improvement to display these results graphically. We'll now show one simple way to do that. As with many of the problems we discuss, we do not intend to show you a fully developed program of production quality. If you do much programming, you'll probably have graphical routines available to you. But we want you to understand the basic ideas involved and to think a little about what is happening when the machine produces complicated graphical displays.

What we're up to is showing a two-dimensional plot of the values of a function of one variable. To keep the programming simple, we'll make some assumptions that you would probably want to reconsider for a production version of the program. First, we'll assume that the x-axis will run vertically in the output; this isn't the most intuitive way to do it, but it is simple. Second, we'll assume precisely the same input parameters as in the program we just developed. This means we don't know the range of values the function takes on in the specified segment of the x-axis. We'll walk along the axis once to determine these values, and then we'll walk along it again to plot the values. This is very inefficient since it means we're computing everything twice. Once you see how to use arrays (Chapter 10), you can improve on this approach.

We won't discuss the program in any more detail. If you examine it, you should have no difficulty understanding all the pieces of it. The output statements are complicated, but do not use anything we haven't discussed previously.

```
*       Program 9.22
*       Illustrates plotting function values using simple output
*
        program ShowMe
        external FOfX
        real FOfX
        call Plot(FOfX, 0., .4, 15)
        stop
        end

        real function FOfX(X)
        real X
        FOfX = sin(X)
        return
        end
*
*       This subroutine plots the function values using
*       information about the line length and function
*
        subroutine Plot(F, Start, Step, Number)
        real F, Min, Step
        integer Number
        integer Width
        parameter (Width = 35)
        real Here, FHere, FMin, FMax
        integer I, N, Over
*       find the bounds
        FMax = F(Start)
        FMin = FMax
        do 100 I = 0, Number
           FHere = F (Start + I * Step)
           if (FHere .gt. FMax) then
              FMax = FHere
           else if (FHere .lt. FMin) then
              FMin = FHere
           endif
100     continue
```

```
*         now plot it
          print 600, FMin, FMax, ('-', N=1, Width + 2)
  600     format (11X, 'Y =', G12.5, ' to ', G12.5,/
     &    1X, 'X =', 9X, 67A1)
          do 200 I = 0, Number
             Here = Start + I * Step
             FHere = F(Here)
             Over = Width * (FHere - FMin) / (FMax - FMin)
             print 610, Here, (' ', N=1, Over), '*'
  610        format (1X, G12.5, '|', 67A1)
  200     continue
          print 620, ('-', N=1, Width + 2)
  620     format (13X, 67(A1))
          return
          end
```

We now show the output produced from the program, that is, plotting the values of sin x that we previously printed.

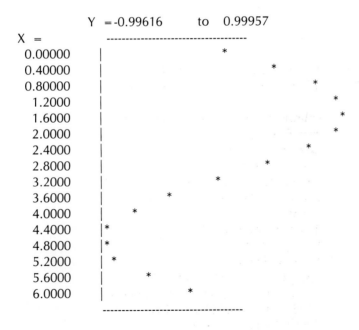

This program is a good demonstration that even simple concepts become much more difficult to program once interesting output is being produced.

We can change this program to plot other functions in a straightforward manner. The output that follows was generated for the function $f(x) = x^2$, using 20 as the value of the Width parameter.

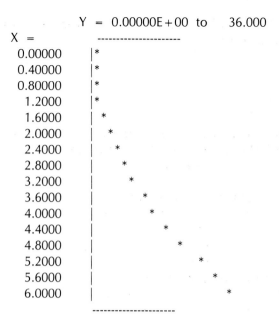

Programming Example

Problem Statement

Write a program to compute the trajectory of a shell fired with given muzzle velocity and angle of elevation, and plot the trajectory.

Inputs

Muzzle velocity and angle of elevation.

Outputs

Plot of vertical height against horizontal range in steps of 1000 m, until the shell returns to Earth.

Discussion

We need to plot the vertical height of the shell during its trajectory, but we don't know what the maximum height will be in advance. The program will compute the height as a fraction of the maximum height attained. This quantity can then be scaled for the plotting.

Program

```
*       Program 9.23
*       This program displays a shell trajectory numerically
*       or graphically. It uses equations of motion to
*       calculate height at each range.
*
*         X, Y - current x and y values
*         YMax, XMax - point of max height
*         MaxRng - maximum range
*         V - muzzle velocity
*         Theta - angle of elevation
*         A, B - useful temporary values
*         Outflg - flag for type of output required
*         IntHgt - height as an integer
*
        program Gun
        real X, Y, XMax, YMax, MaxRng, V, Theta, A, B
        integer OutFlg, I, IntHgt
        parameter (G = 9.8)
        write (*, 100)
100     format('Enter muzzle velocity and elevation (radians)')
        read *, V, Theta
        write (6,200) V, Theta
200     format(F10.2, F7.3)
        write (*, 300)
300     format('Numeric or Graphical Output? (0/1)')
        read *, OutFlg
        YMax = (V * sin(Theta)) ** 2 / (2 * G)
        XMax = V * V * sin(Theta) * cos(Theta) / G
        write (*, 400) YMax, XMax
400     format(' Max height is ', G12.2, ' at range ', G12.2)
        write (*,*)
        MaxRng = 2 * XMax
        X = 0.0
        A = tan(Theta)
        B = G / (2.0 * (V * cos(Theta)) ** 2)
```

Input, Output, and Formatting

```
        while (X.le.MaxRng) do
            Y = A * X - B * X * X
            Rat = Y / YMax
            if (OutFlg.eq.0) then
                write (*, 500) X, Y, Rat
500             format(2F10.2, F9.5)
            else
                IntHgt = 50 * Rat
                write (*,600) X, 'I', (' ',I = 1, IntHgt), '*'
600             format(F9.0,60A1)
            endif
            X = X + 10000
        end while
        stop
        end
```

Testing

Here is the output using the graphical output option:

```
Enter muzzle velocity and elevation (radians)
    1500.00   0.700
Numeric or Graphical Output? (0/1)
  Max height is     0.48E+05 at range       0.11E+06
```

```
      0.I*
 10000.I        *
 20000.I                 *
 30000.I                     *
 40000.I                         *
 50000.I                             *
 60000.I                                 *
 70000.I                                    *
 80000.I                                      *
 90000.I                                        *
100000.I                                          *
110000.I                                           *
120000.I                                           *
130000.I                                          *
140000.I                                         *
150000.I                                       *
160000.I                                     *
170000.I                                  *
180000.I                              *
190000.I                          *
200000.I                      *
210000.I                 *
220000.I           *
```

Here is the output in its numeric form:

```
Enter muzzle velocity and elevation (radians)
    1500.00   0.700
Numeric or Graphical Output? (0/1)
 Max  height  is        0.48E+05  at  range        0.11E+06

          0.00        0.00    0.00000
      10000.00     8050.60    0.16898
      20000.00    15356.64    0.32233
      30000.00    21918.13    0.46006
      40000.00    27735.05    0.58215
      50000.00    32807.41    0.68862
      60000.00    37135.21    0.77946
      70000.00    40718.45    0.85467
      80000.00    43557.13    0.91426
      90000.00    45651.23    0.95821
     100000.00    47000.78    0.98654
     110000.00    47605.77    0.99924
     120000.00    47466.20    0.99631
     130000.00    46582.06    0.97775
```

Input, Output, and Formatting

```
140000.00   44953.38   0.94356
150000.00   42580.13   0.89375
160000.00   39462.38   0.82831
170000.00   35600.00   0.74724
180000.00   30993.06   0.65054
190000.00   25641.56   0.53821
200000.00   19545.50   0.41026
210000.00   12704.88   0.26667
220000.00    5119.69   0.10746
```

Design, Testing, and Debugging

- Using formatted output almost always requires making a guess at how to construct the output appearance that you want and then running the program and seeing how the output looks. The formatting can then be adjusted to iron out any problems. It's usually faster to take this approach than to try and get it exactly right the first time.
- When using formatted input, always check that the values are placed where you expect them in the input line. If you are reading formatted data interactively, then it is very important to inform the user of the expected format in the prompt.

Style and Presentation

- Formatting output is, of course, crucial to making a program readily usable. You should be prepared to put some thought into the ordering and layout of output from your programs to make them as effective as possible.
- Visual data are more readily understood than tables of numbers. Presenting program output using a plot, even a simple one, can drastically improve its impact. You should give some thought to developing your own general-purpose plotting tools (if your system doesn't already have them) and using them as part of the ordinary output of your programs.

Fortran Statement Summary

Format Statement

This statement describes the patterns or templates to be used for input and output.

stmtNumber **format**(*formatCodeList*)

100 **format**(1X, I3, F7.2, E10.3, /, A10)

Write Statement

This statement allows output from the program. It is the most powerful of the Fortran output statements.

write (*unit, stmtNumber*) *expressionList*

write (*unit, formatSpecification*) *expressionList*

write (6, 100) I, J, A, B

Implied Do Loop

The implied do loop is used to repeat the printing or reading of objects without having to list them explicitly. It behaves exactly like an ordinary do loop except that it has more compact syntax.

(*itemList, index* = *start, stop*)

print *, ('?', I = 1, 10)

Chapter Summary

- The most general form of output statement is the **write** statement. Both **print** and **write** statements may have an associated format specification that defines how output is to be printed.
- A format specification is included in an output statement, enclosed by single quotation marks, or is placed in a separate **format** statement. It consists of format codes sufficient to match the items in the associated output list.
- The following format codes can be used: I for integer values, F, E, or G for floating point values, A for character values, L for logical values, X for horizontal space.
- Vertical spacing of output can be arranged by using carriage control. For some output devices, the first character of every output line is used to determine how to space vertically. Common characters are: blank for single spacing, 0 for double spacing, — for triple spacing, + for remaining on the previous line, and 1 for moving to the top of a new page. It is wise to assume

- that carriage control will always occur on output and provide a blank at the beginning of each line for which spacing is not explicitly required.
- The slash format code is used to begin a new output record.
- Any format code can be preceded by a group count indicating how many times it is to be used. A group count is required for the X format code.
- A set of format codes can be repeated by enclosing them in parentheses and placing a repetition count around them.
- The implied do loop is used to repeat a value in the output list of an output statement.
- Both the **read** and **write** statements can be used for input or output from somewhere other than the terminal keyboard and screen. This is done by referencing a unit number in the IO statement. This unit number can be associated with a file or other IO device.
- An input statement can perform some special action if it reaches the end of a file while it is reading. This can be done using the **end=** option on the **read** statement.
- The **open** statement is one way to associate a unit number with a file or other external IO device.
- Simple graphical output can significantly enhance the presentation of output data.
- Format can also be used for input statements but it is usually inconvenient for users. It can sometimes be useful when the input comes from the (formatted) output of some other program.

Define These Concepts and Terms

Format codes
Carriage control
Group counts
Unit descriptors

Standard input
Standard output
Merge

Exercises

1. Write a program to produce a formatted calendar for a single month, given the name of the month, the year, and the day the month begins.
2. Extend the program of the previous exercise to produce a calendar for an entire year, given only the year number and the day on which January 1st falls.
3. Describe the behavior of the program that merges the streams of voltage values and the one that finds the inconsistencies in the streams of voltage values, when either or both of the input files is empty.

4. Modify Program 9.22 to "shade" the area under the curve by putting some other character (perhaps a period) in place of blank in all positions to the left of the asterisk.
5. Write a program to plot using the same parameters as the program given in this chapter, but with the plot oriented so the *x*-axis is horizontal. You will have to do a great deal of extra computation for this. Once you have seen how to use arrays (Chapter 10), this can be done much more easily and efficiently.
6. Rewrite the programs in Chapter 8 so that they format their outputs in a sensible way.
7. Write a program to print an appropriately labeled table in which each row contains an integer, its square, its cube, its fourth power, and its fifth power. Have the integers range from 2 to 10.
8. The binomial theorem states that

$$(x + 1)^n = \sum_{i=0}^{n} \binom{n}{i} x^i$$

The coefficients on the right-hand side form an interesting pattern called Pascal's triangle when they are written with the coefficients for successive values of *n* on successive lines. Here are the first few lines:

```
            1
         1    1
      1    2    1
   1    3    3    1
1    4    6    4    1
```

Write a program to print the triangle in this form, given a maximum value for *n* as input.
9. Modify the program from the previous exercise to print the triangle rotated ninety degrees clockwise. We show the output produced when the input is 4.

```
1
   1
4     1
   3     1
6     2     1
   3     1
4     1
   1
1
```

Input, Output, and Formatting 235

10. Write a subprogram that takes a single integer parameter and prints the indicated number of spaces followed by an asterisk. Use this subroutine to write another subroutine that produces one zigzag.

```
*
 *
  *
   *
    *
   *
  *
 *
```

11. Write a program to print a file, putting some predetermined number of lines on each page (say 60 lines) and putting a line at the bottom of each page giving the filename and the page number.
12. Write a program that will read a file that may contain consecutive identical lines, and produce a new file in which each line appears exactly once.
13. Write a program that will read a file of lines of length 40 and produce, as output, a file of lines of length 80, where the bottom half of the file is moved up beside the top half. Thus the input file

    1
    2
    3
    4

    produces the output file

    1   3
    2   4

14. Write a subprogram that takes a two-dimensional array and its sizes in each dimension as arguments, and prints the array formatted in rows and columns.
15. Write a program that will read an integer and a filename from the standard input and then print the indicated number of lines from the end of the indicated file. This can be useful if you are looking for something that is at the end of a file, and you just want to see the last 10 lines.

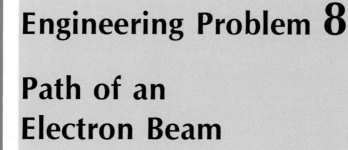

Engineering Problem 8

Path of an Electron Beam

You probably know that an electron gun is at the heart of devices such as television sets and oscilloscopes. Electrons are fired in a beam from a gun and the beam passes between two sets of orthogonal plates, one set that is vertical and the other that is horizontal. By applying voltages to the plates, the path of the electron beam can be altered.

If sinusoidal voltages are applied to each set of plates (not necessarily the same on each), then the path of the electron beam on the screen will follow a curve that can be described parametrically by

$$x = A_x \sin(\omega_x t + \theta_x)$$
$$y = A_y \sin(\omega_y t + \theta_y)$$

where A_i is the amplitude in direction i, ω_i is the angular frequency in direction i, and θ_i is the initial phase angle.

If the frequencies in the two directions are appropriately chosen, then the figure traced on the screen will form a closed curve that will slowly change its orientation.

Programming Problems

1. Write a program that will calculate the curves traced given initial data on amplitudes, frequencies, and phase angles.
2. Write an interface to a printer (or to your terminal screen if you know how) to draw the figures that are traced. Determine pairs of frequencies that give rise to closed curves. Can you determine any patterns in frequencies that work?

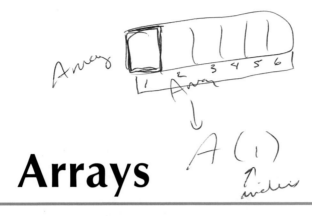

Arrays

10.1 Structured Data
10.2 Using Arrays
10.3 Recalculating the Standard Deviation
10.4 Other Uses of Arrays
10.5 Indexes
10.6 Other Types of Indexes
10.7 Arrays as Vectors
10.8 Polynomials
10.9 Linear Search
10.10 Binary Search
10.11 Sorting
10.12 An Example: The Sieve of Eratosthenes
10.13 Using the Common Statement

10.1 Structured Data

We have seen that there are several types of simple data objects that can be declared and used in Fortran. In this chapter we will look at the most important of the structured data objects that Fortran provides.

Consider the problem of writing a program to read a list of integers and print only those that are within 10 of the largest. It is easy to see that this cannot be done by only looking through the list of numbers once, because it's impossible to tell which numbers are within 10 of the largest until we know which is the largest. Determining the largest means looking at all the numbers because the largest might also be the last. So a program to solve this problem is going to have to examine the list of input values twice—the first time to determine the largest and the second time to print the appropriate numbers.

From the last chapter we already know one way to do this. We can read through the list of input values and then close and reopen the input device. The next read operation will get its data from the beginning of the file and so we will be able to look at the same list of integers twice. The program to implement this algorithm will contain two loops, one after the other. The first loop will find the largest element in the list. The second loop will determine whether or not to print each element. Here is a program that uses this method.

```
*       Program 10.1
*       Prints those numbers in a file that are within 10
*       of the largest. No prompts are used
*       because the input comes from a file.
*
        program Tops
        integer Num, Max
        open (5, file='SOME DATA')
        read (5, *, End=100) Num
        Max = Num
        while (true.) do
            if (Num.gt.Max) then
                Max = Num
            endif
            read (5, *, End=100) Num
        end while
100     close (5)
```

```
      open (5, file='SOME DATA')
      read (5, *, End=200) Num
      while (.true.) do
         if (Max - Num.le.10) then
            print *, Num
         endif
         read (5, *, End=200) Num
      end while
200   stop
      end
```

For this input file

127
45
136
89
117
5

this output is produced

127
136

There are two drawbacks to this kind of solution. It will only work if the input values come from a file of data that can be accessed more than once. The standard input file cannot be reused in Fortran—it wouldn't make much sense since it usually corresponds to the keyboard. You can imagine that a user would not be very happy with a program into which all of the data had to be entered twice. This limits the generality of the solution. Also, the time taken to read a value from an input device is much larger than the time taken to access a memory location, so that reading the data twice from the input file will be slow. A solution in which the data are kept in memory will be much faster. For very long lists of values, differences in speed like this may mean considerable savings in execution time.

This is only one example of a problem that requires using a list of data values more than once. Other examples are: finding the standard deviation using the usual formula, finding the values that are below average, and so on. In this chapter we look at the ways of representing data that are organized as lists and tables. The data structure used to represent them is called an *array*.

10.2 Using Arrays

A list, when thought of as a structured way of organizing data, has a number of properties. Any list contains items, or elements, that are themselves either simple or structured data. For example, a list of telephone numbers has items that are simple pieces of data—seven-digit numbers. A list of mailing labels has items that are structured, since each element of the list contains several pieces of information such as names and addresses. So the first property of a list is the kind of things that it contains.

The second property of a list is the number of things in it. A list may contain 20 elements or 200, but clearly the length of the list is important.

The third property of a list is the way in which the elements are distinguished. Some lists are implicitly numbered, with the first element of the list at the beginning, followed by the second, and so on. Other lists are organized using some kind of label for each entry—for example, a calendar is a list of months labeled by names such as January and April.

A list can also be considered to be ordered if the arrangement of the items is important, or unordered, if it is not. It turns out that all representations of lists in Fortran are ordered (at least implicitly), so we won't consider this aspect further.

An array in Fortran can be used to model lists with all of these properties. The elements of an array can be any of the simple data types we have seen so far—things like integers and reals. There is one major restriction in representing a list by an array—an array must have a fixed size determined at the time the program is compiled. Thus we cannot directly represent a list whose size cannot be determined, or at least bounded, when the program using it is written.

When an array is declared, the compiler has to be told what sort of elements it will contain, how many of them there will be, and how they are going to be referenced. The declaration must also give the array a name. All of the individual elements of an array share the same name and are distinguished by their position. Arrays implicitly order their elements since the elements are referred to by their positions within the array. However, programs do not need to exploit this information where it is not relevant, such as when the array is being used to represent an unordered list.

Here is an array declaration:

integer A(10)

This declares A to be an array of size 10 in which each item is an integer. The items, or elements, are referred to as A(1), A(2), A(3), A(4), A(5), A(6), A(7), A(8), A(9), and A(10). Each of these elements is large enough to hold a single integer value. The whole array occupies the same space as ten simple integer variables.

Entire arrays are not usually manipulated as units. Elements of arrays can be referred to in expressions and assignment statements in exactly the same way as simple variables. We can assign the value of an expression to an element of an array by writing

A(3) = 2 * X + Y

which assigns the value of the expression to the third element of the array. (The expression and the array element must be of compatible types, of course.)

The value of an array element can be used by including it in an expression such as the following:

X = 2 * A(3)

which assigns to X twice the value of the array element A(3). The object that distinguishes the particular element of the array being referenced is the **array index**.

We can now see how to write a program to solve the problem in the previous section without having to read the data values more than once. The data values that are read during the first pass can be kept in an array and accessed from the array during the second pass, which is required to compare them to the largest. Here is a program that uses an array as a list to keep the input values until they are needed for the second phase of the algorithm.

```
*       Program 10.2
*       Prints the numbers within 10 of the largest
*       using an array. It is much faster.
*
        program Tops
        integer Num(10), Max, I
        print *, 'Enter a number'
        read *, Num(1)
        Max = Num(1)
        do 200 I = 2, 10
            print *, 'Enter a number'
            read *, Num(I)
            if (Num(I).gt.Max) then
                Max = Num(I)
            endif
200     continue
        do 400 I = 1,10
            if (Max - Num(I) .le. 10) then
                print *, Num(I)
            endif
400     continue
        stop
        end
```

In this version of the program we have assumed that there are exactly ten values in the list of integers that we are concerned with. The variable Max holds the largest input value at any point. It is initialized to the value of the first element of the list because that is the largest value seen at that point.

The first loop reads the values and assigns them to successive locations in the array. The first value read in is placed in Num(1), the second in Num(2), and so on. As the values are being read, they are also being compared to the largest value seen so far. If a larger value is encountered, it replaces the value in Max. The second loop now looks at each element of the array again, checking to see whether it is within 10 of the largest. If it is, the value is printed. We know that none of the values in the array can be greater than the value of Max (because otherwise Max doesn't contain the largest), so we only need to consider the difference between Max and each array element, and not the absolute value of the difference.

10.3 Recalculating the Standard Deviation

We saw in Chapter 6 how to calculate the standard deviation of a set of data by only looking at each item of data once. The more usual way of calculating the standard deviation is to use a summation of the following form:

$$\sqrt{\frac{1}{n-1} \sum_{i=1}^{n} (x_i - \bar{x})^2}$$

This sum illustrates why the standard deviation is often called the root mean square of the deviation from the mean. Clearly the terms of this sum cannot be easily calculated until the mean itself has been calculated. Therefore this problem also has the property that each item of data must be used twice — once in calculating the mean and once in calculating the standard deviation. So a program that uses this formula to calculate a standard deviation will use an array and will contain two loops. The first loop will calculate the mean by summing the data values and will also place them into an array. The second loop will calculate the preceding summation, using the value of the mean determined in the first loop.

We are forced to specify the size of an array when we declare it, but we also usually want programs to be flexible enough to be used on lists of data values of different lengths. Today we might have a list of 20 values whose standard deviation we want to calculate, but tomorrow we might have 100 values. A program that only allows a fixed-size array in which to store the list is not very useful if the only way to use it with different amounts of data is to change and recompile it for each. Unfortunately, Fortran does not allow the flexibility to allocate an array of unspecified size. You can probably already see why this might be — the compiler has to allocate space for each element of the array and so must know how many elements there are going to be.

The usual solution to this problem is to declare an array of a size large enough to hold the largest set of data that we reasonably expect. Then we only use a part of it — a different amount depending on the particular data we are processing. We might declare our array to contain 200 elements. The program can then be run with a list of 20 values, 70 values, or 200 values. The compiler has allocated all 200 locations for the elements of the array, but we will only have used 20 (or 70, or possibly even all 200) of them. The other array elements are allocated but undefined. That is not a problem as long as we don't try to print their contents or refer to them. Of course, we have wasted some space in memory because we have allocated many locations that are not usually used, but on most modern computers, memory is not a significant limitation. We have to be sensible and appreciate that there is a trade-off involved between the wasted storage space and the difficulties caused by finding a set of data too large for the program to handle. It would be ridiculous to allocate an array of size 3000 for a situation where the usual size of data is 200, on the off-chance that one day we would need such a large array.

Programs must keep track of how many of the array locations are in use during each particular execution so that the undefined elements are not referred to by accident. It wouldn't work, for example, to read 100 values into the first 100 locations of an array with 200 elements and then print the entire array. We can think of the array as having a static bound, declared at compile time, and a dynamic bound marking the part that is actually in use. The dynamic bound must be maintained by the program.

We use this idea in the next version of the standard deviation program. The program reads in data until it encounters the end of file. When it processes the list the second time, it uses only those array locations that had values placed in them by the first loop.

```
*       Program 10.3
*       Calculates standard deviations keeping
*       the values in an array.
*
        program CalcSDV
        real X(200)
        real Sum, SumSq, Mean, StdDev
        integer I, J, N
        Sum = 0.0
        I = 1
        while (I.le.100) do
            print *, 'Enter next number'
            read (*,*, end=100) X(I)
            Sum = Sum + X(I)
            I = I + 1
        end while
100     continue
        N = I - 1
        Mean = Sum / N
        SumSq = 0.0
        do 200 J = 1,N
            SumSq = SumSq + (X(J) - Mean) * (X(J) - Mean)
200     continue
```

```
      StdDev = Sqrt(SumSq / N)
      print *, 'Mean is ', Mean, ' Standard Dev is ', StdDev
      stop
      end
```

Each pass through the data takes a time proportional to n, the number of elements in the list. Two passes therefore take time proportional to $2n$, which is itself proportional to n. The complexity of this program is $O(n)$.

10.4 Other Uses of Arrays

So far we have seen how arrays can be used to represent data in the form of lists. Lists are not the only kind of structured data that arrays can be used to model. In this section we look at arrays as the representation for tables.

Consider the following problem: we are given a list of numbers (say between 1 and 5) and are asked to determine the frequency of occurrence of each of the numbers. If the list is

1 2 3 2 2 3 1 1 2 1 3 2 3 1 2 3 2 3

then the answer is that there are five 1s, seven 2s, six 3s, zero 4s, and zero 5s. A simple-minded approach to this problem might result in the following program:

```
*       Program 10.4
*       Counts the frequencies of numbers in
*       in a list. This is not a good method
*
        program Freq
        integer Count1, Count2, Count3, Count4, Count5
        integer Number
        Count1 = 0
        Count2 = 0
        Count3 = 0
        Count4 = 0
        Count5 = 0
        print *, 'Enter numbers'
        while (.true.) do
           read (*,*,end=100) Number
           if (Number.eq.1) then
              Count1 = Count1 + 1
           elseif (Number.eq.2) then
              Count2 = Count2 + 1
           elseif (Number.eq.3) then
              Count3 = Count3 + 1
```

```
            elseif (Number.eq.4) then
                Count4 = Count4 + 1
            elseif (Number.eq.5) then
                Count5 = Count5 + 1
            else
                print *,'Number must be in 1-5. Try again'
            endif
        end while
100 continue
        print *,'Number of Ones', Count1
        print *,'Number of Twos', Count2
        print *,'Number of Threes', Count3
        print *,'Number of Fours', Count4
        print *,'Number of Fives', Count5
        stop
        end
```

This program works and correctly counts the frequency of the numbers 1 to 5 in any set of input data. However, it is not a very general solution. If the program requirements change so that we want to count the frequency of numbers from 1 to 10, then we could adapt the program to handle that—but it wouldn't be very easy and the resulting program would be both tedious to write and very confusing. If the requirements change so that the program must count the frequencies of the numbers from 1 to 100, then it's clear that this form of program is not going to be very useful.

We can use an array to hold the table of frequencies in the following way: the first element of the array will record the frequency of 1s in the input, the second element of the array will hold the frequency of 2s in the input, and so on. At first this might seem to be just a notational convenience that saves us having to declare lots of variables—and it does do that. The real power, however, of using an array in this way comes because we can use the input value as the array index.

Suppose that we have just read an input value into the variable Number. Then the assignment statement

```
        Count(Number) = Count(Number) + 1
```

increments one of the elements of the array by 1. But the interesting part is which array element is incremented. If the value read was a 1, then the first element of the array is incremented; if the value read was a 2, then the second element is incremented; if the value read was a 37, then the thirty-seventh element is incremented. The element of the array that is recording the frequency of a value is incremented precisely when that value is read in. This eliminates all of the **if** statements determining the correct counter to be incremented.

Here is the new version of the program that uses an array as the frequency counter. You can see that it's even simpler than the program that only counts the frequency of the numbers from 1 to 5.

Arrays

```
*       Program 10.5
*       Counts frequencies of numbers using arrays. Much better.
*
        program Frequency
        integer Count(100)
        integer Number, I
        do 100 I = 1,100
            Count(I) = 0
100     continue
        print *, 'Enter numbers'
        while (.true.) do
            read (*,*,end=200) Number
            if (Number.ge.1.and.Number.le.100) then
                Count(Number) = Count(Number) + 1
            else
                print *,' Number must be in 1-100. Try again'
            endif
        end while
200     continue
        do 300 I = 1, 100
            print *, ' Frequency of ',I,' is ',Count(I)
300     continue
        stop
        end
```

This program can be easily modified to count the frequencies of a larger set of numbers. Only the size of the array needs to be changed. The rest of the program can remain exactly as it is. Well-designed programs should, whenever possible, have the property that they can be altered to meet changing requirements by making small changes in the program.

The last program illustrated one of the ways in which the data from an array can be printed. In general, printing the contents of an array is a little more difficult than just printing simple variables because the array data belong together and we usually want the appearance of the output to indicate that. In the preceding program we printed the frequencies of the numbers as a simple table in which the first column contained the numbers and the second column the frequencies. However, because we didn't specify any formatting, the actual appearance of the table will be unattractive because each number will be placed in a wide field.

Here is a better version of the output part of the program:

```
        do 300 I =1,100
            write (6,250) I, Count(I)
250         format ('Frequency of number ',I4,' is ',I5)
300     continue
```

Because we have specified the number of columns that each value in the output will occupy, we can be sure that the numbers will be neatly arranged in columns, unless they are too large to fit within the format fields we have specified.

A different problem arises when we want to print out all of the elements of an array as a list, particularly when we want to be able to handle lists of different lengths. If we simply write the name of an array in a **write** statement, then it is taken to mean that we want to print the entire contents of the specified array. Thus if we have the declaration

    integer X(20)

then the output statement

    write (6,100) X

results in all twenty elements of the array being printed. However, twenty values will not all fit on a line so we would have to do something in the corresponding **format** statement to allow the numbers to be separated onto two or three lines. We could, for example, write a format statement of the form

100   **format**(10I6 / 10I6)

which, you will recall from the previous chapter, prints the first ten values on the first line of output and the next ten on a second line of output. This method doesn't generalize very well. For instance, if we ever have to change the size of the array, then we need to rethink the format codes we have used.

Fortunately this is where the rule about reusing format codes is useful. You'll remember that if there are more data values in the output list of the **write** statement than there are format codes, then the format codes are reused, and a new output line is begun. This is exactly the property that we want for output of arrays where we don't want to be constrained by the exact length of the array. If we write a **format** statement like this

100   **format**(10I6)

the first ten data values will be printed on the first line. Then the codes will be reused, starting from the beginning of the **format** statement, and a new line begun so that the next ten data values will appear on the second line. Thus, for the particular array X, the preceding **format** statement is exactly equivalent to the first one. It causes output values to be printed ten per line, regardless of how many values are printed.

We have talked about declaring an array to be one size and only using a part of it. The technique for printing arrays described earlier will obviously not handle this very well because we don't want to print the whole array—only the part that has been used. The solution to this is an **implied do loop**.

Suppose that we have declared an array in the following way

Arrays   **249**

>   **integer** X(100)

but we will only use the first N elements of it during a particular execution of the program. N may be an input variable and can be anything up to and including 100. When we want to print the defined contents of the array, then we want only to print the first N elements. The following statements do this

>   **write** (6,100) (X(I), I=1, N)
> 100 **format** (10I6)

The output list of the **write** statement is an implied **do** loop. I is the index variable and this loop says that I takes on values from 1 up to N successively. Thus the list in the **write** statement is logically equivalent to

>   **write**(6,100) X(1), X(2),...,X(N)

except that there's no way to write "..." in Fortran. The **format** statement behaves exactly as it did in the previous example—it causes the values, however many there are, to be printed ten per line.

Thus printing of arrays uses the power of **format** statements to produce output that is well arranged and clear.

10.5 Indexes

In this section we look more closely at index sets for arrays. When we have declared arrays up to now we have only said how many elements they have, not how the elements are indexed. The assumption has been that the elements are numbered from 1 up to the size of the array. This does not always have to be the case. The indexes of an array can be any contiguous set of integers.

To declare an array with some other contiguous set of integers as its index set, we simply include the range as part of the declaration. Here is an example

>   **integer** A(100), B(0:99)

A and B both contain the same number of elements, but the difference is that the elements of A are numbered from 1 to 100 while the elements of B are numbered from 0 to 99. The reason that this provision is made in Fortran is that it is often useful to have the index set represent more directly what is happening in the problem domain. For example, model numbers are often arranged so that they have four digits. Thus, a model number is in the range 1000 to 9999, and an array used to store the number of each model kept on hand should be declared as

>   **integer** Count(1000:9999)

In the same way, an array that was indexed by average temperature might be indexed from -40 to $+45$ like this

character*10 Place(-40:45)

10.6 Other Types of Indexes

Let's consider a variation of the frequency counting problem described in an earlier section. Suppose that, instead of counting the frequency of numbers in the input, we were concerned with counting the frequency of characters. This is an interesting and important problem because the frequency of characters plays an important part in analyzing text.

Clearly our program will be very similar to the previous version except that we will be reading characters, not integers, and we will have a counter for each character rather than one for each number. The general outline of the program will be

> while there is more input
>   read the next character
>   increment the appropriate counter

The difficulty comes in the step of incrementing the right counter. We need a counter for "a's", a counter for "b's", and so on. We've already seen that a solution with an individual counter for each possible input doesn't work very well, so obviously we need to use an array of counters. But then we need some way of determining easily that if the input is an "a", then we must increment the "a" counter and so on.

You'll remember that there is a representation for each character as a string of bits stored in a byte. Fortran provides two functions that allow you to switch between the representations used for characters and the integers. There are 256 possible bit strings representable in a byte so that these functions are mappings between the set of characters and the set of integers between 0 and 255. The functions are called Ichar and Char.

The Ichar function takes a character as argument and returns the integer value corresponding to it. The Char function takes an integer (in the range 0–255) and returns the character with that integer as its representation. The two functions are inverses since

> Ichar(Char(I)) = I, for all integers in [0-255]
> Char(Ichar(C)) = C, for all characters

We can use the Ichar function to map the characters that are read into integers that can be used as array indexes. The next program does this.

Arrays

```
*       Program 10.6
*       Counts frequencies of characters
*
        program CharF
        integer Freq(0:255), I
        character*1 C
        do 100 I = 0,255
           Freq(I) = 0
100     continue
        print *, 'Enter characters, one per line'
        while (.true.) do
           read (*, '(A1)', end=300) C
           Freq(Ichar(C)) = Freq(Ichar(C)) + 1
        end while
300     continue
        print '(A)', ' Character      Frequency'
        do 500 I = 0,255
           print '(5X,A3,5X,I5)', Char(I), Freq(I)
500     continue
        stop
        end
```

When we read a character as input, say the character "a", then we increment the array element whose index is Ichar('a') (that is element 97 in the ASCII representation). Of course, any of the 256 input characters have their frequency counted so that the program counts how many of the unusual characters are read by the program, as well as the ordinary alphabetic and numeric characters.

Notice how the Char function is used to handle the output. We want to print the entire contents of the array and to label the entries so we know what characters they correspond to. The Char function allows us to determine which character's frequency was recorded at position I in the array. Of course, this method does have one drawback—many of the 256 characters are nonprinting so that nothing will appear when they are printed out.

10.7 Arrays as Vectors

In a coordinate system, every point can be represented by a set of coordinates. If the system has dimension n, then n coordinates are required. Many problems are naturally formulated within coordinate systems. It is therefore important to be able to write programs to manipulate coordinates in sensible ways. Manipulating coordinate systems computationally is part of numerical linear algebra. If each point in the coordinate space is regarded as equivalent to a vector from the origin to that point, then a coordinate system can be thought of as a vector space. Because they are equivalent, no careful distinction is usually made between vectors and points. Transformations in the space either transform one vector into another or move one point to another. In coordinate systems, many useful transformations can be represented by matrices.

Arrays are a natural way to represent sets of coordinates. In an *n*-dimensional space, an array with *n* elements can hold the coordinates of a single point. Such a set of coordinates can be thought of as representing either a point or a vector. The transformations that are usually applied to points or vectors in the *n*-dimensional space can also be applied to the array.

Suppose that we are concerned with 3-dimensional space and we have declared two vectors like this

  real X(3), Y(3)

Then the length of such a vector (or the distance of the point from the origin) is given by calculating

$$\sqrt{\sum_{i=1}^{n} x_i^2}$$

The next program segment does this calculation for our 3-dimensional example.

```
          Sum = 0.0
          do 100 I = 1,3
             Sum = X(I) * X(I)
  100     continue
          Length = sqrt(Sum)
```

where the variables Sum and Length are assumed to have been declared appropriately as reals. This piece of code has been written to make it easy to generalize it to higher-dimensional spaces, even though we could have just added up the three terms required in a single expression.

Other common operations that are applied to vectors are rotation, stretching, and finding the dot product of two vectors. General subprograms can be written to do these things and they should be useful in any larger program involving the use of vectors.

Before we discuss the implementation of pieces of code to do these operations, we must discuss the way in which arrays are passed as arguments to subprograms. When a subprogram has an array as an argument, the parameter that is declared in the subprogram heading must be of the same type and size as the one that appears at the point of invocation. Thus if we have a function that is to calculate the length of a vector, it will have to be passed the vector as an input. We next show one way to implement the function.

```
      real function Length(X)
      real X(3), Sum
      integer I
      Sum = 0
      do 100 I = 1,3
         Sum = Sum + X(I) * X(I)
100   continue
      Length = sqrt(Sum)
      return
      end
```

The parameter X appears in the heading but without any size mentioned. It is then declared in the body of the function just as if it were an array that was local to the function.

In an invoking program there would have to be declarations and an invocation something like this:

```
*     Program 10.7
      program UseSub
      real Y(3), Length
      print *, 'Enter vector'
      read *, Y
      print *, 'Length of vector is', Length(Y)
      stop
      end
```

The array Y has been declared to be the same size (3) and same type (**real**) as the dummy array declared within the function. Because each subprogram in a Fortran program is compiled separately, the compiler will not produce error messages if the sizes or types of the respective arrays do not match, but the linker may catch problems of this kind before the program is run. Nevertheless, this kind of mistake is easy to make and can sometimes lead to frustrating searches before the error is found.

We have already seen how to declare an array to be of one size, but only use a part of it in a particular situation. We can use this technique to develop a Length function that is independent of the number of dimensions (up to some fixed maximum).

The next function assumes that it will be passed an array of up to 10 elements. It is also passed an integer specifying how many of the entries are meaningful.

```
          real function Length(X, N)
          real X(10), Sum
          integer I, N
          Sum = 0.0
          do 100 I = 1,N
              Sum = Sum + X(I) * X(I)
100       continue
          Length = sqrt(Sum)
          return
          end
```

This function can be invoked by the previous program (with one small change) to calculate the length of a vector in a 3-dimensional space, but it can also be used to calculate the length of a vector in a 6-dimensional space or a 9-dimensional space. The next program uses the function to calculate the length of a vector in 3-dimensional space:

```
*         Program 10.8
*         Calculates length of a vector of dimension <= 10
*
          program UseSub
          real Y(10), Length
          integer I
          print *, 'Enter vector (dimension 3)'
          read *, (Y(I), I=1,3)
          print *, 'Length is ', Length(Y, 3)
          stop
          end
*
*         Calculates the length of a vector passed in
*
          real function Length(X, N)
          real X(10), Sum
          integer I, N
          Sum = 0.0
          do 100 I = 1,N
              Sum = Sum + X(I) * X(I)
100       continue
          Length = sqrt(Sum)
          return
          end
```

In fact, we can build even more general array passing mechanisms than this. We can write a function or subroutine in which we don't say how big the array is to be—we don't even fix an upper limit on the size of the array.

```
*       Program 10.9
*       Calculates length of a vector of any dimension
*
        program UseSub
        real Y(10), Z(5), Length
        integer I
        print *, 'Enter vector (dimension 10)'
        read *, (Y(I), I=1,10)
        print *, 'Length is ', Length(Y, 10)
        print *, 'Enter vector (dimension 5)'
        read *, (Z(I), I=1,5)
        print *, 'Length is', Length(Z, 5)
        stop
        end
*
*       Calculates the length of a vector passed in
*
        real function Length(X, N)
        integer N
        real X(N), Sum
        integer I
        Sum = 0.0
        do 100 I = 1,N
            Sum = Sum + X(I) * X(I)
100     continue
        Length = sqrt(Sum)
        return
        end
```

When the subprogram is invoked, a parameter is passed giving the size of the array; this is used in the declaration where we would expect the array's size to be given. The variable that gives the size, in this case N, must have been declared before it appears as a placeholder for the size of the array. That is why we changed the order of the declarations for this example. In fact, because it would be too inefficient, arrays are not copied wholesale to and from subprograms. Instead, the location of the array is passed to the subprogram and the subprogram operates directly on the invoking program's copy. This explains how it is possible to declare arrays in subprograms without explicitly describing their sizes.

As a further example, we show how to compute the dot product of two vectors. The dot product of two vectors is given by the summation

$$\sum_{i=1}^{n} x_i y_i$$

The next program does this computation. We will present several different functions to do it.

```
*       Program 10.10
*       Compute dot product using several variations on
*       passing the information to a function
*
        program Prodex
        integer A(10), B(10), P(5), Q(5), Dot1, Dot2, Dot3, Dot4
        integer I, J
        print *, ' Enter 10 elements for A, and 10 for B:'
        read *, (A(I), I=1, 10), (B(I), I=1, 10)
        print '(10I5)', (A(I), I = 1, 10), (B(I), I = 1, 10)
        print *, ' Enter J<=5, and J values for P, J values for Q:'
        read *, J, (P(I), I=1, J), (Q(I), I=1, J)
        print *, ' Length is ', J
        print '(5I5)', (P(I), I=1, J)
        print '(5I5)', (Q(I), I=1, J)
        print '(/, A)', ' Dot products:'
        print *, ' Using first two vectors:          ', Dot1(A, B)
        print *, ' Using first 5 elements of these:', Dot2(A, B, 5)
        print *, ' Using first two vectors again:    ', Dot3(A, B, 10)
        print *, ' Using last two vectors:           ', Dot4(P, Q, J, 5)
        stop
        end
*
*       Dot product of full vectors of fixed size
*
        integer function Dot1(X, Y)
        integer X(10), Y(10)
        integer Sum
        Sum = 0
        do 100 I = 1, 10
            Sum = Sum + X(I) * Y(I)
100     continue
        Dot1 = Sum
        return
        end
*
*       Dot product of partial vectors of fixed size
*
        integer function Dot2(X, Y, M)
        integer X(10), Y(10)
        integer M
        integer Sum
```

```
        Sum = 0
        do 100 I = 1, M
            Sum = Sum + X(I) * Y(I)
100     continue
        Dot2 = Sum
        return
        end
*
*       Dot product of full vectors of variable size
*
        integer function Dot3(X, Y, N)
        integer N
        integer X(N), Y(N)
        integer Sum
        Sum = 0
        do 100 I = 1, N
            Sum = Sum + X(I) * Y(I)
100     continue
        Dot3 = Sum
        return
        end
*
*       Dot product of partial vectors of variable size
*
        integer function Dot4(X, Y, M, N)
        integer M, N
        integer X(N), Y(N)
        integer Sum
        Sum = 0
        do 100 I = 1, M
            Sum = Sum + X(I) * Y(I)
100     continue
        Dot4 = Sum
        return
        end
```

Notice how the input is handled in the main program. The arrays A and B are filled to capacity, but P and Q are partly filled. The size of P and Q is read in just before the arrays themselves. The variable J contains the number of elements that will go into the arrays P and Q. The index I can be used for all the implied do loops because they are independent in time; the first implied loop must have finished before the second one can start, and so on. So there is never any possibility of confusion about the value of I. We now show the output produced by this program.

```
Enter 10 elements for A, and 10 for B:
   1    2    3    4    5    6    7    8    9    0
   2    3    4    5    6    7    8    9    0    1
Enter J<=5, and J values for P and J values for Q:
Length is            4
   1    3    5    7
   2    4    6    8
Dot products:
Using first two vectors:               240
Using first 5 elements of these:        70
Using first two vectors again:         240
Using last two vectors:                100
```

We have shown four different versions of the dot product function. In the first version the size of the vectors passed as parameters is known at compile time (it is 10), and thus is not passed as a parameter. Thus, Dot1 only works for vectors of size 10, and only when those vectors have exactly ten valid elements in them.

The second version is slightly more general. It works only for vectors of size 10, but it allows for the possibility that only some of the elements might be valid. This is a useful way to write subprograms if we have a main program with structures of a fixed size, but we want to use the program to solve problems of varying size. For example, we could have a different main program than the one we have here that would read in a variable number of elements into the vectors A and B, and then invoke Dot2 to find the dot product of these partially filled vectors (our main program actually behaves this way with P and Q). To make Dot2 work, we have to tell it how many valid elements there are in each vector, and it uses this value as the limit on its iteration statement. In the example we have "pretended" that A and B contain only five valid elements to test Dot2.

The third version of the function is generalized in a different manner. It accepts vectors of arbitrary size—they do not have to be ten elements long. To make this work, we have to tell the function (using a parameter) how long the vectors are and then use this parameter in the declaration of the vector parameters. This approach is useful in a very obvious way, since Dot3 can be used to find dot products of vectors of many different declared sizes. But notice that this version assumes that these vectors, no matter what size they are declared to be, are full of valid elements.

The fourth version of the function is generalized in both the ways used in the second and third versions. It can work with vectors of arbitrary declared size, and it can also work with any number of valid elements in these arrays. Accordingly, we have to pass it parameters that specify how large the vectors have been declared to be in the calling program and how many elements are actually used. It uses one parameter to declare the array sizes and the other to control its iteration statement.

These different functions illustrate a general point about Fortran arrays. They are very powerful mechanisms, and extremely general routines can be written to manipulate them. The last version of the dot product routine could be used for any two integer vectors (assuming their declared sizes are the same). But to take advantage of this available generality, we have to pass more information as parameters to subprograms and be careful that this information is used correctly.

10.8 Polynomials

Another common class of algebraic objects that may be represented by arrays are polynomials. In this section we will discuss ways of representing polynomials and manipulating them. Let us begin with a polynomial of the form

$$a_n x^n + a_{n-1} x^{n-1} + \cdots + a_0$$

The obvious way to represent such a polynomial is to use a vector of size $n + 1$ (because a polynomial of degree n has $n + 1$ coefficients). The vector will contain the coefficients a_i.

Using this representation, we can write programs that manipulate polynomials. For example, multiplying two polynomials requires us to take all possible products of their coefficients. This corresponds to taking the cross product of the two vectors that represent the polynomials and then summing appropriate terms. Suppose that we want to form the product of two polynomials whose coefficients are (in the order of the previous example):

$$2x^2 + x + 3$$
$$5x^2 + 4x + 6$$

Taking the product of these two polynomials means taking all possible products (2*5, 2*4, 2*6, 1*5, 1*4, 1*6, 3*5, 3*4, 3*6) and calculating the power of x to which each product belongs. For example, the product 2*4 corresponds to $2x^2 \times 4x$, so the result is $8 \times x^3$. We can determine the appropriate power of x by summing the powers of x corresponding to the coefficients we are multiplying.

Suppose that we want to write a program to carry out this multiplication. We will want to limit the maximum size of polynomial that the program will handle (say x^{10}) and then use zero coefficients for the higher-order terms for polynomials of smaller degree. The program will then have to form all possible products of coefficients (which requires a double nested loop) and add the product of the coefficients to the appropriate coefficient of the result, which is determined by adding the powers of x corresponding to the multiplicands.

We will use the convention that the coefficient of x^i will be stored in the ith position of the vector representing the polynomial. The program reads the coefficients of the polynomials to be multiplied. For convenience, the order of the polynomials (the highest power of x) is read first, followed by the coefficients in increasing order of coefficient power. The program fills in the remaining coefficients to zero.

```
*         Program 10.11
*         Multiplies two polynomials represented as arrays
*
          program PolMul
          real X1(0:10), X2(0:10), Y(0:10), Prod
          integer M, N, I, J, Power
          print *, 'First polynomial'
          print *, 'Enter degree and coefficients (ascending)'
          read *, M, (X1(I), I = 0,M)
          do 200 I = M+1, 10
             X1(I) = 0.0
      200 continue
          print *, 'Second polynomial'
          print *,' Enter degree and coefficients (ascending)'
          read *, N, (X2(J), J = 0, N)
          do 300 J = N+1, 10
             X2(J) = 0.0
      300 continue
          do 400 I = 0, 10
             Y(I) = 0.0
      400 continue
          do 600 I = 0, M
             do 500 J = 0, N
                Prod = X1(I) * X2(J)
                Power = I + J
                Y(Power) = Y(Power) + Prod
      500    continue
      600 continue
          print '(A,5F7.2)', 'Coefficients of product ', (Y(I), I = 0, 10)
          stop
          end
```

We can also differentiate polynomials by manipulating coefficient vectors. When a polynomial is differentiated, each coefficient is multiplied by the power of x to which it belongs and then the result is divided by x. This last operation corresponds to shifting the coefficients one place down. The next program reads a polynomial and prints its derivative.

```
*         Program 10.12
*         Carries out symbolic differentiation of polynomials
*
          program Diff
          real P(0:10), DP(0:10)
          integer I, N
          print *, 'Enter degree and coefficients (ascending)'
          read *, N, (P(I), I = 0, N)
```

```
      do 200 I = 1,N
         DP(I - 1) = P(I) * I
200   continue
      print *, 'Coefficients of derivative are:'
      print '(5F7.2)', (DP(I), I = 0, N-1)
      stop
      end
```

Another important operation for polynomials is evaluating them. We want to take a polynomial in its representation as a vector and evaluate it for a specific value of x. We can, of course, do this by writing a large expression of the form

Value = P(10) * X**10 + P(9) * X**9 + ... + P(0)

but this is tedious and not very general. It is also extremely inefficient because it involves calculating powers of x. Furthermore, it takes no advantage of the fact that if the value of x^8 is known, then it only takes one more multiplication to calculate x^9.

A better way (computationally) to evaluate polynomials is to express them in a form known as **Horner's rule**. This involves taking out common factors of x wherever possible. For example, a polynomial of degree three can be written as

$$[[ax + b]x + c]x + d$$

and evaluating it using this formulation requires many fewer operations than in the straightforward way. More usefully, we can express the operations that are required to evaluate a polynomial of this form in an iterative way. At every stage of the evaluation, working from the innermost nested brackets outward, we take the product of what we have so far with x and add in the next coefficient. This next function evaluates a polynomial at a given point, with the polynomial as an argument.

```
*     Program 10.13
*     Evaluates a polynomial written as an array at a point
*
      program UseEval
      real P(0:10), X, Eval
      integer N, I
      print *, 'Enter degree, X value and coefficients'
      read *, N, X, (P(I), I = 0, N)
      print *, 'Value of polynomial is'
      print '(F10.4)', Eval(P, N, X)
      stop
      end
```

```
*
*       This function sums the values of each term of the polynomial
*
        real function Eval(P, N, X)
        integer N, J
        real P(0:N), X, Sum
        Sum = 0.0
        do 300 J = N, 0, -1
            Sum = Sum * X + P(J)
300     continue
        Eval = Sum
        return
        end
```

We can now put these pieces together to construct larger useful programs. For example, we can now write a general root-finding package that uses the **Newton-Raphson method** on any polynomial. The polynomial can be provided as input to the program. We can differentiate the polynomial and we can evaluate both the polynomial and its derivative at any point. Here is the complete set of subprograms to do this:

```
*       Program 10.14
*       Finds the roots of a polynomial equation
*       using the Newton-Raphson method
*
        program Newton
        real P(0:10), DP(0:10), Num, Denom, X, Eps, Eval
        integer I, N
        Eps = 1.0e-6
        print *, 'Enter degree, coefficients and initial point'
        read *, N, ( P(I), I = 0, N), X
        print '(A, I5)', 'Polynomial has degree ', N
        print '(A, (3F12.4))', 'Coefficients are', (P(I), I = 0,N)
        print '(A, F10.4)', 'Initial point       ', X
        call Diff(P, DP, N)
        Num = Eval(P, N, X)
        Denom = Eval(DP, N-1, X)
        while ( Abs(Num) .gt. Eps ) do
            X = X - Num / Denom
            Num = Eval(P, N, X)
            Denom = Eval(DP, N-1, X)
        end while
        print '(A, F12.4)', 'Root is at X = ', X
        stop
        end
```

```
*
*       This subroutine symbolically differentiates the polynomial
*
        subroutine Diff(P, DP, N)
        integer N
        real P(0:N), DP(0:N)
        integer I
        do 100 I = 1,N
            DP(I - 1) = P(I) * I
100     continue
        return
        end
*
*       This function evaluates the polynomial
*
        real function Eval(P, N, X)
        integer N, J
        real P(0:N), X, Sum
        Sum = 0.0
        do 100 J = N, 0, -1
            Sum = Sum * X + P(J)
100     continue
        Eval = Sum
        return
        end
```

The output produced by this program looks like this for a sample quadratic.

```
 Enter degree, coefficients and initial point
Polynomial has degree       2
Coefficients are      3.0000      -4.0000      1.0000
Initial point         0.0000
Root is at X  =       1.0000
```

10.9 Linear Search

We very often need to use standard data in calculations. When these calculations are done by humans, the relevant data can usually be found in reference books. When the data are used by computer programs, they are usually placed in files and can then be accessed by programs just like any other data.

However, there are some major differences between files of, say, data from an experiment that has just been run, which are being analyzed by a program, and files that contain reference data. For example, a program analyzing experimental data will almost certainly use all those data in its calculations. However, the proportion of the reference data that a particular program will require is usually very small. The difference is like that between reading a novel and looking up the population of Upper Volta in an encyclopedia.

The second major difference is that the reference data files are likely to be very large compared to program-specific files. Again, the comparison between the size of a novel and an encyclopedia is relevant. In this section we will look at ways of accessing data when we are only interested in a small number of them.

Suppose that we are given a long list of chemical names and densities and are asked to look up the density of a particular chemical. How we go about this depends on the way in which the list is organized. If the list is organized in some useful way, we may be able to look in an organized manner (see the next section). However, if the list is not organized, then there is really no other algorithm than to start at the beginning of the list and look through it until we find the chemical we want.

Lists in the real world are often not organized into any particular order — usually because they change so often that it's not practical to do so. We need therefore to be able to write programs that will handle lists that are not sorted. Here is a program that uses the algorithm: look at each element in turn and see whether or not it is the one required. If the value is present, then it prints the position at which the value appears. As usual, the list is implemented as an array.

```
*       Program 10.15
*       Uses a linear search to match a key value
*
        program Search
        real List(1000), Key
        integer Posn
        logical Found
        print *, 'Enter list and key'
        read '(10F7.2)', List
        read '(F7.2)', Key
        Found = .false.
        Posn = 1
        while (Posn .le. 1000 .and. .not. Found) do
           if (List(Posn).eq.Key) then
              Found = .true.
           endif
           Posn = Posn + 1
        end while
```

Arrays

```
        if (.not. Found) then
            print *, ' Key not found '
        else
            print *, ' Key found at position ', Posn - 1
        endif
        stop
        end
```

Notice that there are two reasons to stop searching through the array—we may have found the value for which we are searching, but we may also have run out of array to search without finding the value we are looking for. We need to know, after the loop has completed, which of the two possible reasons caused the loop to finish because this tells us whether or not the value was found. Notice also that the value of the variable Found must be used as the test deciding whether the value has been found or not. If we attempt to test whether Posn contains the value 1001 and conclude that if it does, then the value is not present in the list, then we will be wrong in the situation where the value appears in position 1000. In this case, Found will be true and Posn will equal 1001—so the test in our program will work where the other one won't.

It's tempting to simplify the loop by using a definite iteration that looks at each item. This is an expensive change as we can see by considering the complexity of this algorithm. If the value that is being looked for is, in fact, present in the list, then sometimes it will be near the beginning and sometimes near the end of the list. If there is no particular bias in the way the list is created (items are placed in the list randomly), then we will have to examine about half the items in the list on average before we find the one we're looking for. Thus the number of array elements of an n-element array that are examined will be about $n/2$. We call this an **average case analysis**. If the value we are looking for is not in the list, then we will have to look all of the way through the list before we can safely conclude that it is not there. Thus we have to search all n elements to discover that a value is not present in the list. If the fraction of possible values that are in the list is f, then the fraction that aren't in the list is $1 - f$ and the number of array elements examined over all possible inputs is $f(n/2) + (1 - f)n$. This is clearly linear in n, and so this algorithm, called **linear search**, is $O(n)$.

Now this does not completely solve the problem because the position at which the sought value appears is not usually interesting by itself. We can distinguish two different parts for each item in the list. The first part is called the **key**; it is the part that distinguishes this item from others and is used when we search for it. In the preceding example, the chemical name is the key. The second part is the actual information in which we are interested—this might be called the **record**. In the preceding example, the record was the density of each chemical.

The preceding programs really only tell us which position in the list the key occupies. However, once we know a position, we can look in the list consisting of the records and find the matching information.

Here is a program that uses two arrays. The first array holds the keys and the second holds the records. The arrays are parallel in the sense that the key at position *i* in the first array matches the record at position *i* in the second array. We can use the same method as in the previous program to find the position corresponding to the key and then print the record that belongs with that key.

```
*       Program 10.16
*       Uses a linear search on a more complicated list
*       (artificial method of reading list)
*
        program Linear
        real List(1000), Record(1000), Key
        integer Place, Search
        print *, 'Enter lists followed by key'
        read *, List, Record, Key
        Place = Search(List, Key)
        if (Place.ne.-1) then
            print '(A, F9.3)', 'Value requested is ', Record(Place)
        else
            print *, ' Item not present in list '
        endif
        stop
        end

        integer function Search(List, Key)
        real List(1000), Key
        integer Posn
        logical Found
        Found = .false.
        Posn = 1
        while (Posn.le.1000.and..not.Found) do
            if (List(Posn).eq.Key) then
                Found = .true.
            endif
            Posn = Posn + 1
        end while
        if (.not.Found) then
            Search = -1
        else
            Search = Posn - 1
        endif
        return
        end
```

Arrays

10.10 Binary Search

Linear search is so time-consuming that for lists that don't change rapidly, it soon becomes cost effective to maintain the list in a sorted order. As we shall see in this section, searching through a sorted list is much faster than searching through an unsorted one.

As an example, imagine the largest list with which you come into regular contact—the telephone directory. If it were maintained in an unsorted order, with the names of new subscribers added at the end, it would be useless except for very small communities. Because it is sorted, it is possible to look up a name and telephone number without looking at even a small fraction of the entries. In fact, the algorithm that you use to look up names in the phone book is (pretty much) the one that we will discuss.

Suppose then that you have been asked to look up the name Moriarty in a telephone directory. Generally, you would open the book at about the halfway point, reasoning that Moriarty begins with M and M is near the middle of the alphabet. There's a hidden assumption in this reasoning—that the number of names that begin with each letter of the alphabet is about the same. This assumption is close enough to being true for names, but cannot be generalized to other kinds of lists with other kinds of keys. At this point in your search, you look at the names that appear on the page you have opened and decide whether Moriarty should appear after this page or before it. If you're very lucky, you may have opened it at the correct page right away. Assuming you're not that lucky, you probably turn over a chunk of pages in the direction in which you expect the name to appear. Then you examine the names on the newly opened page to see whether or not you've overshot the page with Moriarty on it. Continuing in this fashion, you very quickly find the correct page. Now the same procedure is followed to select the correct column of names. Most people then use a short linear search when they have the correct area of the column located.

When we want to turn this into an algorithm that can be used on many different kinds of lists, we need to make it simpler. In particular, we cannot make assumptions about the distribution of keys in the list, as we did when we assumed the names starting with M would be near the middle of the book. When we decide which part of the list we want to examine first, we always look at the middle element. By comparing the key belonging to the middle element with the key we are searching for, we can determine in which half of the list the record we want is located. Looking at the middle element is as sensible as looking at any other if we don't know the distribution of keys, and it has the advantage that it reduces the search space by one half. By that we mean that the possible places in the list where the record we want could be located are reduced by 50 percent simply by examining a single element of the list.

Now we repeat this step, this time starting with the half of the list we selected after the first probe. We choose the middle element of this half list and examine it. We can then tell whether the record we want is before or after it in the list.

The search space is now reduced to one quarter of the whole list. If we continue to work on smaller and smaller sections of the list, then eventually we will have a list consisting of a single element. If this element has a key matching the key we are searching for, then we have found the record. If it does not match, then the record we are looking for cannot be in the list because, if it were, then it would be at this spot.

An example will illustrate how the algorithm works. Suppose we have this list:

1 2 4 5 6 7 9 10 12 14 15 17 18 19 22

and we want to see if the value 4 is present in the list. We know that the elements in the list are sorted into ascending order, but we don't know what their distribution is. So we look first at the middle element of the list. The middle element has the value 10; 10 is larger than 4, so if 4 is in the list it must be to the left of the middle. We now look at the middle element of the left-hand sublist. The middle element has the value 5; 5 is larger than 4, so 4 must be to the left, if it is present. We look at the middle element of the left-hand list—it has the value 2; 2 is smaller than 4, so we know that 4 must be in the right-hand sublist if it is present. The middle element of this sublist has the value 4, so we have found the record we were looking for.

If one of the elements that we examine in the list happens to be the one we're looking for, then obviously we should stop immediately. The algorithm, however, still works without this refinement.

We can easily calculate the complexity of this algorithm. Suppose that we have a list of N elements. Then, examining one element reduces the search space to $N/2$, examining two elements reduces the search space to $N/4$, and so on. When the search space has been reduced to one, then the search is finished. Therefore the number of elements that must be examined is the smallest number n such that

$$\frac{N}{2^n} \leq 1$$

Rearranging, we can see that n is the smallest number such that

$$2^n \geq N$$

Therefore n is the logarithm of N, taken to the base 2. This is usually written as $\lg N$.

It is interesting to see how the number of elements to be examined behaves as N increases. If N is 4, then n is 2 and the **binary search** algorithm is about the same complexity as the linear search; if N is 16, then n is 4; if N is 256, then n is 8; if N is 4096, then n is 12. You can see that n increases very slowly compared to N and that n is much smaller than $N/2$, which is the number of elements that the linear search would need to examine. We say that binary search is $O(\log N)$. Notice that we don't need to specify the base to which we mean the logarithm to be taken since logarithms to different bases are proportional.

The next program uses this algorithm to search a list of up to 1000 elements. To keep it simple, it finds only the location of the search key.

```
*       Program 10.17
*       Searches an ordered list using binary search algorithm
*
        program ShowBS
        real Number(1000), Key
        integer N, I, Place, Posn
        print *, 'Enter key, size of list and then list'
        read *, Key, N, (Number(I), I= 1,N)
        Place = Posn(Number, N, Key)
        if (Place .ne. -1) then
            print *, ' Found at position ', Place
        else
            print *, ' Not present in the list '
        endif
        stop
        end
*
*       This function carries out the binary search
*
        integer function Posn (List, Size, Key)
        integer Size
        real List(Size), Key
        integer Left, Right, Midpt
        Posn = -1
        Left = 1
        Right = Size
        while (Right - Left .ge. 1) do
            Midpt = (Left + Right) / 2
            if (Key.lt.List(Midpt)) then
                Right = Midpt - 1
            else if (Key.gt.List(Midpt)) then
                Left = Midpt + 1
            else
                Right = Left
                Posn = Midpt
            endif
        end while
        return
        end
```

The variables Left and Right indicate the part of the array in which the value may lie. As the algorithm proceeds, this part shrinks. That is, Left and Right get closer together. The loop terminates when the remaining part of the array is less than one element. As we saw in the linear search, when we have finished the search, we must decide whether or not it was successful. In this program the third part of the **if** statement is executed when a match is found between the Key and an element of the array. Rather than using a check after the loop terminates to decide on success, this function begins with the assumption that the search will be unsuccessful and so assigns -1 to Posn. If the search does in fact succeed, then this is changed to indicate the position at which the value was found. This technique often makes programs simpler. It is often easier to assume one outcome and then correct if necessary than to try and decide between two outcomes.

10.11 Sorting

We saw in the last section how sorting a list of values could improve by an enormous amount the speed at which the list could be searched. In many other situations, the presentation of data is made easier by sorting them. It is therefore clear that algorithms to sort lists are very important.

Sorting is such a common operation that most computer systems provide some tools for sorting as part of the operating system. It is unlikely that you will ever have to write your own sort as a working engineer, and the fastest sorting algorithms are not easy to understand. In this section we will present some of the simple algorithms for sorting and show how they can be used. All of these algorithms should be understood to be for demonstration purposes only—they are much slower than the best algorithms and should not be used in more than trivial situations.

The first sort that we will look at is called a **selection sort**. It uses the straightforward idea that we should attempt to place each element of the list into its sorted position. In general, of course, we can't tell what the final position of an element in the list will be since it depends on the values of the other elements. However, we can tell the final position of the largest element in the list—it will finish up at the end of the list. The second-largest element will finish up second from the end of the list, and so on.

The sorting algorithms we will consider are called "in-place" algorithms because they do not use more memory than the array itself occupies (except for a few extra variables). If two arrays are used, one with the initial arrangement and the other with the sorted array, then some small changes are required in the algorithm. We assume that the sorting is done within the array itself by rearranging, because it is very expensive to use twice the amount of memory that the array needs if the array is large.

The selection sort begins by examining the unsorted list to find the largest element. This element is then moved to the end of the array—which means that the element that is presently at the end of the array must be moved somewhere else. It is natural to move it to the place where the largest element was. So the largest element and the element at the end are swapped.

Now the element at the end of the list is in its correct place, so the unsorted part of the list is now smaller (by one) than it was to begin with. We now repeat the search, looking for the largest among the unsorted part of the list (which is the second largest in the whole list). This element is then swapped with the element at the end of the unsorted part. Now two elements are in their correct places.

This sequence is repeated, with each pass through the list placing one more element in its correct place and shrinking the unsorted part of the list by one. After $n - 1$ passes, the list must be sorted (only $n - 1$ passes are required because if $n - 1$ of the elements of the list are in their correct places, then the remaining element only has one place left and it must be the correct one).

We can easily calculate the complexity of this algorithm. It requires $n - 1$ passes through the list and each pass examines, on average, half the list. Thus the complexity is $O(n^2)$.

```
*       Program 10.18
*       Illustrates the selection sort
*
        program SSort
        integer List(1000), N, I
        print *, 'Enter size and list'
        read *, N, (List(I), I = 1,N)
        call Sort(List, N)
        print *, ' Sorted list is'
        print '(5I10)', (List(I), I = 1,N)
        stop
        end
*
*       This subroutine sorts by finding the largest element
*       and moving it to the end of the list
*
        subroutine Sort (List, Size)
        integer Size, List(Size)
        integer I, Last, Maxpos, Max, Temp
        do 200 Last = Size, 2, -1
           Max = List(1)
           Maxpos = 1
           do 100 I = 1, Last
              if (List(I).gt.Max) then
                 Max = List(I)
                 Maxpos = I
              endif
100        continue
           Temp = List(Last)
           List(Last) = List(Maxpos)
           List(Maxpos) = Temp
200     continue
        return
        end
```

The inner loop searches the unsorted part of the array, looking for the largest element remaining. We have seen this algorithm before. After the largest remaining element has been found, it is swapped with the element at the end of the unsorted part, using the three statements after the inner loop. It is a common mistake to write statements like

```
A = B
B = A
```

to swap the contents of two variables. However, if you work out what happens using these two assignments, you see that this loses one of the values. To swap two values properly, we have to make room at one location by copying its value to a temporary location, then copying the first value into the space created, and finally copying the value from the temporary back into the first location. Because the steps require creating a space for each value before moving it, it's easy to check the steps — they are cyclic, with the right-hand side of each assignment becoming the left-hand side of the next.

The second sort that we will discuss is called the **bubble sort** because the values "bubble" up through the array. The selection sort placed one more value in its correct final place after each pass through the data — the bubble sort can, for some initial arrangements of the list, do considerably better.

The bubble sort goes as follows: compare the elements in the first two positions; if they are out of order, then swap them. Then compare the elements in the second and third positions, swapping them if they are out of order. This continues for each adjacent pair of locations (notice that the pairs of locations are overlapping). By the time we reach the end of the list, the largest element must have been placed in the last entry, because as soon as we encounter the largest element in the list, it will always be out of order with any element in the further location and so will always be swapped. Thus, the largest element has bubbled up to the end of the list.

The benefit of the bubble sort is that other large elements will tend to move along the list — not just the largest. For example, if we begin with the list

```
3  1  2  6  5  4
```

then, as we go through the list for the first time, the list goes through these intermediate stages:

```
1  3  2  6  5  4
1  2  3  6  5  4
1  2  3  5  6  4
1  2  3  5  4  6
```

After the first pass through the list, the value 6 has bubbled up to the last position in the list. But the value 3 has also bubbled up past the 1 and the 2 and is therefore closer to its final position (in fact in its final position in this example).

The idea behind the bubble sort is that you can improve the general order of the list by going through it in this fashion. This compensates (one hopes) for the extra work involved in swapping elements extra times. If the list is in exactly reverse order, then the bubble sort will place exactly one element in its correct position in each pass and so will be no faster than the selection sort.

This next program carries out a bubble sort:

```
*         Program 10.19
*         Illustrates the bubble sort
*
          program BSort
          integer List(1000), N, I
          print *, 'Enter size and list'
          read *, N, (List(I), I = 1,N)
          call Sort(List, N)
          print *, 'Sorted list is'
          print '(5I10)', (List(I), I = 1,N)
          stop
          end
*
*         This subroutine sorts by passing through the array
*         swapping pairs that are out of order
*
          subroutine Sort (List, Size)
          integer Size, List(Size)
          integer I, Last, Final
          integer Temp
          Last = Size
          while (Last .gt. 1) do
             Final = 1
             do 100 I = 1, Last-1
                if (List(I).gt.List(I+1))then
                   Temp = List(I)
                   List(I) = List(I+1)
                   List(I+1) = Temp
                   Final = I
                endif
100          continue
             Last = Final
          end while
          return
          end
```

The termination condition for the outermost loop is interesting. It may happen that after a certain point in the list, no more reversals will take place during a particular pass through the data. If this happens, then there cannot be any values out of place between the point at which the last reversal occurred and the end of the unsorted part. If there were, a larger value would be bubbling up to the end of the list and there would be swaps all the way to the end. The program makes use of this by recording the place where the last swap occurred during each pass in the variable Final. This value is the highest point required for the next pass and so is assigned to Last before the next pass.

We need to show that the outer loop will in fact terminate. The inner loop takes on values of I up to and including Last $- 1$. So either Final is assigned the value 1 before the loop or it is assigned a value in the set

$$\{1, 2, ..., Last - 1\}$$

All of these values are smaller than the current value of Last so that Final always contains a smaller value than Last does after the inner loop completes. Therefore, the assignment of Final to Last always reduces the value of Last and hence Last must eventually become smaller than 2.

The complexity of this algorithm is harder to determine than for the selection sort because the complexity depends on the particular arrangement of the list. For example, if the list is in order except that the largest element is at the front, then one pass is sufficient to sort the list into order (the program will actually take two since it can't detect that the list has been sorted by the first pass without making a further pass). We can compare the two algorithms by asking what their worst-case behavior is. The worst possible arrangement for the bubble sort is when the list is in order but exactly backward. Then one pass is required to move each element into its proper place. When this happens, $n - 1$ passes are required, with each one processing about half the list. Thus the worst-case complexity of the bubble sort is the same as the selection sort. In actual efficiency, the bubble sort is worse because each pass involves large numbers of exchanges. However, the average behavior of the bubble sort is better than the average behavior of the selection sort.

The best sorting algorithms known have complexity $O(n \log n)$ rather than $O(n^2)$. We won't show them here since they are complex both to understand and to program.

10.12 An Example: The Sieve of Eratosthenes

Prime numbers are of considerable theoretical and practical interest. A number is prime if it is only divisible by itself and 1. By convention, 1 is not considered a prime.

If we are given a number n and asked if it is prime, then we can find out by taking all of the numbers up to the square root of n and checking whether any divide n. We only need to consider possible divisors up to the square root of n because if n is the product of a and b, then at least one of a and b must be less than or equal to the square root. This method works reasonably well if we are only asked whether or not a single number is prime. However, if we are interested in preparing a list of primes, then it is much more efficient to approach the problem in another way. For example, if we write down all of the numbers from 1 to 1000, then we can immediately cross off all of the numbers that are multiples of other numbers, because they can't possibly be prime. The algorithm using this idea is called the **Sieve of Eratosthenes**.

Suppose that we have written down a list of the numbers from 1 to N. Then we can start from the beginning (we ignore 1 for obvious reasons and actually start at 2) and go along the list crossing off all of the multiples of 2, because none of them are prime. All of the multiples of 2 are two places apart in the list. Then we return to the beginning of the list and cross off all of the multiples of 3 (but not 3 itself). The multiples of 3 are three spaces apart in the list. We repeat this, crossing off all of the multiples of 4, then of 5, and so on. We stop when we get to the first number larger than the square root of N (say M) for the following reason: multiples of the form $2M$, $3M$, and so on have already been crossed off because they are also multiples of 2, 3, ... The first new multiple that hasn't already been crossed off is $M \times M$ and this number is beyond the end of the list (by definition).

The program therefore contains two loops. The outermost loop controls the starting point of the crossing off activity and runs from 2 up to the square root of the list length. The innermost loop crosses off all multiples of the starting element. We use an array in a slightly new way as the list itself. We don't need to keep the numbers—we use position i in the array to indicate whether or not the value i has been crossed off. The array is an array of flags indicating whether the number is composite (that is, not prime).

```
*       Program 10.20
*       Finds primes by crossing off multiples of
*       numbers up to the square root of the list size
*
        program Sieve
        logical Compos(2:1000)
        integer Start, I, Sroot
        do 100 I = 2,1000
            Compos(I) = .false.
100     continue
        Sroot = sqrt(real(1000))
        do 300 Start = 2, Sroot
            do 200 I = 2 * Start, 1000, Start
                Compos(I) = .true.
200         continue
300     continue
```

```
          print *, 'Prime numbers:'
          do 400 I = 2,1000
              if (.not. Compos(I)) then
                  print *, I
              endif
400       continue
          stop
          end
```

There are several easy optimizations that can be made to this algorithm. For instance, every multiple of 4 is also a multiple of 2, so we do not need to cross off multiples of 4 because they will all have been crossed off already. We can detect this by checking whether the start value has been crossed off before beginning a new inner loop. The revised version is much quicker because almost all numbers are composite.

```
*         Program 10.21
*         Does not consider multiples of composites
*
          program Sieve
          logical Compos(2:1000)
          integer Start, I, Sroot
          do 100 I = 2,1000
              Compos(I) = .false.
100       continue
          Sroot = sqrt(real(1000))
          do 300 Start = 2, Sroot
              if (.not.Compos(Start)) then
                  do 200 I = 2 * Start, 1000, Start
                      Compos(I) = .true.
200               continue
              endif
300       continue
          print *, 'Prime numbers:'
          do 400 I = 2,1000
              if (.not. Compos(I)) then
                  print *, I
              endif
400       continue
          stop
          end
```

Another simple optimization is to observe that all even numbers are composite, except for 2. Therefore we can save a great deal of space in the array, and also save the first pass through the data, by not including the even numbers in the list at all. When this is done, the list still has the useful property that all of the multiples of x are separated by x places in the array. The array element at position i represents the state of the number $2i + 1$. The next version exploits this property.

```
*       Program 10.22
*       Does not include even numbers in the list
*
        program Sieve
        logical Compos(1:1000)
        integer Start, I, Sroot
        do 100 I = 1,1000
            Compos(I) = .false.
100     continue
        Sroot = sqrt(real(2 * 1000 + 1))
        do 300 Start = 1, (Sroot-1)/2
            do 200 I = 3 * Start + 1, 1000, 2 * Start + 1
                Compos(I) = .true.
200         continue
300     continue
        print *, 'Prime numbers:'
        print *, 2
        do 400 I = 1,1000
            if (.not. Compos(I)) then
                print *, 2 * I + 1
            endif
400     continue
        stop
        end
```

10.13 Using the Common Statement

The Fortran mechanism we've described for invoking subprograms requires all data values from the calling environment to be explicitly passed as parameters. This can be tedious, especially when large arrays are to be passed. Fortran provides a **common** mechanism that is a second means of making values available in the called subprogram.

The idea is that variables can be declared and placed in a block, and this block can be seen in several subprograms. The statement that does this has the form

   common /id/ variableList

The **common** declaration gives the name of the block (there can be many in a large program) set off by slashes, followed by a list of the variables in the block. The same declarations appear in all program segments that use the block. Although the variable names used do not have to be the same, the locations in the block appear identical to any program segment. Therefore, the types of the variables must match in all program segments. When the subprogram is invoked, the variables in the common block are available to be used, without having to be passed as parameters. This next program shows a simple example.

```
*       Program 10.23
*       Illustrates the use of common blocks
*
        program Lookup
        common /Table/ Size, Words
        character*5 Words(10)
        integer Size
        character*5 Target
        integer I
        print *, 'Enter length of list'
        read *, Size
        print *, 'Enter list of words and target word'
        read '(A5)', (Words (I), I=1, Size), Target
        call Find(Target)
        stop
        end
*
*       This subroutine does a linear search of the word list
*
        subroutine Find(W)
        character*5 W
        common /Table/ Size, Words
        character*5 Words(10)
        integer Size
        integer I
        do 100 I=1, Size
            if (Words(I) .eq. W) then
                print *, W, ' found at position ', I
                return
            endif
100     continue
        print *, W, ' not found'
        return
        end
```

Some variables are declared and placed in a common block in the main program. We now show a sample input file.

```
10
print
write
check
value
there
often
limit
array
count
order
value
```

We now show the output produced with this input.

```
value found at position          4
```

The use of **common** blocks requires some caution and they should only be used when they really do save some parameter passing overhead. More details of using **common** can be found in Appendix C.

Programming Example

Problem Statement

A Caesar cipher is a method of encryption that shifts the letters of the alphabet by some amount (the key) in order to encrypt them. For example, if the key is 3, then "a" becomes "d" and so on. This method of encryption is easy to do using a computer, but it is also easy to break using a computer. Because there is always one cipher letter for each plaintext letter, a frequency analysis will usually reveal which letters are which. Write a program to encrypt text using a provided key.

Inputs

A key in the range 1 to 26. Lines of text of maximum length 60 characters.

Outputs

Encrypted versions of the input lines.

Discussion

We need to read in the characters of the input line, one by one, and convert them to their corresponding integer value. This value must be incremented by the key and care must be taken that values greater than 26 are "wrapped around" so that they remain in the range 1 to 26. The encrypted integer value must then be converted back to a character and printed.

Program

```
*       Program 10.24
*       Encrypts lines of text using
*       a Caesar cipher. A Key must be supplied
*
*           Line - current line of text
*           Char - current character from Line
*           Key - shift to apply to alphabet
*           CInt - integer representation of character
*
        program Caesar
        character*1 Line(60), Char
        integer Key, CInt, Ichar
        print *,' Enter key followed by lines of text'
        read *, Key
        print *, Key
        while (.true.) do
            read (*, '(60A1)', end=200) Line
            write (*, '(60A1)') Line
            do 100 I = 1,60
                if (Line(I).ge.'a'.and.Line(I).le.'z') then
                    CInt = Ichar( Line(I) )
                    CInt = CInt + Key
                    if (CInt.gt.Ichar('z')) then
                        CInt = CInt - 26
                    endif
                    Line(I) = Char(CInt)
                endif
100         continue
            print '(A)', 'Encrypted version of line is'
            print '(60A1)', Line
        end while
200     stop
        end
```

Testing

```
 Enter key followed by lines of text
     5
aabbccddeeffgghhiijjkkllmmnnooppqqrrssttuuvvwwxxyyzz
 Encrypted version of line is
ffgghhiijjkkllmmnnooppqqrrssttuuvvwwxxyyzzaabbccddee
```

Discussion

The output of this program depends on the particular character sequence used on your machine. For an ASCII machine, it performs as you would expect, mapping alphabetic characters to alphabetic characters. However, for an EBCDIC machine there are nonalphabetic characters that come logically in the middle of the ones from the alphabet. Thus the encrypted output can be a little surprising. Some letters are mapped to punctuation, such as the double quotation mark character. Notice that the program still works. An equivalent decryption program for the same type of machine will produce the correct cleartext again.

Design, Testing, and Debugging

- Always ensure that the array index of any array is within the bounds of the array. When the array reference is within a loop and the array index depends on the loop index, check that the loop index can never exceed the array index.
- Choose the array index range to be meaningful. Most arrays will be indexed from one to some upper bound. But if it makes sense for your problem to index it from -10 to -1, then go ahead and do it.
- We've seen in this chapter that we can write small pieces of code that do quite complicated things. Try to maintain an awareness of the potential execution complexity of what you write when using arrays.
- Be careful when passing arrays to subprograms. If you have explicitly dimensioned the array in the subprogram, then check to make sure that the array that is passed in will fit.
- Don't change an array in a subprogram unless that is the explicit purpose of the subprogram. Nothing is worse than having an array altered by a subprogram that didn't need to do so. We recall one student program that found the second largest element of an array by sorting it and printing the second element from the end. It works, but it's not a good idea.

Style and Presentation

- Arrays allow programs to manipulate large amounts of data easily. This means that output from such programs must be carefully thought out and well laid out. It's tempting to bury the user in data just because it's easy to produce. Always give thought to which values the user will want to see and how to present them effectively.

Fortran Statement Summary

Declaring Arrays

An array declaration names the array and its type and specifies the number and index range of its elements.

    type variable(start: stop)

    integer L(10), I(0:15)
    real X(-100:100)

Assignment to and Reference of Arrays

Array elements can be assigned and referenced in exactly the same way as ordinary variables.

    arrayName(index) = *expression*
    variable = *arrayName(index)*

    A(I) = 3 * J + 2
    A(2 * I + 1) = 54
    X = A(J) / A(K)

Passing Arrays to Subprograms

Arrays can always be passed to subprograms when the invoking environment and the subprogram understand the array to be the same size. Arrays can be passed to subprograms in which their dimension has not been made explicit. The variable describing the array size must be declared before the array itself is declared.

```
real A(10)
call Examp(10, A)
    .
    .
subroutine Examp(N, A)
integer N
real A(N)
```

Chapter Summary

- An array is a way of representing structured data. Arrays can represent lists and tables.
- An array is declared like a variable, except that the number of items it contains must be specified, along with its name and type.
- An array consists of a contiguous set of memory locations. These are referenced using the name of the array and the number of the specific member of the set. This last value is called the array index.
- The individual locations within an array are called its elements. An array element can be assigned, or its value used in an expression, just as an ordinary variable can.
- An array index can be a constant, a variable, or an expression. Its value must fall within the range of legal index values specified when the array was declared. Most compilers do *not* check this.
- An array cannot be manipulated as a unit except when it is passed to or from a subprogram, or when it is read or written.
- Arrays with only one possible index are called one-dimensional arrays. They can be used to model vectors or polynomials.
- There are two main algorithms for searching for a value in a list: linear search and binary search. Linear search is used when the list is unsorted and binary search is used when it is sorted.
- Sorting algorithms arrange elements of a list in order. Selection sort does this by successively finding the largest remaining element and moving it to the end. Bubble sort rearranges pairs of values that are out of order on a series of passes through the list.
- The **common** statement allows blocks of storage that are not local to any particular subprogram to be defined. These can be used to store large amounts of data that are used in many subprograms. This avoids the overhead of specifying that they are arguments at each invocation and also avoids the overhead of moving the values between subprograms. Multiple common blocks can be defined, since each has its own name. The association of names with memory locations in common blocks is done positionally.

Define These Concepts and Terms

Index of an array
Implied do loop
Horner's rule
Newton-Raphson method
Average case analysis
Linear search

Key
Binary search
Selection sort
Bubble sort
Prime numbers
Sieve of Eratosthenes

Exercises

1. Write a program to count the frequency of letters in text and print them out in decreasing order of frequency. What are the ten most frequent letters in English text? What are the ten most frequent letters in Fortran programs?
2. Write output statements that will print a character 6 times on the first line, 5 on the second, and so on.
3. Write a program to read sets of three real numbers representing coordinates in a 3-dimensional space and print the coordinates of the point closest to the origin.
4. Write a program to symbolically integrate a polynomial and evaluate it between two limits given as input. Assume that the constant of integration is given as input.
5. Rewrite Program 10.5 to check for a zero denominator.
6. Often a list is searched to check whether a particular value is present, and if it isn't, the value is inserted. In this situation there is an improvement to the linear search technique. The value being searched for is inserted at the end of the list. Now the linear search is made, but the termination condition is much simpler because we know that the value is present in the list (either at the end or earlier). When the search terminates, having found the value, we examine the location at which it was found. If it is at the end of the list, then the value was not already present and we can leave it in the list. If it is found before the end, then it is actually present twice, so we delete the copy at the end. Write a program to implement this linear-search-and-insert algorithm.
7. Write a selection sort that sorts an array of integers into descending order.
8. Write a bubble sort that sorts an array of integers into descending order.
9. Write a bubble sort that sorts an array of characters into ascending order.
10. Write a program that will read a string and print a banner showing the letters in the string in large formats, say 10 or 12 lines high. Use an array to store pieces of the representation of each character.

11. Using the Sieve of Eratosthenes, generate as many primes as you can and count them. Produce a table showing the number of primes between 1 and 1000, between 1001 and 2000, and so on.
12. Write a program to merge the contents of two arrays into a third array.
13. Write a program that simulates a simple calculator with ten memory locations. The commands come from the standard input and are of the form

    <operand> <operand> <operator> <target>

    where an <operand> is either an integer or a memory reference, an operator is one of +, −, *, /, and where a <target> is either a memory reference or is absent, in which case the result is printed immediately. A memory reference is the character ":" followed by an integer in the range 0 to 9.
14. Write a function that accepts an array of **logical** values and returns an integer indicating how many of the values were *true*.
15. Write a subroutine that plots the values in an array. The parameters should be the array, the number of values, a number to label the left-hand end of the *x*-axis, and a number to label the right-hand end of the *x*-axis. The values in the array indicate the *y* values of equally spaced *x* values. Plot the array with the *x*-axis running horizontally. Assume that the function is positive in the interval.

Engineering Problem 9

Building a Custom Resistor

Resistors come in standard resistance values. When a resistor of another value is needed, it can be built by using a set of the standard resistors in parallel. You may recall that when two resistors R_1 and R_2 are connected in parallel, the combined resistance R is given by

$$\frac{1}{R} = \frac{1}{R_1} + \frac{1}{R_2}$$

Suppose that we are given a box of resistors with resistances chosen from a set

$$\{R_1, R_2, \ldots, R_n\}$$

(with a potentially infinite number of each) and we are asked to construct a resistor with resistance R. One way to approach this is to take the reciprocal of R and then try to express it as a sum of reciprocals of the R_i's. We can choose the smallest resistors first. This gives a good solution that uses a small number of resistors, although there will be situations in which some other choice would do better.

As an example, suppose that we are given resistances of 1, 2, 5, and 10 ohms, and that there are an infinite number of each kind. Suppose further that we want to build a resistance of 1.25 ohms. Then the reciprocal of 1.25 is 0.8, and we want to express this as a sum of 1, 0.5, 0.2, and 0.1 (the reciprocals of the resistors we are given). By inspection, we see that

$$0.8 = 0.5 + 0.2 + 0.1$$

Therefore, we can build a 1.25-ohm resistor by placing 2-, 5- and 10-ohm resistors in parallel.

Programming Problems

1. Write a program to implement this algorithm. The input to the program will be a list of resistor denominations, in no particular order, and a required resistance. The output of the program should indicate the number and type of resistors required to produce the required resistance using the smallest possible number of resistors.
2. Modify the program of the previous part to handle the case where there are only a finite number of resistors of each kind. The input will include the number of resistors of each kind available. Where it is not possible to exactly construct the required resistance, construct the closest one and indicate the percentage error.

Engineering Problem 10

Motion of Particles

Many problems are concerned with the movements of sets of particles that interact with each other. Here we consider a computational approach to this problem. A completely analytic approach is extremely difficult for more than trivial numbers of particles.

Consider a cloud of charged particles in 3-dimensional space. Each particle repels the others according to an inverse square law

$$F = k \frac{q_i q_j}{d^2}$$

where q_i and q_j are the charges of the particles and d is the distance between them. If the charges are expressed in coulombs, then the constant k has the value 8.9874×10^9 nm²/coulombs². The charge on an electron is 1.602×10^{-19} coulomb.

Each particle has associated with it a position given as a set of coordinates in 3-dimensional space. It also has a velocity that is a vector and can be described by a triple of coordinates.

Given that a particle is moving with velocity v and is being acted upon by a force F, we can calculate an approximation for its position after a small time interval Δt by calculating the acceleration applied to it. Newton's second law tells us that the acceleration (also a vector) is calculated by dividing the force by the mass of the particle.

$$a = \frac{F}{m}$$

This acceleration can be used to calculate the new velocity:

$$New\ v = v + a$$

and the new position

$$new\ position = old\ position + at$$

The force acting on each particle is the vector sum of the forces due to all of the other particles.

Programming Problems

1. Write a program that reads the initial positions and velocities of a set of n electrons, each with mass 9.11×10^{-31} kg. The program should also read a value for a time step and a number of steps. It should calculate and print the position of the electrons after each time step.
2. Modify your program so that the particles are assumed to be in a gravitational field at the Earth's surface.

Engineering Problem 11

Separation of Components

Multicomponent chemical mixtures can be separated in a flash drum, a device of the general form shown in Figure P11.1.

A mixture of components flows into the flash drum through a reducing valve. Because the pressure within the flash drum is lower than in the feed stream, the volatile components of the mixture will tend to vaporize while the less volatile will remain as liquids. If two output streams are connected to the top and bottom of the flash drum, then the mixture in the top stream (the vapor stream) will contain a greater percentage of the lighter components, while the bottom stream (the liquid stream) will contain a greater percentage of the liquid components.

Suppose that the incoming mixture contains c components. Then we use the following notation

- F is the flow rate of the feed
- V is the flow rate of the vapor stream
- L is the flow rate of the liquid stream
- z_i is the mole fraction of component i in the incoming stream
- y_i is the mole fraction of component i in the vapor stream
- x_i is the mole fraction of component i in the liquid stream

Typically we would be given the incoming mole fractions and would be interested in calculating the mole fractions of each of the components in the two outgoing streams, as well as the flow rates of the liquid and vapor streams. For example, if we are dealing with two components and we calculate that $y_1 = 1.0$ and $x_1 = 0.0$ for the first component, while $y_2 = 0.0$ and $x_2 = 1.0$ for the second, then we will have separated the two substances completely. We will restrict our attention to a simple case, known as TP flash separation, in which the temperature and pressure inside the drum are known *a priori*.

Because of the law of conservation of mass

$$F = V + L \qquad (1)$$

and
$$Fz_i = Vy_i + Lx_i \qquad (2)$$

It is also true by definition that the sum of the mole fractions in each stream equals 1 so that

$$\sum z_i = 1 \quad \sum y_i = 1 \quad \sum x_i = 1 \qquad (3)$$

We define K-values, or distribution coefficients, K_i by the following equations:

$$K_i = \frac{y_i}{x_i} \qquad (4)$$

Using equations (1), (2), (3), and (4) we can derive

$$x_i = \frac{z_i}{1 - (\frac{V}{F}) + (\frac{V}{F})K_i} \qquad (5)$$

and

$$y_i = \frac{z_i K_i}{1 - (\frac{V}{F}) + (\frac{V}{F})K_i} \qquad (6)$$

Now, using (3) we can write

$$\sum x_i - \sum y_i = 0.0$$

Figure P11.1
Flash Drum

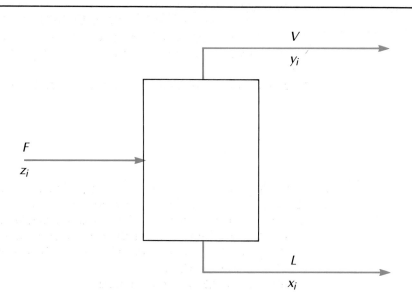

so that, substituting, we get

$$\sum_{i=1}^{i=c} \frac{z_i(1 - K_i)}{1 + (\frac{V}{F})(K_i - 1)} = 0.0 \tag{7}$$

This is called the *flash equation*.

The K_i's are functions of temperature, pressure, and of the compositions x_i and y_i. However, a simplified or shortcut approximation known as Raoult's law is:

$$K_i = \frac{p_{i*}}{P} \tag{8}$$

where p_{i*} is the vapor pressure of component i at temperature T and pressure P. The vapor pressure of each component can be calculated using *Antoine's equation*:

$$\ln p_{i*} = a_1 + \frac{a_2}{(T + a_3)} + a_4 T + a_5 \ln T + a_6 T^{a_7} \tag{9}$$

where the a_i's are coefficients for each specific component.

In this problem we are given z_i for each component, the temperature T, and the pressure P in the drum. We can calculate the K_i's from Antoine's equation and Raoult's law (equations (8) and (9)) and then use the flash equation (7) to determine the ratio V/F. Once we know this we can substitute into equations (5) and (6) to find the mole fractions of each component in the output streams.

Here are the steps to follow:

1. Read in the coefficients for Antoine's equation, the temperature and pressure, and the feed fractions z_i.
2. Find values of V/F for which the left-hand side of the flash equation is positive and negative. Note that V/F must lie between 0 and 1.
3. Use the bisection method to find the value of V/F that satisfies the flash equation.
4. Calculate the mole fractions x_i and y_i, and V and L as fractions of F.

Programming Problems

1. Using the steps shown earlier, write a program that will find the vapor and liquid mole fractions for each component, given the input mole fractions, a temperature, and a pressure. Your program should handle up to six components. You may find it easier to read the Antoine coefficients from a file. See Chapter 9 for details of how to do this. Use the following data:
   - Temperature = 372°K
   - Pressure = 0.20673×10^7 pascals

- Mole Fractions:
  - $C_2H_4 = 0.02$
  - $C_2H_6 = 0.03$
  - $C_3H_6 = 0.05$
  - $C_3H_8 = 0.1$
  - $i\text{-}C_4H_{10} = 0.6$
  - $n\text{-}C_4H_{10} = 0.2$

The following table contains the values of the Antoine coefficients for the preceding compounds, in order.

Antoine Coefficients

1	2	3	4	5	6	7
60.2591	-2517.89	0.0	1.23597E-2	-7.00865	2.97139E-16	6
54.9873	-2636.62	0.0	8.50757E-3	-5.89314	2.06619E-16	6
52.7642	-3243.44	0.0	4.29984E-3	-5.13649	8.59519E-17	6
63.5869	-3550.19	0.0	8.18977E-3	-7.09227	7.13799E-17	6
66.1499	-4301.38	0.0	6.53669E-3	-7.24760	3.44859E-17	6
62.5348	-4038.33	0.0	5.35525E-3	-6.64216	4.63607E-17	6

Computing with Characters

11.1 An Algebra of Strings
11.2 Converting Numeric Representations
11.3 More Program Tracing
11.4 Text Editing
11.5 Text Formatting Techniques
11.6 A Framework for Formatting

11.1 An Algebra of Strings

So far we have been computing mostly with numeric values (although we have made some use of Boolean values). That is, we have been using numeric values as the operands in expressions and producing numeric values as the results of expressions. This is the most "natural" kind of computation for most of us, since we encountered it when we were very young and have used it extensively since then. But you know from algebra (secondary school or post-secondary) that there's nothing special about numbers; algebraic structures like groups, rings, fields, and so on can be defined for many object-operator combinations, as long as the operators obey some simple rules.

Another very useful domain for computing on modern machines is character strings and their associated operations. In Fortran there are string variables, constants, assignments, and expressions. Variables are declared like this

    character*10 S, T

where S and T are declared to be strings of ten characters each. Strings are of a fixed length.

String constants are written enclosed in single quotation marks. To represent a single quotation mark *within* a string, we write two successive single quotation marks.

'This is a simple string'
'This one''s not so simple'

Strings can be assigned to variables.

    S = 'Hello, Jo!'
    T = 'This will be truncated'

If the string on the right-hand side of the assignment is shorter than the string to which it is being assigned, then the destination is padded on the right with blanks. Character strings can be read and written using the standard approach we've seen so far. Strings to be read must be enclosed in single quotation marks.

There are several operations that can be applied to character strings, including asking how long a string is (which uninterestingly returns the size a variable was declared to be, rather than something more useful like the number of characters explicitly used in the last assignment to the variable), catenating two strings together, and extracting or replacing a substring of a string.

A **substring** is specified by writing the left and right character positions of the desired piece, separated by a colon.

```
character*9 S
S = 'computing'
print *, S(4:6)
```

will print "put". It's also possible to assign to a substring.

```
S(5:6) = 'ar'
print *, S
```

results in the output of the word "comparing". We can write the same value for the left and right ends of the substring and thus specify a single character. We can also omit either endpoint; omitting the left means start at the beginning, omitting the right means go to the end, and omitting both means take the entire string.

```
print *, S(:4), S(6:), S(:)
```

will print "comp", "ring", and "comparing".

Catenation can be used to put pieces of things together in interesting ways. The symbol for catenation is "//" written between the two strings.

```
character*18 S, T
S = 'Robert G. Crawford'
T = S(11:) // ', ' // S(1:1) // '.' // S(8:9)
print *, T
```

will print "Crawford, R.G.". The strings used on the left and right sides must not overlap. The following assignment would be illegal.

```
         character S(10)
*        illegal
         S = S(2:10)
```

Another useful tool in dealing with strings is the intrinsic function named *index*. It tells at what position in one string a second string appears. For example,

```
index ('Hi there.', 'the')
```

has the value 4. If the second string is not present in the first string, the value 0 is returned.

Comparing strings (as opposed to single characters) works in the way you would probably expect it to: strings are equal if all the characters in them are equal, and S precedes T precisely when they differ in some position and the corresponding character in S precedes the corresponding character in T. If necessary, a shorter string is extended with blanks on the right to make it long enough for the comparison to be carried out to the end of the longer string.

11.2 Converting Numeric Representations

As an example of string manipulation, consider the problem of converting number representations from one base to another. For example, the binary (base 2) number 1101 has the decimal (base 10) representation 13. We can write functions to perform this conversion. In fact, the most convenient way to do this is to have two functions: the first will take a string representing a number, together with a base, and compute the value of the number. This is a bit of a slippery concept; it's sometimes hard to separate the idea of a number from a representation, but think about the number "four" being represented either by the digit 4 or by the number of fingers on your left hand. Now it should be possible for you to generalize and think about "four" without having to see the representation 4. We'll build a function to convert number representations into the machine's internal conceptualization of number values. We know that the machine is using binary representations internally, but that doesn't matter; we also know that if we try to print values, they'll come out as decimal representations (since that's what we normally use, it is what machines normally print), but that doesn't matter either. We'll work with the internal **integer** type that Fortran has.

So, the first thing we need is to convert a string representing a number to a value. The way to think about the problem is this: read the string representation left to right, a digit at a time; at each step we have a partially formed value and the next digit; then we form the new partial value by multiplying by the base of the representation and adding in the value of the new digit.

Here's how it works with 1101 in base 2. We start with a partial value of 0. Now look at the leftmost digit of 1101, which is, of course, 1. The value of 1 in base 2 is 1, so the new partial value is the old partial value, 0, times the base, 2, plus the value of the new digit, so $0 \times 2 + 1 = 1$ is our result. The problem is now one digit smaller. Consider the leftmost digit of 101, which is again 1. Now we form a new partial value as the old partial value times the base, plus the value of the new digit, so $1 \times 2 + 1 = 3$ is our result. The problem is now one digit smaller. Consider the leftmost digit of 01, which is 0. Now $3 \times 2 + 0 = 6$ is the new partial value. The problem is now one digit smaller. The last digit is 1. The new partial value is $6 \times 2 + 1 = 13$ and since that was the last digit, this partial value is the final value. It's easy to build this general approach into a Fortran function.

The inverse transformation is a bit different. Now we have an internal representation (a value), and want to convert it to a string representation in some specified base. This time it's easiest to work right to left, producing the representation a digit at a time.

Consider producing the base 3 representation of the decimal value 113. The rightmost digit in the final representation will indicate how many *units* are in the value (that is, the amount less than the base). We can find this by taking the modulus of the number with respect to the base. In this case 113 mod 3 = 2 so the rightmost digit, the partial representation generated thus far, is 2. We now divide the value by the base, keeping only the integer part of the result. In this case 113/3 = 37. Now we have a smaller problem, the 37 that is left, and a partial result—the 2. The general step is to find the next digit by taking the modulus of what is left, put that digit in front of the partial representation, and reduce the problem by dividing by the base. So, for the next step, we find the next digit as 37 mod 3 = 1, put it in front of the existing partial solution to form 12, and reduce the problem by dividing 37 by 3 to get 12. Similarly, we now find 12 mod 3 = 0, put it in front to form 012, and reduce the problem by dividing 12 by 3 to get 4. Next we find 4 mod 3 = 1, put it in front to form 1012, and reduce the problem by dividing by 3 to get 1. The final representation is 11012, and there is nothing left to work with (the result of dividing 1 by 3 and retaining only the integer component is the integer 0). We now know that 11012 is the base 3 representation of the value 113. It is possible to build this into a Fortran function, too.

```
*       Program 11.1
*       Converts numbers from string representations in
*       various bases to internal values, and vice versa
*
        program NumBas
        character*10 StrIn, StrOut, SVal
        integer BasIn, BasOut, Value, IVal
        print *, 'Enter input base and output base'
        read *, BasIn, BasOut
        while (BasIn .ne. 0) do
            print *, 'Enter string to be converted'
            read *, StrIn
            Value = IVal(StrIn, BasIn)
            StrOut = SVal(Value, BasOut)
            print 100, BasIn, StrIn
100         format(' Input base ', I5, ' Input string ', A)
            print 200, BasOut, StrOut
200         format(' Output base ', I5, ' Output string ', A)
            print 300, Value
300         format(' Output as integer is ', I6)
            print *
            print *, 'Enter input base and output base'
            read *, BasIn, BasOut
        end while
        stop
        end
```

Computing with Characters

```
*
*       given a string and base, find the value the string represents
*

        integer function IVal(S, Base)
        character*10 S
        integer Base
        character*11 T
        integer I, V
*       add a blank at the end to stop the scan below
        T = S // ' '
*       skip leading blanks
        I = 1
        while (T(I:I) .eq. ' ') do
            I = I + 1
        end while
        V = 0
        while (T(I:I) .ne. ' ') do
*           multiply by Base to shift left, add next digit value
*           the digit value is one less than its position
            V = V * Base +
*               Index ('0123456789ABCDEF', T(I:I)) - 1
            I = I + 1
        end while
        IVal = V
        return
        end

*
*       given a value and base, construct a string representation
*

        character*10 function SVal(I, Base)
        integer I, Base
        character*11 S, T
        integer Remain, DigPos
        character*16 Digits, Blanks
        S = ' '
        Digits = '0123456789ABCDEF'
        Blanks = ' '
```

```
        if (I .eq. 0) then
            S = '0'
        else
            Remain = I
            while (Remain .gt. 0) do
*               the rightmost digit of what's left is found
*               by taking the modulus with respect to the base
                DigPos = mod(Remain, Base) + 1
*               now divide to effectively shift right
                Remain = Remain / Base
                T = Digits(DigPos:DigPos) // S
                S = T
            end while
        endif
        SVal = Blanks(1: 11 - index (S, ' ')) // S
    return
    end
```

In the function IVal, the assignment statement inside the main **while** loop is a bit tricky. It computes the new value of V by multiplying by the Base (thus shifting the result one column to the left in the Base representation), and then adding in the value corresponding to the current digit; this latter value is one less than the position of the Ith digit in the string of digits (digit 0 is in position 1, and so on).

The tricky statement in SVal is the last assignment, which does a right justification of the computed string in a field of the appropriate width. This is done by finding how many valid characters there are (looking past these for a blank), and prefixing blanks in front of them.

Here is some sample output from the program. The output shows the input values, so we don't need to show them separately. The prompts have been removed except at the beginning to save space.

```
Enter input base and output base
Enter string to be converted
Input base      2  Input string  100
Output base     4  Output string            10
Output as integer is       4

Input base      8  Input string  465
Output base     2  Output string  100110101
Output as integer is     309

Input base      16 Input string  1F
Output base     4  Output string           133
Output as integer is      31
```

Input base   2   Input string 1101
Output base   16   Output string   D
Output as integer is   13

Input base   9   Input string 1555
Output base   4   Output string   102200
Output as integer is   1184

11.3 More Program Tracing

We have already seen how to trace programs when they are quite small and simple. To illustrate tracing on a larger and more complicated example, we'll trace the number representation program using these lines of input:

4 5
'13'
0 0

So, when the NumBas program starts, we produce the following table corresponding to the point immediately prior to executing the first input statement.

Main:
  StrIn: undefined
  StrOut: undefined
  BasIn: undefined
  BasOut: undefined
  Value: undefined

There are as yet no values associated with any of the variables. After executing the input statements, the table becomes:

Main:
  StrIn: '13
  StrOut: undefined
  BasIn: 4
  BasOut: 5
  Value: undefined

Notice that we have padded the input value for StrIn on the right with blanks, just as the machine would do. Now we encounter the **while** statement. Since the condition is true (BasIn is not zero, which is the sentinel value we use in this program to indicate the end of the input list), the body of the statement is executed.

We want to assign a value to the variable Value, but must first trace the execution of the **function** IVal. We now create a section of the table for the variables declared in IVal.

Main:
    StrIn: '13   '
    StrOut: undefined
    BasIn: 4
    BasOut: 5
    Value: undefined

IVal:
    S: undefined
    Base: undefined
    T: undefined
    I: undefined
    V: undefined

The parameters of IVal are S and Base, and these take on the values of the corresponding arguments in the main program. We can fill them in immediately.

Main:
    StrIn: '13   '
    StrOut: undefined
    BasIn: 4
    BasOut: 5
    Value: undefined

IVal:
    S: '13   '
    Base: 4
    T: undefined
    I: undefined
    V: undefined

Now we carry out the assignment to T. Its sole purpose in the program is to allow us to extend the input parameter with a blank. In this way we can guarantee later in the subprogram that a search for a blank in the string will terminate successfully. We'll also incorporate the assignment to I in this next version of the table.

Main:
    StrIn: '13   '
    StrOut: undefined
    BasIn: 4
    BasOut: 5
    Value: undefined

IVal:
    S: '13   '
    Base: 4
    T: '13   '
    I: 1
    V: undefined

Now, since I is 1, the Ith character of T is the first character, and that is not a blank, so the body of the **while** is not executed. This loop allows for the input value being padded on the left with blanks, rather than on the right. So we come to the assignment to V.

Main:
    StrIn: '13   '
    StrOut: undefined
    BasIn: 4
    BasOut: 5
    Value: undefined

IVal:
    S: '13   '
    Base: 4
    T: '13   '
    I: 1
    V: 0

Now the Ith character of T is the first character and this is not a blank. So the body of the **while** is executed. We compute the value of the expression on the right side of the assignment to V. The product of V and Base is zero because V is zero. The Ith character of T appears in the second position of the string of digits. The value of the expression is 2 − 1, or 1, which we assign to V. We also include the result of incrementing I in this version of the table.

Main:
  StrIn: '13   '
  StrOut:
  BasIn: 4
  BasOut: 5
  Value: undefined

IVal:
  S: '13   '
  Base: 4
  T: '13   '
  I: 1̸ 2
  V: 0̸ 1

Now we're at the end of the **while**, so we go back to the beginning of it. The Ith character of T is now the second character, which is 3 and thus not blank, so the body of the **while** is executed again. This time the product of V and Base is 4. We find 3 at the fourth position in the string of digits. We subtract 1 from the position of the Ith character to give 3, and add this to the 4 to get 7. We also include the result of incrementing I in this version of the table.

Main:
  StrIn: '13   '
  StrOut:
  BasIn: 4
  BasOut: 5
  Value: undefined

IVal:
  S: '13   '
  Base: 4
  T: '13   '
  I: 1̸ 2̸ 3
  V: 0̸ 1̸ 7

This time when we look at the condition on the **while**, it fails, since the third character of T is a blank. We now come to the assignment of V to IVal and the subprogram terminates. The value of the function IVal is assigned to Value in the main program.

Main:
  StrIn: '13   '
  StrOut: undefined
  BasIn: 4
  BasOut: 5
  Value: 7

When a subprogram is invoked, we have to keep track of where the invocation occurs. You can do that by recording the name of the invoking routine (which will be the main program or some other subprogram) and the line number from which it was invoked, if the program is complicated enough to warrant that. This one is straightforward because there's only one point of invocation for each subprogram. Another thing that you may have to do that is not needed here is to copy back values of arguments to the invoking routine. If the values of the arguments have been changed within the subprogram, they are copied back. In this case, there were no assignments to the parameters.

So, now we're back in the main program, and the next thing to happen is the invocation of SVal to get a value to assign to StrOut. The table of values is expanded again.

Main:
    StrIn: '13   '
    StrOut: undefined
    BasIn: 4
    BasOut: 5
    Value: 7

SVal:
    I: undefined
    Base: undefined
    S: undefined
    T: undefined
    Remain: undefined
    DigPos: undefined
    Digits: undefined
    Blanks: undefined

The values of the arguments supplied in the main program are copied into the arguments in the subprogram. This is straightforward, as are the assignments to S, Digits, and Blanks, so we will incorporate all of these changes into the next version of the table.

Main:
    StrIn: '13   '
    StrOut: undefined
    BasIn: 4
    BasOut: 5
    Value: 7

SVal:
    I: 7
    Base: 5
    S: '   '
    T: undefined
    Remain: undefined
    DigPos: undefined
    Digits: '0123456789ABCDEF'
    Blanks: '   '

Now we encounter the **if**, and since I is not zero, execute the **else** part. Remain is set to I.

Main:
    StrIn: '13   '
    StrOut: undefined
    BasIn: 4
    BasOut: 5
    Value: 7

SVal:
    I: 7
    Base: 5
    S: '   '
    T: undefined
    Remain: 7
    DigPos: undefined
    Digits: '0123456789ABCDEF'
    Blanks: '   '

The condition on the **while** is true since Remain is greater than zero, so we compute DigPos as the modulus of 7 with respect to 5, which is 2, and add 1 to get 3. Remain is the integer part of dividing 7 by 5, which is 1. T is the old value of S with the third character of Digits, which is 2, stuck on the front. And finally, this value is stored in S.

```
Main:                        SVal:
  StrIn: '13        '          I: 7
  StrOut: undefined            Base: 5
  BasIn: 4                     S: '//////////' '2         '
  BasOut: 5                    T: '2        '
  Value: 7                     Remain: 7̶ 1
                               DigPos: 3
                               Digits: '0123456789ABCDEF'
                               Blanks: '          '
```

Back at the top of the **while**, the condition still holds, so we execute the body again. This time DigPos gets the value 2 (because the modulus of 1 with respect to 5 is 1). Remain becomes zero, which is the integer part of the result of dividing 1 by 5. T becomes the old value of S with the second digit, which is 1, stuck on the front. S gets this value as well.

```
Main:                        SVal:
  StrIn: '13        '          I: 7
  StrOut: undefined            Base: 5
  BasIn: 4                     S: '//////////' '2//////////' '12       '
  BasOut: 5                    T: '2//////////' '12       '
  Value: 7                     Remain: 7̶ 1̶ 0
                               DigPos: 3̶ 2
                               Digits: '0123456789ABCDEF'
                               Blanks: '          '
```

Once more at the top of the **while**, we examine the condition, which fails. We skip to the next statement. The purpose of prefixing the substring of Blanks to S is to push the computed representation string to the right of the variable's storage. So, knowing that the nonblank characters of S are at its left end, we find the first blank. We now know how many nonblank characters there are, and insert enough blanks in the beginning to push these nonblanks to the right end. In this case, the first blank in S is at position 3, so we insert 8 blanks before the nonblank characters. This value is returned as the result of SVal, and stored in StrOut in the main program.

Main:
  StrIn: '13   '
  StrOut: '   12'
  BasIn: 4
  BasOut: 5
  Value: 7

The output statements print the values 13, 4, 7, 12, and 5. The next statement in the main program is the input statement that will read the second set of input values. The **while** condition will fail, leaving this final table of values.

Main:
  StrIn: '13   '
  StrOut: '   12'
  BasIn: 4̸ 0
  BasOut: 5̸ 0
  Value: 7

11.4 Text Editing

We have mentioned several times that programs are often designed to work together to solve interesting problems. In a situation like this, it is often the case that a file produced by one program either has to be modified to reflect additional work that has been done by another program, or have its format changed to be compatible with the input format required by another program.

One way to bring about these changes is with a **stream editor**. This is a program that reads a file and produces a new one containing modified copies of the lines in the original file. The modifications can include replacing a string by another string, or perhaps deleting lines that contain a specified string.

We will write a simple version of such a program. It will read one line of input that will contain a target string to look for and a replacement string. The target string will be compared to all lines in the given file. Any lines not containing the target are copied to the output file; any lines containing the target are copied to the output file with the replacement string in place of the target.

The line specifying the target and replacement strings will begin with an arbitrary separator character that must not be included in either string. This is immediately followed by the target string, then another separator, then the replacement string, and finally another separator. Here are some examples.

/e/EE/
;one;1;
,semi-colon,;,

The first example indicates that the lowercase letter "e" is to be replaced by two uppercase letters. The second example indicates that the word "one" is to be replaced by the digit 1. The third example indicates that the word "semi-colon" is to be replaced by the semicolon character, which was the separator used in the second example.

The program we now show can accept a single line of this type as its command input.

```
*      Program 11.2
*      Edits an input stream producing an output stream with
*      a specified target string modified to a specified replacement
*
       program Stream
       character*80 Pat, Old, New, InLin, OutLin
       character*20 InFil, OutFil
       character*1 Sep
       integer OldSiz, NewSiz, Where
       print *, ' Enter editing pattern'
       read '(a)', Pat
       print *, ' Enter input file name'
       read '(a)', InFil
       print *, ' Enter output file name'
       read '(a)', OutFil
       open (1, file=InFil)
       open (2, file=OutFil)
       Sep = Pat(1:1)
       OldSiz = Index(Pat (2:), Sep) - 1
       Old = Pat (2: OldSiz + 1)
       NewSiz = Index(Pat (OldSiz + 3:), Sep) - 1
       New = Pat(OldSiz + 3: OldSiz + NewSiz + 2)
       while (.true.) do
          read (1, '(a)', end=100) InLin
          Where = Index(InLin, Old (1: OldSiz))
          if (Where .eq. 0) then
             OutLin = InLin
          elseif (Where .eq. 1) then
             OutLin = New(1: NewSiz) // InLin (OldSiz + 1:)
          else
             OutLin=InLin(1:Where-1)//New(1:NewSiz)//
     *             InLin(Where+OldSiz:)
          endif
          write (2, '(a)') OutLin
       end while
100    continue
       stop
       end
```

You have been using a text editor to prepare programs and possibly documentation as you have worked through this book. Preparing files using editors is one of the most frequent applications of computers. In this section we develop a program to do simple editing; you might find it (or perhaps an extended version of it) to be a useful tool for some applications; use it if your computing system does not provide an editor (these are, admittedly, rare).

We'll keep the editor simple by making it a **line editor**; that is, we'll expect the user to type commands to effect all changes to the file rather than displaying the contents of the file, allowing the user to make changes on the screen, and having these changes implicitly incorporated into the file, as a screen-oriented editor does. So we'll have to keep track of the contents of the file being edited, which will involve both the lines themselves and the number of lines, and we'll have to keep track of a position in the file—the current line—where the editing operations are being performed.

Our editor will respond to a set of commands. To make the editor simple and also to minimize the number of keystrokes required of the user, we'll make the commands single characters. We need to have commands to add a line to the file (preferably to add any number of lines to the file), delete a line from the file, save the results of an editing session, and terminate the session. These would be enough, in theory, to produce any required output file, but we'll want some other things to make the program more usable. We should also have a way to display the current line pointer (the one the editor's attention is focused on) and possibly several lines in that vicinity, a way to read in an existing file (so we can use the editor to change files and not just to create them), a way to change the current line (so we can, for example, move to a specific line and delete it), and a way to look for a particular string in the file.

Just as the input is simple and terse, so is the output. We will have our program print a single question mark as a prompt for command input. There will be no prompt for text input (the program will just sit and wait for it) and there will be no messages printed to indicate errors.

When working with interestingly large problems like this one, it's extremely important to choose data structures that let you produce programs with the features you want. For example, we could maintain the lines of the file being edited either in a temporary file on disk, or in memory. The former has the advantage (on most machines) that there is more space in the file system than in memory, so larger files can be edited. The latter has the advantage that it is faster to access main memory than disk and we might expect to have simpler programs to manipulate memory than disk. Even once we have decided to use main memory, there are several possibilities. We could maintain the lines in a 1-dimensional array; when a line is deleted, we just shift the remaining lines down, and when a line is inserted, we shift the existing lines up to make room for it. Alternatively, we could maintain the lines in a 1-dimensional array, but not try to keep them in order; thus when we added a line we could just add it at the end of the array, which should be faster than shifting and making room. But this scheme has the disadvantage that we would have to maintain another set of information saying what the *logical* order of the lines is. Now, if we were writing a production-quality editor, we might consider this extra complexity worthwhile to get the speed advantages it might give. For our present purposes, we'll stick with the simpler approach that maintains the lines in an array and shifts them when necessary.

The commands that the editor will understand are these:

s	show current line
r<file>	read file adding after current line
w<file>	write all lines into file
l	list next 10 lines
a	append typed lines after current line
.	end append mode
d	delete current line
n	print current line number
:<line>	set current line pointer to line
-<some>	set current line pointer back some lines
+<some>	set current line pointer ahead some lines
/<pat>/	search for pattern

The structure of the program will be like this. First, in the main program we'll simply read command lines. A separate subprogram will be used as a "router" to decode the command and invoke the appropriate subprogram to handle it; any commands that can be handled with a single line of Fortran code will be done directly. We'll have individual subprograms for the commands that read lines from a file, write lines to a file, print several lines, add several lines, delete a line, and search for a string in the file. We'll also include a (somewhat simplified) version of the routine developed earlier to convert a string into a number; this is needed when we deal with the operands to the commands that move the current line forward and backward.

We'll maintain the state of the editor as the array of lines, the current command line, an integer indicating how many lines are in the array, and an integer indicating the current line. These can be passed to subroutines in a **common** block.

Most of the subprograms are straightforward and we won't say more about them. However, both the insertion and reading commands require making room in the array for lines that are to be added. This is done by a loop that moves all lines from the current position to the end of the file up one position, and then inserts the new line. Perhaps the most interesting algorithm is the one that handles searching for a string. We arrange to stop on the line where the string is found, and we always stop at the last line in any case; you should look at both the condition on the inner **while** loop and the computation of the variable Found, to be sure that you understand them.

```
*       Program 11.3
*       A simple line editor
*       The state of the program is represented by an array of lines,
*       the last input line, a number of lines, and a current line
*
        program Editor
        common /Edit/ Lines, L, Size, P
        character*80 Lines(3000), L
        integer Size, P
        logical Done
        Size = 0
        P = 0
        Done = .false.
        while (.not. Done) do
            print *, '?'
            read '(A80)', L
            call Handle(Done)
        end while
        stop
        end
*
*       Router for commands; handle the simple ones here
*
        subroutine Handle(Done)
        logical Done
        common /Edit/ Lines, L, Size, P
        character*80 Lines(3000), L
        integer Size, P
        integer IVal
        if (L .eq. 'q') then
            Done = .true.
        elseif (L .eq. 'h') then
            call Help
        elseif (L .eq. 's') then
            print *, Lines(P)
        elseif (L(1:1) .eq. 'r') then
            call Input
        elseif (L(1:1) .eq. 'w') then
            call Output
        elseif (L .eq. 'l') then
            call List
        elseif (L .eq. 'a') then
            call Append
```

```
      elseif (L .eq. 'd') then
          call Delete
      elseif (L(1:1) .eq. '/') then
          call Search
      elseif (L .eq. 'n') then
          print *, P
      elseif (L(1:1) .eq. '-') then
          P = P - IVal (L(2:11))
      elseif (L(1:1) .eq. '+') then
          P = P + IVal (L(2:11))
      elseif (L(1:1) .eq. ':') then
          P = IVal (L(2:11))
      else
          print *, 'Unrecognizable'
      endif
      return
      end
*
*     Display help information on the user's screen
*
      subroutine Help
      print 100
100   format (
     &    ' h            help',/
     &    ' q            quit',/
     &    ' s            show current line', /
     &    ' r<file>      read file adding after current line',/
     &    ' w<file>      write all lines into file',/
     &    ' l            list next 10 lines',/
     &    ' a            add typed lines after current line',/
     &    ' .            end append mode',/
     &    ' d            delete current line',/
     &    ' n            print current line number',/
     &    ' :<line>      set current line pointer to line',/
     &    ' -<some>      set current line pointer back some lines',/
     &    ' +<some>      set current line pointer ahead some lines',/
     &    ' /<pat>/      search for pattern',/
     &    )
      return
      end
```

```
*
*       Read lines from a file and add to the array
*
        subroutine Input
        common /Edit/ Lines, L, Size, P
        character*80 Lines(3000), L
        integer Size, P
        integer I
*       The filename is in the input line after the command char
        open (1, file=L(2:))
        while (.true.) do
            read (1, '(A80)', end=100) L
*           make room
            Size = Size + 1
            if (Size .gt. 1) then
                do 200 I = Size, P+1, -1
                    Lines(I) = Lines(I - 1)
200             continue
            endif
*           insert the new line
            P = P + 1
            Lines(P) = L
        end while
100     continue
        close (1)
        return
        end
*
*       Write the array of lines to a file
*
        subroutine Output
        common /Edit/ Lines, L, Size, P
        character*80 Lines(3000), L
        integer Size, P
        integer I
*       The filename is in the input line after the command char
        open (2, file=L(2:))
        do 100 I = 1, Size
            write (2, '(A80)') Lines(I)
100     continue
        close (2)
        return
        end
```

```
*
*       Display next 10 lines on the user's screen
*
        subroutine List
        common /Edit/ Lines, L, Size, P
        character*80 Lines(3000), L
        integer Size, P
        integer I
        do 100 I=P, min (Size, P + 9)
            print *, Lines(I)
100     continue
        P = min (Size, P + 9)
        return
        end
*
*       Accept input lines from terminal and add to array
*
        subroutine Append
        common /Edit/ Lines, L, Size, P
        character*80 Lines(3000), L
        integer Size, P
        integer I
        read '(A80)', L
        while (L .ne. '.') do
*           make room
            Size = Size + 1
            if (Size .gt. 1) then
                do 200 I = Size, P+1, -1
                    Lines(I) = Lines(I - 1)
200             continue
            endif
*           insert the new line
            P = P + 1
            Lines(P) = L
            read '(A80)', L
        end while
        return
        end
*
*       Delete a line from the array
*
        subroutine Delete
        common /Edit/ Lines, L, Size, P
        character*80 Lines(3000), L
        integer Size, P
        integer I
```

```
*       Move the lines down one slot
        do 100 I = P, Size - 1
            Lines(I) = Lines(I + 1)
  100   continue
        Size = Size - 1
*       did we just delete the last one?
        if (P .gt. Size) then
            P = P - 1
        endif
        return
        end
*
*       Search for a particular string in the file
*       quitting when it's found, or the last line is reached
*
        subroutine Search
        common /Edit/ Lines, L, Size, P
        character*80 Lines(3000), L
        integer Size, P
        character*80 Target
        logical Found
        integer Len, I
*       The target is in the input line after the command char
        Target = L (2:)
*       There's a / at the end of the target string
        Len = index (Target, '/') - 1
        Found = P .eq. Size
        while (.not. Found) do
            P = P + 1
            I = 1
            while (I .le. 80 - Len .and.
     &         Target(1:Len) .ne. Lines(P)(I:I+Len-1)) do
                I = I + 1
            end while
            Found = P .eq. Size .or.
     &         Target(1:Len) .eq. Lines(P)(I:I+Len-1)
        end while
        print *, Lines(P)
        return
        end
```

```
*
*       Convert a string to an integer value
*
        integer function IVal (S)
        character*10 S
        character*11 T
        integer I, V
        T = S // ' '
        V = 0
        I = 1
        while (T(I:I) .ne. ' ') do
            V = V * 10 + Index ('0123456789', T(I:I)) - 1
            I = I + 1
        end while
        IVal = V
        return
        end
```

An editor like this can be used as a "batch" editor: it uses the standard input to find its commands and thus it normally would be finding its command input from the keyboard. However, rather than running it interactively, we could put commands in a file, cause the standard input to be from that file (using the appropriate operating system command), and then run the editor. For example, here is a command file that will replace the second line of a file with a new line given in the command file.

```
rdata file
d
a
Report has been generated.
.
wdata file
q
```

If you had a series of programs that were logically related, you could use this editor script to cause this file (stored as DATA FILE)

System started.
Report not yet generated.

to be transformed to this one

System started.
Report has been generated.

and thus use the editor as a mechanism for maintaining a file showing the status of the system as it's running; if the system of programs failed, you could look in the status file to see how much had been successfully completed. Of course, this particular application is not a difficult one, but you can probably see how to generalize from this example to others where having such a batch editor would be more useful.

11.5 Text Formatting Techniques

One of the other textual applications that computers have made commonplace is text formatting. Until computers became ubiquitous, only text that had been printed could be arranged on the page so that the right margin was straight (right-justified) and words and letters were spaced pleasingly. It's very difficult to avoid a ragged right margin using a typewriter. To illustrate some of the more sophisticated uses of string variables, we will construct a simple **text formatter**. Your system may already have one. If it doesn't, even the one we will develop should improve the appearance of your writing.

We will begin by considering some of the common operations that need to be performed to carry out text formatting. The simplest involve recognizing words in text and being able to count them. We will suppose, for the time being, that we are concerned only with each line of text separately. The current line of text will be kept in a string declared like this:

character*1 Line(80)

Counting the number of words on the line involves scanning the characters of the line in turn and watching for a string of alphabetic characters surrounded by blanks and/or punctuation. To avoid having always to refer to the different possible characters, let us divide them into white space (which includes punctuation as well as blanks) and nonwhite space, which we will assume includes only alphabetic characters. Thus a word is a sequence of nonwhite-space characters surrounded by white space.

Now as we look at each individual character of the line, we must remember something about what happened earlier in the line. For example, if we see a nonwhite-space character that was preceded by white space, then we have found the beginning of a word. However, if it was preceded by another nonwhite-space character, then it is somewhere in the middle of a word. Many of the difficulties of writing programs to handle text arise from the necessity to remember previous information and in deciding what needs to be remembered.

To handle this difficulty we introduce a concept that you will find useful in other programs, as well as being a powerful way of thinking about the world. This concept is that of *state*. In the problem that we have been discussing, the key piece of information that must be remembered is whether or not, as we scan the line of characters, we are inside a word. There are thus two possible states: IN and OUT. For each state, we describe an action that depends on the next character we encounter. For example, if we are IN a word and we encounter a nonwhite-space, then we remain IN the word. On the other hand, if we encounter a white space, then we move OUTside a word. Similarly, if we are OUT and we encounter a white space, then we stay OUT. If we encounter a nonwhite-space, then we become IN. For each of these possible *transitions*, we can also carry out some other action. The transition from IN to OUT represents discovering the end of a word. So if there is some action to be performed at the end of each word, then this is the time to perform that action. The transition from OUT to IN occurs at the beginning of a word. The operations and changes of state are illustrated in Figure 11.1.

We can easily write a program to behave in this way. During each iteration of a loop, we examine the next character of the input line and carry out the appropriate actions and change of state. Suppose, for example, that we wish to count the number of words on each line. We can do this by counting either the beginnings or endings of the words. The first version of the program counts the beginnings of words.

```
*       Program 11.4
*       Count words in a line
*
        program Words
        character*1 Line(80), Currt
        character*3 State
        integer I, Word
        logical White
        read '(80A1)', Line
*       a word is a sequence of non-white characters
        State = 'OUT'
        Word = 0
        do 200 I = 1, 80
            Currt = Line(I)
            if (State.eq.'IN ') then
                if (White(Currt)) then
                    State = 'OUT'
                else
                endif
            else
                if (White(Currt)) then
                else
                    State = 'IN '
                    Word = Word + 1
                endif
            endif
200     continue
        print *, ' Number of words =', Word
        stop
        end
```

```
*
*          punctuation and blanks are not in words
*
       logical function White(C)
       character*1 C
       White = C.eq.' '.or.C.eq.'.'.or.C.eq.','.or.C.eq.';'
       return
       end
```

Although this program can be simplified, all of the **if** statements have been left in the program so that you can see what happens in each case. The transitions that do not change the state do not have any actions associated with them.

Of course, the initial state is OUT because we clearly are outside a word when we have not yet looked at the line. This program makes a number of assumptions that are not very realistic, for the sake of simplicity. For instance, it assumes that all lines are exactly 80 characters long. We could have written the program using the ends of words instead.

```
*          Program 11.5
*          Count words in a line
*
       program Words
       character*1 Line(80), Currt
       character*3 State
       integer I, Word
       logical White
       read '(80A1)', Line
*          a word is a sequence of nonwhite characters
       State = 'OUT'
       Word  = 0
```

Figure 11.1
Scanning Words

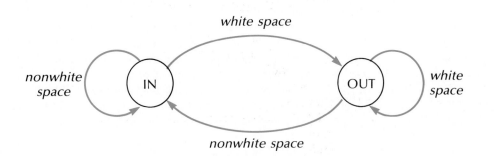

```
      do 200 I = 1, 80
         Currt = Line(I)
         if (State.eq.'IN ') then
            if (White(Currt)) then
               State = 'OUT'
               Word = Word + 1
            else
            endif
         else
            if (White(Currt)) then
            else
               State = 'IN '
            endif
         endif
200   continue
      if (State.eq.'IN') then
         Word = Word + 1
      endif
      print *, ' Number of Words = ', Word
      stop
      end
*
*     punctuation and blanks are not in words
*
      logical function White(C)
      character*1 C
      White = C.eq.' '.or.C.eq.'.'.or.C.eq.','.or.C.eq.';'
      return
      end
```

It turns out that counting ends of words rather than beginnings is slightly more complicated. This is because of the situation in which the last word on the line finishes in the last position on the line. Since we are counting the occurrences of a white space following a word, we do not count such words. Thus we have to add a test to the end of the program to make sure that we finish in the state OUT. If we reach the end of a line and are not in the OUT state, then we add 1 to the count of the number of words.

Now if we want to format lines, we will want to move the words around so that, for instance, the last word ends at the right margin. To do this we need to know not only how many words are on the line, but also how much other space there is. We enhance the preceding program so that it will count the number of words, the number of letters, and the number of blanks. It turns out that punctuation is easily treated as if it belongs to the word that immediately preceded it, so that in this version the only white spaces are blanks.

```
*       Program 11.6
*       count words, blanks, and letters in a line
*
        program Words
        character*1 Line(80), Currt
        character*3 State
        integer I, Word, Letter, Blank
        logical White
        read '(80A1)', Line
        State = 'OUT'
        Word = 0
        Letter = 0
        Blank = 0
*       a word is a sequence of nonblank characters
        do 200 I = 1, 80
            Currt = Line(I)
            if (State.eq.'IN ') then
                if (White(Currt)) then
                    State = 'OUT'
                    Blank = Blank + 1
                else
                    Letter = Letter + 1
                endif
            else
                if (White(Currt)) then
                    Blank = Blank + 1
                else
                    State = 'IN '
                    Word = Word + 1
                    Letter = Letter + 1
                endif
            endif
200     continue
        print *, ' Number of Words = ', Word
        print *, ' Number of Letters = ', Letter
        print *, ' Number of Blanks = ', Blank
        stop
        end
*
*       blanks are not in words
*
        logical function White(C)
        character*1 C
        White = C.eq.' '
        return
        end
```

Now suppose that we are dealing with an 80-character line that contains w words, n letters, and b blanks. The number of gaps between words is $w - 1$ because the first word begins at the left margin and the last word ends at the right margin. Therefore, to create the formatted line, we have to place the w words so that the number of blanks in the gaps between words is enough to make this happen. We have b blanks to be placed in $w - 1$ gaps so that the average number in each gap is $b / (w - 1)$. However, this will not usually be an integer, and we can't place 3.5 blanks in each gap, so clearly we will have to put more blanks in some gaps than in others. The result of the integer division of b by $w - 1$ gives the minimum number of blanks required in each gap (call this m). There are $b - m$ blanks left over. The sensible thing to do with these is to insert one into each gap as long as they last. Thus the inter-word gaps will be slightly larger for gaps toward the beginning of lines. We will discuss a more effective, but more complicated, way of handling this problem in the next section.

Now the program to carry out the formatting makes two passes through each line. On the first pass, the number of words and blanks is calculated as we did in the preceding program.

In the second pass, the words are copied from the input line to an output line, and the correct number of blanks is inserted into each inter-word gap. The normal number of blanks required in each inter-word gap is kept in the variable Regblk. The variable Extblk holds the total number of extra blanks that have to be inserted. This program uses the strategy of putting the extra blanks in the early gaps on each line.

Once again, for simplicity, this program only handles a single line of input. It is easily extended to handle many lines by embedding the body of the program in an iteration. We now show the program.

```
*       Program 11.7
*       Adjust a line of text to get a specified right margin
*
        program Words
        character*1 ILine(80), OLine(80), Currt
        character*3 State
        integer I, J, K, Word, Letter, Blank, Regblk, Extblk
        logical White
        print *, 'Enter line to be formatted'
        read '(80A1)', ILine
        State = 'OUT'
*       a word is a sequence of nonblank characters
        Word = 0
        Letter = 0
        Blank = 0
```

```
*       determine number of words, blanks, and letters
        do 200 I = 1, 80
            Currt = ILine(I)
            if (State.eq.'IN ') then
                if (White(Currt)) then
                    State = 'OUT'
                    Blank = Blank + 1
                else
                    Letter = Letter + 1
                endif
            else
                if (White(Currt)) then
                    Blank = Blank + 1
                else
                    State = 'IN '
                    Word = Word + 1
                    Letter = Letter + 1
                endif
            endif
200     continue
        print *, ' Number of Words = ', Word
        print *, ' Number of Letters = ', Letter
        print *, ' Number of Blanks = ', Blank
        Regblk = Blank / ( Word - 1 )
        Extblk = Blank - Regblk * (Word - 1)
        J = 1
*       distribute the extra blanks in the line
        do 700 I = 1, 80
            Currt = ILine(I)
            if (State.eq.'IN ') then
                if (White(Currt)) then
                    State = 'OUT'
                    Word = Word -1
                    if (Word.ge.1) then
                        do 600 K = 0, Regblk-1
                            OLine(J+K) = ' '
600                     continue
                        J = J + Regblk
                        if (Extblk.ne.0) then
                            OLine(J) = ' '
                            J = J + 1
                            Extblk = Extblk - 1
                        endif
                    endif
```

```
              else
                  OLine(J) = Currt
                  J = J + 1
              endif
          else
              if (White(Currt)) then
              else
                  State = 'IN '
                  OLine(J) = Currt
                  J = J + 1
              endif
          endif
700   continue
      print *, 'Formatted line is'
      print '(80A1)', OLine
      stop
      end

      logical function White(C)
      character*1 C
      White = C.eq.' '
      return
      end
```

The main drawback of this approach to text formatting is that we have treated each line as a separate entity. However, a much better effect can be achieved if the program can move words from one line to another to create a more even distribution of gaps. For example, if one line has only a few words while the next has many, then the formatted version of the first line will have large gaps between words while the second line will look cramped by comparison.

11.6 A Framework for Formatting

You've now seen several basic text-handling techniques, and we can move on to a more interesting approach to text formatting. We'll present a framework that has several useful capabilities and can be extended in obvious ways to do more complicated things.

The major advance over the previous section is decoupling input lines and output lines. This is done by treating the input stream as a sequence of *words*, and using a subprogram to find the next word in the input as required. This subprogram hides all the details of lines that are made up of characters, reading from the input as required and "chunking" the input into words. The remainder of the program can then behave as though the input actually occurs that way; this is the way good abstractions should contribute to the programming process.

The main loop of the program, then, has to process a single word. A word is simply a contiguous sequence of nonblank characters. We allow text words and command words. We assume that punctuation characters will be attached to the word they follow or precede. Let's deal with text words first. A partial output line is kept as part of the program's state. Processing a text word involves seeing if the word will fit on the current partial line; if not, the current partial line is printed. The new word is then added to the partial line that has just been newly emptied. This is all there is to it.

The complexity of the program arises from adjusting lines (making the right margin straight, as in the previous section) and handling commands. Just before a line is printed, we may want to adjust it. The algorithm here does not need to count words and blanks since we keep track of the number of words, the size of the partial line, and the desired size as we add words. Thus we know how many blanks to add to do the adjusting. We use essentially the same idea as in the previous section, but the setting is a bit different. We examine the line moving right to left, and at each inter-word gap we insert additional spaces as required. Some inter-word gaps may be wider than others, and these will always be at the left end of the line using the algorithm given here; a better approach would be to alternate from line to line, putting the extra spaces at the left of one, then at the right of the next, and so on. We have not separately kept track of required blanks and extra blanks. The required blanks are in the line as it is built and only the extra blanks are inserted.

We allow commands in the input stream. These are words that begin with a period. We expect that these will be written on input lines by themselves, but if you examine the details of the program, you'll see that we don't require this. These are the commands we handle.

- *.ad* — Turn adjusting on.
- *.na* — Turn adjusting off.
- *.ll n* — Set line length to *n* characters.
- *.sp n* — Leave *n* blank lines.

The program requires some study, but the remainder of the details should be understandable if you've read the preceding parts of the book.

```
*       Program 11.8
*       Simple Text Formatter
*
        program Pretty
        character*82 InLine
        logical Adjust, Done
        integer WordSz, LineSz, MaxLin, NumWds, InPtr
        character*40 Word
        character*80 Line
```

```
        LineSz = 0
        InLine = '~'
        Line = ' '
        InPtr = 1
        Done = .false.
        call NextWd(InLine, InPtr, WordSz, Word, Done)
        MaxLin = 60
        NumWds = 0
        Adjust = .false.
        while (.not. Done) do
*           Commands begin with a period
            if (Word(1:1) .eq. '.') then
                call Commnd(LineSz,Line,NumWds,Word,Adjust,MaxLin,
     &               InLine,InPtr,WordSz,Done)
            else
                call Text(WordSz,LineSz,MaxLin,Adjust,Line,NumWds,Word)
            endif
            call NextWd(InLine, InPtr, WordSz, Word, Done)
        end while
        call Out(Line, NumWds, LineSz)
        stop
        end
*
*       Turn the input stream into words, next one goes in Word
*
        subroutine NextWd(InLine, InPtr, WordSz, Word, Done)
        character*82 InLine
        integer WordSz, InPtr
        character*40 Word
        logical Done
        integer First
*       a word starts with a nonblank, ends before the next blank
        while (InLine(InPtr:InPtr) .eq. ' ' .or.
     &         InLine(InPtr:InPtr) .eq. '~') do
            if (InLine(InPtr:InPtr) .eq. '~') then
                read (*, '(a80)', end=100) InLine
                InLine (81:82) = ' ~'
                InPtr = 1
            else
                InPtr = InPtr + 1
            endif
        end while
```

```
            First = InPtr
            WordSz = Index(InLine(InPtr + 1:), ' ')
            InPtr = InPtr + WordSz + 1
            Word = InLine(First: InPtr - 2)
            return
    100     continue
            Done = .true.
            return
            end
*
*       Force out the current line and start again
*
            subroutine Out(Line, NumWds, LineSz)
            character*80 Line
            integer NumWds, LineSz
            print *, Line(1:LineSz)
            NumWds = 0
            LineSz = 0
            return
            end
*
*       Insert interword blanks to get a solid right margin
*
            subroutine DoAdj(LineSz, MaxLin, Line, NumWds)
            integer LineSz, MaxLin, NumWds
            character*80 Line
            integer I, J, K
            character*80 Blanks, Other
            Blanks = ' '
            LineSz = LineSz - 1
*           I moves on existing line, J on new line
            I = LineSz
            J = MaxLin
            while (I .lt. J) do
                while(Line(I:I) .ne. ' ') do
                    I = I - 1
                    J = J - 1
                end while
```

```
*         How many spaces are there to insert blanks?
          NumWds = NumWds - 1
*         How many blanks go here?
          K = (J - I) / NumWds
          if (K .ge. 1) then
              Other = Line(1:I) // Blanks(1:K) // Line(I+1:)
              Line = Other
          endif
          J = J - K - 1
          I = I - 1
      end while
      LineSz = MaxLin
      return
      end
*
*     Deal with command words in the input stream
*
      subroutine Commnd(LineSz,Line,NumWds,Word,Adjust,MaxLin,
     &    InLine,InPtr,WordSz,Done)
      integer LineSz,NumWds,MaxLin,InPtr,WordSz
      character*80 Line
      character*40 Word
      character*82 InLine
      logical Adjust, Done
      integer I
      if (LineSz .gt. 0) then
          call Out(Line, NumWds, LineSz)
      endif
      if (Word(2:3) .eq. 'na') then
          Adjust = .false.
      elseif (Word(2:3) .eq. 'ad') then
          Adjust = .true.
      elseif (Word(2:3) .eq. 'll') then
          call NextWd(InLine, InPtr, WordSz, Word, Done)
          MaxLin = IVal(Word, WordSz)
      elseif (Word(2:3) .eq. 'sp') then
          call NextWd(InLine, InPtr, WordSz, Word, Done)
          print '(a)', (' ', I=1, IVal(Word, WordSz))
      endif
      return
      end
```

```
*
*      Add a text word to the output stream
*
       subroutine Text(WordSz,LineSz,MaxLin,Adjust,Line,NumWds,Word)
       integer WordSz, LineSz, MaxLin, NumWds
       logical Adjust
       character*80 Line
       character*40 Word
*      if it won't fit, force out the current line
       if (WordSz + LineSz .gt. MaxLin) then
           if(Adjust) then
               call DoAdj(LineSz, MaxLin, Line, NumWds)
           endif
           call Out(Line, NumWds, LineSz)
       endif
*      now we know there's room, so add it
       NumWds = NumWds + 1
       Line(LineSz+1:LineSz+WordSz+1) = Word // ' '
       LineSz = LineSz + WordSz + 1
       return
       end
*
*      Convert a string to an integer value
*
       integer function IVal(S, L)
       character*40 S
       integer L, I, V
       V = 0
       do 100 I = 1, L
           V = V * 10 + Index('0123456789', S(I:I)) - 1
100    continue
       IVal = V
       return
       end
```

We next show a sample of text interspersed with commands, in a form suitable to be used as input to this program.

.ll 45
Many of the things that we do in the normal course of life could be written down as a specific set of steps to follow. Some examples are:
the routine you follow between awakening and actually getting out of the house each morning;
.ll 50
the instructions for cooking a cheese omelet;
the procedure for doing the laundry;
the directions to follow to get from your house to City Hall;
and so on.
.sp 1
.ad
12345678 91357 924 680 1471025 811 36912 15102611 3790 138657429 17963 110
12 1324 1538 7563 35547 3465265654
14789 0987647 8 8749 74 9382704981 9249 7893 63 2436.
.na
(for example, you're lying in bed),
a list of "material" to work with
(for example, three eggs, 1/4 pound of cheese, etc.),
and a list of things to do, in order.

We now show the output that would be produced from this input.

Many of the things that we do in the normal
course of life could be written down as a
specific set of steps to follow. Some
examples are: the routine you follow between
awakening and actually getting out of the
house each morning;
the instructions for cooking a cheese omelet; the
procedure for doing the laundry; the directions to
follow to get from your house to City Hall; and so
on.

12345678   91357 924 680 1471025 811 36912 15102611
3790   138657429   17963 110 12 1324 1538 7563 35547
3465265654 14789 0987647 8 8749 74 9382704981 9249
7893 63 2436.
(for example, you're lying in bed), a list of
"material" to work with (for example, three
eggs, 1/4 pound of cheese, etc.), and a list of
things to do, in order.

In the part of the input that is to be formatted with adjusting turned on, we have used strings of digits since they all have the same width in the font that is used for examples in this book, and the adjusted text can be clearly seen.

This program could be extended in a number of useful ways, some of which are suggested in the exercises. If there is no other formatter on the computer system you use, you might find it worthwhile to build a system based on these ideas.

Programming Example

Problem Statement

Extend the programming example of Chapter 10 so that it will work for any alphabet.

Inputs

A key and lines of text to be encrypted.

Outputs

Lines of encrypted text.

Discussion

We pointed out in discussing the program at the end of the previous chapter that the program would only work if we assumed that the alphabetic characters are contiguously represented. We want to produce a general version that will work for any alphabet, so we include the alphabet as part of the program.

Program

```
*       Program 11.9
*       Encrypts lines of text using a Caesar cipher. A Key
*       must be supplied. Works with any alphabet.
*
*       Line - line currently being encrypted
*       Letters - letters in proper sequence
*       Key - amount to shift for encryption
*       CInt - integer value of each character
*
```

```
              program Caesar
              character*1 Line(60)
              character*26 Letters
              integer Key, CInt
              Letters = 'abcdefghijklmnopqrstuvwxyz'
              print *,' Enter key'
              read *, Key
              print *, 'Key is', Key
              print *, 'Enter lines of text'
              while (.true.) do
                  read (*, '(60A1)', end=200) Line
                  write (*, '(A, / 60A1)') 'Plain text', Line
                  do 100 I = 1,60
                     if (Index(Letters, Line(I)) .ne. 0 ) then
*
*                      select the Ith character, find its position,
*                      minus 1 to number from zero, add in shift,
*                      wrap around if necessary, plus 1 to number from 1
*
                       CInt = mod(Index (Letters, Line(I))-1+Key, 26) + 1
                       Line(I) = Letters(CInt:CInt)
                     endif
  100             continue
                  write (*, '(A, / 60A1)') 'Encrypted text', Line
              end while
  200         stop
              end
```

Testing

We next show the output from the program on an EBCDIC machine. As you can see, it correctly maps the characters.

```
 Enter key
Key is           7
 Enter lines of text
Plain text
aabbccddeeffgghhiijjkkllmmnnooppqqrrssttuuvvwwxxyyzz
Encrypted text
hhiijjkkllmmnnooppqqrrssttuuvvwwxxyyzzaabbccddeeffgg
```

Design, Testing, and Debugging

- Input and output of strings can be more confusing than when other data types are used. Make sure you understand when an input string has to be written in single quotes and when it need not be.
- Because the ordering of characters is machine-dependent, you usually have to know what kind of machine your program will run on to be sure of how character comparisons will behave. If you want to write a portable program that will run on any machine, then you have to work harder to make sure that it does not make assumptions about character ordering.
- Check your programs to make sure that you don't assign a part of a string to itself. It's not always possible for the compiler to detect this and it can lead to apparently strange behavior from the program. So it's especially important that you check it yourself.

Style and Presentation

- Strings can significantly enhance the appearance of your output. Learn to use them well.
- Some of the examples in this chapter are longer than any you have seen before. Take note of how we upgrade the documentation for these programs. Extra comments are used, but sparingly. Most of the structure of the program is conveyed by indenting and breaking the program up into appropriately sized subprograms.

Fortran Statement Summary

String Declaration Statement

Character declarations require an extra field specifying the size of character string each variable is to contain.

    character*size *variableList*

    character*1 C
    character*60 Line, Table(10)

Catenation Operator

This operator catenates two strings.

$$\text{variable} = \text{variable} \mathbin{/\!/} \text{variable}$$

$$A = B \mathbin{/\!/} \text{'extra stuff'}$$

Chapter Summary

- A string is a sequence of characters that are treated as a unit. Strings can be stored in **character** variables. String constants are formed by enclosing the characters in single quotation marks.
- A substring can be selected by referencing a character variable and providing left and right character positions. Strings can be assigned to character variables. If the string expression is shorter than the character variable, then the string is padded on the right with blanks. If it is longer, then it is truncated on the right.
- The catenation operator pastes two strings together to form a single string.
- The index intrinsic function provides the position of one string within another.
- Character variables and string manipulation operations allow us to write programs to handle interesting applications like text editing and formatting.

Define These Concepts and Terms

Substring
Catenation
Tracing execution

Stream editor
Formatter

Exercises

1. What is printed by the following program?

```
program What
character*1 A
character*3 E
character*4 B, D, F, G
character*7 C
character*10 X, Y, Z, T, U
A = 'Interrupt'
B = 'That'
C = 'example'
D = 'Ju' // C(4:5)
E = 'put'
F = 'ears'
G = 'Cool'
X = A
Y = B(2:4) // C
Z = ' ' // G(1:2) // D(3:4)
T = Z(:4) // E // F
U = T(1:8) // F(3:4)
print *, X(1:2), ' ', Y(1:4), U
stop
end
```

2. Write a program to determine the percentage of letters in some input lines that are equal to a given letter. Use it to find the percentage of "e's" in some input text. What happens if there is no input text? What if there are no "e's"?
3. Write a function to compute the Roman numeral representation of a given value for values from 0 to 999.
4. Write a function to remove duplicate blanks from a string of characters.
5. Trace the execution of NumBas when the input representation is a string of blanks. Trace the execution of NumBas when the input representation is 512, the input base is 10, and the output base is 2. Should the program's behavior be modified in these cases? If so, how?
6. Write a function to compute the reverse of a given string. Use this to write a function that determines if a string is a palindrome (reads the same forward and backward). Extend your function so that capitals, blanks, and punctuation are ignored in checking for the palindrome property. Thus the statement attributed to Napoleon "Able was I, ere I saw Elba," will be a palindrome. Use an additional subprogram that removes all extraneous characters from a given string.
7. Write a function that takes a single parameter that is a string representing a time in the form *hh/mm/ss* and returns a string of the same form representing the next second.
8. Write a function to determine the first position at which a string occurs in another string. Return the value zero if it does not occur. Thus, you are to produce the same results as the intrinsic function Index, but of course without using Index to do so.

9. Write a program that will edit a file in the following way. The given file will consist of lines, each containing an integer and a string. The integers at the beginning of the lines are in ascending sequence through the file. Changes to the file are specified from the standard input. They can be: a deletion, indicated by a "D" followed by the integer labeling the line to be deleted; an insertion, indicated by an "I" followed by the integer and string to be added to the file; a change, indicated by a "C" followed by an integer indicating the line to be changed, and a new string value. The changes are also ordered so that the lines they reference are in ascending order. The output of the program will be an appropriately modified copy of the input file, together with a list of changes on the standard output.

10. Write a simple spelling checker. The program first reads an array of words, as strings, from a file. No word is longer than ten letters. It then reads text from the standard input and checks whether or not each word is present in its word list. A list of incorrect or missing words is listed on the standard output.

11. Extend the previous problem so that the text is also read from a file. If a word is found in the text but not in the word list, the program should ask the user whether it is a misspelling or simply a word omitted from the word list. If it is misspelled, then the user should be asked to enter the correct version and the program should make the correction in the file. If it is correctly spelled, the user should be asked if it should be added to the word list.

12. Write a program that reads a list of words into an array. Words may be repeated in the list. The program prints the words out in decreasing order of frequency of appearance. The most frequent word appears first, then the second most frequent, and so on.

13. The simple text editor presented in this chapter makes many assumptions and is thus not robust. For example, it assumes that line numbers given in the commands to change the current line pointer are valid ones. Identify all such assumptions and add appropriate program statements to protect against errors. Your goal should be that no input can cause the program to abort or get its internal data into an inconsistent state.

14. Extend the text editor presented in this chapter by adding a command that will print all the lines containing a given string, together with their associated line numbers.

15. Extend the text editor presented in this chapter by adding a command to write a specified number of lines (starting with the current line) into a specified file.

16. Modify the text formatter to handle the command "ce", which causes the next input line to be centered on an output line.

17. Modify the text formatter to handle the commands "ti" (followed by a number), which causes an indentation of the indicated number of spaces for one line, and "in" (followed by a number), which causes a permanent indentation of the indicated number of spaces. You will probably want to allow a minus sign before the numbers, so you may want to modify the routine that transforms strings to integers in an appropriate way to handle this.

18. Modify the text formatter to handle the command "ne" with a single integer parameter indicating a number of lines (as in the "sp" command). This indi-

cates that the formatter should ensure that there is room for the indicated number of lines on the current page. A new page will be started if not. You will have to keep track of lines per page to do this.

19. Modify the formatter to allow the user to define named sequences of formatting commands (often called macros). The command "de" followed by a two-character name indicates the start of a definition. The command "ed" indicates the end of a definition. Between these two commands, the user can specify any sequence of commands. Once the definition has been made, the name can be used as a regular command. For example, you might define a macro for the beginning of a paragraph:

   .depp
   .sp1
   .ti5
   .ed

   and later use it by writing ".pp" where a paragraph should begin, to get a blank line and an indent of 5 for one line.

20. Modify the stream editor to allow the target string to include "^" as its first character, thus indicating that the remainder of the target string has to be found at the beginning of the line of the input file.

21. Modify the stream editor to allow the target string to include "$" as its last character, thus indicating that the previous characters in the target string have to be found at the end of the line of the input file.

22. What happens to the stream editor if the replacement string is empty, as in this next example?

   /delete me//

23. Modify the stream editor to allow three strings to be specified. The first is a string that must be found in the input line before the replacement can take place, the second is the string to be replaced, and the third is the replacement string. Thus, the command

   /Street/Brown/Green/

   would cause this input file

   Adele Brown,
   of 221 Brown Street,
   was also present.

   to be transformed to the following output file.

Adele Brown,
of 221 Green Street,
was also present.

24. Modify the program resulting from the previous exercise to allow lines to be deleted. The previous command could be written

    /Street/c/Brown/Green/

    and a deletion command could be written

    /Street/d

25. Modify the result of the previous exercise to allow the program to handle multiple commands during a single execution.
26. Write a program that will read a string and a filename, and then read lines from the indicated file, and print those that contain the indicated string, together with their sequence number within the file.
27. Extend the program of Exercise 26 to work with any number of filenames. When a match is found the program must print the filename, the line number, and the line contents.
28. Write a program that will print a Fortran program. Each subroutine or function should begin on a new page. The comments that might precede a subroutine or function should appear on the page with the subprogram, not at the bottom of the previous page. Put a line at the bottom of each page giving the program name and the page number.
29. Assume you have a text file consisting of lists of things to do for each of several days. The entries for a day begin with a line containing only the date in the form mm/dd/yy, which is followed by any number of lines with arbitrary content. Write a program that will read a date in the specified format from the standard input, and then print the lines for that day from the calendar file. There may be no lines for the given day, of course. The program can ignore the semantics of dates and then read the entire file until it finds a match or the end of file.
30. Modify the program produced for Exercise 29 so that it understands the semantics of dates, and stops as soon as it has found a date later than the target.

Engineering Problem 12

File Compression

The continual decrease in price of computer storage devices is offset by an even greater increase in the legitimate demands for storage. There is often good reason, then, for making careful use of file storage space. If you examine Fortran programs as written in this book, you will see that there is a great deal of blank space in them, and thus a large number of blanks must be stored in the files for such programs. This is certainly true of the blanks used at the left end of lines (because we indent our programs) and may be true at the right end of lines (if the system being used requires all lines in a file to be of the same fixed length and pads them with blanks). Of course, there are other types of files that could be effectively compressed.

Many techniques have been used for encoding files. Some are applicable to files of any type, and others depend on particular features of specific types of files to work well. We will apply a general technique that will work with many files. It is especially useful with program files.

The basic idea is this: a run of k copies of a specific character can be replaced by three characters. The first is some special character, called an *escape character*, that would otherwise not appear in the file. You should think of it as a signal to "escape" out of the normal "alphabet" used in the file. The second is the character that was represented in the original run of characters. The third is a single character encoding the number of copies, k, of characters in the original run. Since characters can be mapped to integers, and vice versa, this encoding is easy to produce.

We can, in fact, go further. The character that occurs most frequently in a run is the blank, so we could encode runs of blanks using only two characters. The first would be another escape character, and the second would be an encoding of the number of blanks in the original run of characters.

Another thing to do to make files more compact (at least on some systems) is to compress lines that are not full (that have blanks on the end of the line) into a smaller number of full lines. To do this it must be possible to identify yet another escape character that can serve to indicate the end of a line in the compressed file.

Consider using all three compression techniques (packing lines, encoding runs of arbitrary characters, and encoding runs of blanks) in a single program. This next program is to be compressed.

    program xxx
    print *,'hi there.'
    stop
    end

If we assume that compressed text will go into lines of length 80, the compressed version would be this next single string.

B6**program** Cx3AB6**print** *,'hi there.'AB6**stop**A**end**A

The original program contains no uppercase characters and no digits. In the compressed version we have used A as the escape character for the end of line indication, B as the escape character for a run of blanks, C as the escape character for a run of some nonblank character, and the single digits as the characters that encode the indicated number of characters. If we were actually encoding Fortran programs, we could not use these escape characters since they appear in legal programs. We do not encode a single blank as a run, just as we do not for any other character; this would actually increase the file size.

Programming Problems

1. Write a subprogram that reads a file and compresses it using the three techniques shown, producing the compressed version in another file. You might find it convenient to put an explicit end-of-file indicator in the last line of the compressed file, rather than having to handle the last line specially. This would require a fourth escape character. You can use the characters with integer equivalents 0 through 3 as your escape characters, since it is (reasonably) safe to assume they will not appear in English text or Fortran programs.
2. Write a subprogram that can read a file produced by the subprogram of the previous part and expand it back to its original form. Combine these two into a single program that will compress a file and then expand it again. You should either get an exact copy of the original back or something with systematic differences (perhaps in the treatment of the blanks at the right of lines) that you can explain and that does not affect the use of the files.
3. Use the program to compress and expand a copy of itself. How big is the compressed file as a percentage of the original and final files? Compile the final file to show that compression and expansion do not lose information.

Numerical Linear Algebra

12.1 2-Dimensional Arrays
12.2 IO with 2-Dimensional Arrays
12.3 Matrix Operations
12.4 Gaussian Elimination
12.5 Inverting a Matrix
12.6 LU Decomposition
12.7 Eigenvalues and Eigenvectors
12.8 Plotting Functions of Two Variables
12.9 Linear Regression

12.1 2-Dimensional Arrays

So far we have looked at arrays in which the array elements were simple data objects such as integers or reals. Array elements can also themselves be structured objects such as arrays. This allows us to represent two kinds of sets of data: lists of lists and tables.

We have already seen how to represent vectors as arrays. Now we are interested in representing matrices as arrays so that we can write programs to carry out transformations.

In Fortran we declare the array type and give the size of the array in each dimension. For example, if we want to declare a 3 by 3 array, then the declaration looks like this:

    real X(3,3)

The type of the simple data objects is to be **real**.
A declaration like this

    real Y(3,5)
    integer N(2,4)

declares Y to be an array with 3 rows, each row consisting of 5 elements. The array N consists of 2 rows, each with 4 elements.

An element of a 2-dimensional array is referred to by giving its position in a row and column. Therefore, to refer to the element in the second row and third column of the array Y, as declared above, we say Y(2,3). Such a reference may appear on the left-hand side of an assignment statement such as

    Y(2,3) = 2 * C

or in an expression such as

    C = 2 * Y(2,3)

It's important to remember which index position refers to which dimension, because Y(2,3) is not the same array element as Y(3,2).

As an example of the usefulness of 2-dimensional arrays, let us consider some applications in elementary linear algebra. We begin by writing a function to multiply a 3 by 3 matrix by a 3 by 1 column vector. The operation of multiplying a matrix by a vector is the same as taking the dot product of the first row of the matrix with the vector to get the first element of the product, then multiplying the second row of the matrix by the vector to get the second element, and so on. As we have already shown how to calculate the dot product, this program is not a very difficult extension.

The inputs to the function are the matrix and the vector. The function returns the 3 by 1 column vector resulting from the multiplication. Because Fortran does not allow functions to return anything but single values, we'll have to actually implement this as a subroutine with two input parameters and one output parameter.

```
*       Multiplies a vector by a matrix
*
        subroutine Mult(A, X, Y)
        real A(3, 3), X(3), Y(3), Sum
        integer I, J
        do 200 I = 1,3
            Sum = 0.0
            do 100 J = 1,3
                Sum = Sum + A(I,J) * X(J)
100         continue
            Y(I) = Sum
200     continue
        return
        end
```

The operations involved in matrix multiplication are not very much more complicated. Multiplying two *n* by *n* matrices is equivalent to finding n^2 dot products. The *ij*th element of the product matrix is the result of computing the dot product of the *i*th row of the first matrix and the *j*th column of the second. Alternatively, the *j*th column of the product matrix is the product of the first matrix with the *j*th column of the second. Thus matrix multiplication is a simple extension of the preceding programs.

```
*       Multiplies two 3 by 3 matrices
*
        subroutine Mult(A, B, C)
        real A(3, 3), B(3,3), C(3,3), Sum
        integer I, J, K
        do 300 I = 1,3
            do 200 J = 1,3
                Sum = 0.0
                do 100 K = 1,3
                    Sum = Sum + A(I,K) * B(K,J)
100             continue
                C(I,J) = Sum
200         continue
300     continue
        return
        end
```

The inner loop in this program is calculating a dot product (compare it with the program earlier). The result of the dot product is then placed into the result array C. The two outer loops choose the row of A and column of B whose dot product is to be calculated. We can also allow the sizes of the arrays to be defined by the parameters, as we saw for 1-dimensional arrays. The next version is a general matrix multiplication that multiplies matrices of any size.

```
*       Multiplies two N by N matrices
*
        subroutine Mult(A, B, C, N)
        integer N
        real A(N,N), B(N,N)), C(N,N), Sum
        integer I, J, K
        do 300 I = 1,N
           do 200 J = 1,N
              Sum = 0.0
              do 100 K = 1,N
                 Sum = Sum + A(I,K) * B(K,J)
100           continue
              C(I,J) = Sum
200        continue
300     continue
        return
        end
```

Just as before, the parameter defining the size of the matrices must be declared before the arrays themselves.

12.2 IO with 2-Dimensional Arrays

When 2-dimensional arrays are printed, it is particularly important to lay them out well because otherwise their structure is hidden. Fortunately, the way that format codes are used was designed to make this relatively easy.

Suppose that we wish to print out the contents of a 4 by 4 array in the natural way with four lines, each of which contains four values. We can do this in the following way:

```
        integer P(4,4), I, J
        ....
        write (6,100) ((P(I,J), J = 1,4), I = 1,4)
100     format(4I5)
```

The output list in the write statement contains a nested implied do loop. This behaves exactly as you might expect from the syntax if you work through it in detail, but it may be a bit confusing at first sight. The implied do loop with the index variable I contains 4 objects to be printed. Each of these objects is itself a list created by the implied do loop with index J. The net result is to print the elements of the array in the order P(1,1), P(1,2), P(1,3), P(1,4), P(2,1), P(2,2), P(2,3), P(2,4), P(3,1), P(3,2), P(3,3), P(3,4), P(4,1), P(4,2), P(4,3), and P(4,4). This corresponds to the way in which we want them to appear with the first row first, then the second row and so on. The nested implied do loop is logically equivalent to

```
              do  300  I = 1,4
                do  200  J = 1,4
                  write  (6,100) P(I,J)
100           format(I5)
200         continue
300       continue
```

except that each **write** statement in the explicit nested **do** loops begins on a new line, whereas the **write** statement with the implied loops places all of the values on a single output line (format permitting). The **format** statement (4I5) causes the values from the array to be placed four per line because after the first four values have been printed, the format codes are reused from the beginning of the format code list and a new line is started.

Fortran has a nasty default behavior in the situation where implied do loops are not used. We have already seen that it is not necessary, when printing a 1-dimensional array, to use an implied do loop. The same is true when printing a 2-dimensional array, but it turns out not to be very useful.

Suppose we have declared an array like this:

   integer Q(2,3)

An output statement like this

   write (6,400) Q

is perfectly legal Fortran but it causes the array elements to be printed in the following order: Q(1,1), Q(2,1), Q(1,2), Q(2,2), Q(1,3), and Q(2,3). The array elements are printed in the order in which they appear in the columns—this is called column major order. Another way to remember this is that the first array index changes faster. Unfortunately, this order makes it impossible to print the array in the shape in which we think about it. Printing arrays of more than one dimension without explicitly stating the order in which you want the elements to appear should be avoided whenever possible.

The same problem occurs with array input. If you write

   read *, Q

then the six values read will be placed in the array in the order described earlier. If this is not what you intended (and it almost certainly isn't), then the array does not look like you think it does and the remainder of the program will probably not work properly. Errors of this kind are very difficult to catch because the array has data in all of its elements—but the data are scrambled so that the answers will be incorrect.

12.3 Matrix Operations

In this section we will examine several common transformations that are applied to matrices. These transformations are used in the algorithms that we will discuss in the remainder of this chapter.

Suppose that we multiply any matrix A by a matrix of the form

$$\begin{bmatrix} 0 & 1 & 0 & 0 & \ldots \\ 1 & 0 & 0 & 0 & \ldots \\ 0 & 0 & 1 & 0 & \ldots \\ \ldots & & & & \end{bmatrix}$$

that is, by the identity matrix of the appropriate size with the first two rows switched with each other. It is not hard to see that the result of multiplying A by this matrix will be a matrix exactly like A except that the first two rows of A will have been switched.

Many common matrix operations can be expressed as a multiplication by a matrix of a particular form. For example, if we multiply any matrix A by the matrix

$$\begin{bmatrix} 1 & 1 & 0 & 0 & \ldots \\ 0 & 1 & 0 & 0 & \ldots \\ 0 & 0 & 1 & 0 & \ldots \\ \ldots & & & & \end{bmatrix}$$

then the resulting matrix is identical to A except that the second row of the result is the sum of the first and second rows of A. In general, by premultiplying by an appropriate matrix we can produce a row in the new matrix that is a linear combination of the rows of the original matrix.

Now that we are aware that transformations of this kind really represent matrix multiplications, we can discuss algorithms that have the same effect without actually carrying out the matrix multiplication. For example, since we know that two rows of a matrix can be permuted by multiplying it by another matrix, we could implement a row-permuting algorithm by constructing the appropriate special matrix and using the matrix multiplication algorithm we have already seen. However, it is much more efficient to simply permute the rows of the matrix directly. The following program does this.

```
*       Program 12.1
*       Permutes the rows of a matrix
*
        program Perms
        real A(10, 10)
        integer I, J, Row1, Row2
        print *, 'Enter 10 by 10 matrix'
        read *, ((A(I,J), J = 1,10), I = 1,10)
        print *, 'Enter rows to be permuted'
        read *, Row1, Row2
        call Prows(Row1, Row2, A, 10)
        print *, 'New matrix is'
        print '(10(F5.1, 2X))', ((A(I,J), J=1,10), I=1,10)
        stop
        end
*
*       This subroutine permutes RowA and RowB of the matrix it
*       is passed
*
        subroutine Prows(RowA, RowB, B, N)
        integer N
        real B(N, N), Temp
        integer RowA, RowB, J
        if (RowA.le.N.and.RowB.le.N) then
            do 100 J = 1,N
                Temp = B(RowA, J)
                B(RowA, J) = B(RowB, J)
                B(RowB, J) = Temp
100         continue
        else
            print *, ' Prows: Row out of range '
        endif
        return
        end
```

This program generalizes the technique for swapping using a temporary value that we used in the last value. Here we swap the pair of values in the same column and repeat this for all the columns of the matrix.

Notice that the error message not only prints out the actual reason for the error but also the name of the subprogram where it is printed. When programs are very large it can be very tedious to discover where an error message was produced, particularly if the message is apparently produced in error. It is therefore a good idea to label errors in this way to make their meanings clearer.

Another matrix operation we could consider is swapping two columns of a matrix. This corresponds to postmultiplying the original matrix by an identity matrix in which the first two columns have been permuted. The algorithm to do this operation differs from row permuting only because the reversal uses the second coordinate of the arrays rather than the first. Here, for example, is the program to do column permutation.

```
*      Program 12.2
*      Permutes the columns of a matrix
*
       Program Perms
       real A(10, 10)
       integer I, J, Col1, Col2
       print *, 'Enter matrix'
       read *, ((A(I,J), J = 1,10), I = 1,10)
       print *, 'Enter columns to be permuted'
       read *, Col1, Col2
       call Pcols(Col1, Col2, A, 10)
       print *, 'New matrix'
       print '(10(F5.1,2X))', ((A(I,J), J=1,10), I=1,10)
       stop
       end

       subroutine Pcols(ColA, ColB, B, N)
       integer N
       real B(N, N), Temp
       integer J, ColA, ColB
       if (ColA.le.N.and.ColB.le.N) then
           do 100 J = 1,N
               Temp = B(J, ColA)
               B(J, ColA) = B(J, ColB)
               B(J, ColB) = Temp
100        continue
       else
           print *, ' Pcols: Col out of range '
       endif
       return
       end
```

Operations that manipulate rows are called elementary row operations (EROs), whereas those that manipulate columns are called elementary column operations (ECOs). Elementary row operations correspond to premultiplications by suitable matrices, while column operations correspond to postmultiplications.

Several important transformations can be expressed as compositions of elementary row and column operations. The two most important are upper triangularization and lower triangularization. A matrix is said to be upper triangular if it contains only zeros below the main diagonal and lower triangular if it contains only zeros above the main diagonal. A matrix can be transformed into an upper or lower triangular matrix by applying suitable EROs.

Let us consider the algorithm to transform a matrix into upper triangular form. We begin with the element in the upper left-hand corner of the matrix. Suppose that this element is a and that the element immediately below it is b. Then subtracting b/a times the first row from the second row leaves the second row with zero in place of b. All of the other elements of the second row have new values about which we can't say anything in general. Now, if the element below a in the third row has the value c, then subtracting c/a times the first row from the third row creates a new third row beginning with zero. Continuing in this fashion, we can reduce the first column to zeros except for the element a in the first row.

Now we move down the diagonal to the element in the second column and second row. We can create zeros in place of the elements in the second column and third and subsequent rows by subtracting multiples of the second row from the other rows. Continuing down the diagonal, we can reduce all of the elements below the diagonal to zero in all columns. Thus the matrix is reduced to upper triangular form.

The algorithm can be described more formally like this:

```
for i from 1 to n-1 select diagonal element (i,i)
    zero all values (j,i) for j > i
```

where "zero all values (j,i) for j > i" is:

```
for element (j,i) with j from i+1 to n
    select appropriate multiple of row i to subtract
    for all remaining elements in row j subtract
        corresponding multiple of the element in row i
```

Here is a program that executes the algorithm. You can see that the program contains three nested loops: the outermost selects the diagonal element, the next the row being worked on, and the innermost the subtraction in each column.

```
*          Program 12.3
*          Changes a matrix to upper triangular form
*
           program UpperT
           real A(10,10), Factor
           integer I, J, K
           print *, 'Enter size and matrix (max 10 by 10)'
           read *, N, ((A(I,J), J = 1,N), I = 1,N)
           do 500 I = 1, N - 1
               do 400 J = I + 1, N
                   Factor = A(J,I) / A(I,I)
                   do 300 K = I, N
                       A(J,K) = A(J,K) - Factor * A(I,K)
300            continue
400        continue
500 continue
```

```
          print *, 'Upper triangular matrix is'
          print '(10(F5.2,2X))', ((A(I,J), J=1,N), I=1,N)
          stop
          end
```

There is one obvious problem with this program. If a value on the diagonal is already zero, then the attempt to calculate Factor will cause an execution time error. We will deal with the solution to this problem in the next section.

To transform a matrix into lower triangular form, the same basic algorithm can be applied with a minor modification. The diagonal element must be subtracted from elements in the rows above it instead of those below it.

12.4 Gaussian Elimination

We turn in this section to one of the most important algorithms involving matrices—**Gaussian elimination**. Gaussian elimination is an algorithm used to solve sets of simultaneous **linear equations**. Simultaneous equations are very common in engineering problems. Usually each equation represents a constraint on a system, and the simultaneous solution to all of the constraints determines a wanted state of the system.

We can state the general problem of solving simultaneous equations as follows: we are given a set of n equations

$$a_{11}x_1 + a_{12}x_2 + \cdots + a_{1n}x_n = b_1$$
$$a_{21}x_1 + a_{22}x_2 + \cdots + a_{2n}x_n = b_2$$
$$\cdots$$
$$a_{n1}x_1 + a_{n2}x_2 + \cdots + a_{nn}x_n = b_n$$

in which the a_{ij}'s and the b_i's are given and we wish to find the x_i's. This problem can be formulated in matrix terms as the vector solution X to the matrix equation

$$AX = B$$

where A is the n by n matrix of the a_{ij}'s and B is the n by 1 column vector of the b_i's. A unique solution exists if and only if the matrix A is invertible (that is, has an inverse). If this is the case, we say that A is non-singular. If A does not have an inverse, then it is said to be singular. Clearly, if we know A^{-1}, then we can premultiply both sides of the equation by it and deduce that

$$X = A^{-1}B$$

We will postpone discussion of how to find the inverse of a matrix to the next section and consider how to solve the equations directly. If we could arrange for the matrix A to be in upper triangular form, then the solution is straightforward. The bottom row of the triangularized A has only a single nonzero entry and thus corresponds to an equation involving only x_n. Thus the last equation can be used to find the value of x_n. The second last equation consists only of terms involving x_n and x_{n-1} so that, once we know x_n, we can solve for x_{n-1}. Continuing in this way, we can solve for each of the x_i's in turn. This procedure is called **back substitution**.

We saw in the previous section how to transform a matrix into upper triangular form. However, the matrix A is part of an equation, and so any transformation applied to A must also be applied to the right-hand side of the equation. We have already seen that the transformations used to change a matrix into upper triangular form are really matrix multiplications, so we can apply them to the right-hand side of the equation as matrix multiplications applied to the vector B.

If we do this, we see that the effect on B is exactly the same as the effect on A — if we multiply by a matrix that interchanges two rows of A, then multiplying B by the same matrix interchanges the two corresponding entries of B. Gaussian elimination consists of two steps: transforming A into upper triangular form (performing the same transformations on B) and using back substitution to solve the resulting matrix equation.

We need to modify the program we have written to triangulate the matrix slightly to take into account the need to perform the same operation on the vector B. The next program shows the new version:

```
*       Does triangulation on its inputs
*
        subroutine Triang(A,B,N)
        integer I, J, K, N
        real A(N,N), B(N), Factor
        do 300 I = 1, N - 1
           do 200 J = I + 1, N
              Factor = A(J,I) / A(I,I)
              do 100 K = I, N
                 A(J,K) = A(J,K) - Factor * A(I,K)
100           continue
              B(J) = B(J) - Factor * B(I)
200        continue
300     continue
        return
        end
```

The second part involves taking an upper triangular matrix and a vector and carrying out the back substitution. If we have a set of equations corresponding to an upper triangular matrix, then they look like this:

$$a_{11}x_1 + a_{12}x_2 + \cdots + a_{1n}x_n = b_1$$
$$\cdots$$
$$a_{nn}x_n = b_n$$

Rearranging, we get that

$$x_n = \frac{b_n}{a_{nn}}$$

Numerical Linear Algebra

$$x_{n-1} = \frac{b_{n-1} - a_{n-1,n}x_n}{a_{n-1,n-1}}$$

and in general

$$x_i = \frac{b_i - \sum_{i+1}^{n} a_{ij} x_j}{a_{ii}}$$

Therefore, the back-substitution step consists of calculating these formulas using the known values of A and B. The next subroutine does this:

```
*       This subroutine carries out back-substitution
*
        subroutine BackSub(A, B, X, N)
        integer I, J, N
        real A(N, N), B(N), X(N), Sum
        do 200 I = N, 1, -1
          Sum = 0.0
          do 100 J = I+1, N
            Sum = Sum + A(I,J) * X(J)
100       continue
          X(I) = ( B(I) - Sum ) / A(I,I)
200     continue
        return
        end
```

The next several programs that we develop work on matrices or systems of equations. To demonstrate the algorithms, we will work with this system of equations as an example and use the matrix of coefficients on its left-hand side as required.

$$\begin{aligned}3x_1 + 2x_2 + 1x_3 &= 1 \\ -1x_1 + 5x_2 + 2x_3 &= 10 \\ 6x_1 + 1x_2 + 2x_3 &= -1\end{aligned}$$

We can put the pieces of program just developed together to build a complete program to solve sets of simultaneous equations. The main program needs only to read the matrices A and B and print the vector X.

```
*       Program 12.4
*       Carries out Gaussian elimination
*
        program Gauss
        integer Size
        parameter (Size = 10)
        real A(Size, Size), B(Size), X(Size)
        integer N, I, J
        print *, 'Enter size and coefficient matrix'
        read *, N, (( A(I,J), J = 1,N), I = 1,N)
        print *, ' Enter constant vector'
        read *, (B(I), I=1,N)
        call Triang(A, B, Size, N)
        call Backsb(A, B, X, Size, N)
        print '(5F12.4)', (X(I), I=1, N)
        stop
        end
*
*       Triangulates the matrix A and the vector B
*
        subroutine Triang(A, B, N, M)
        integer I, J, K, N, M
        real A(N,N), B(N), Factor
        do 300 I = 1, M - 1
           do 200 J = I + 1, M
              Factor = A(J,I) / A(I,I)
              do 100 K = I, M
                 A(J,K) = A(J,K) - Factor * A(I,K)
100           continue
              B(J) = B(J) - Factor * B(I)
200        continue
300     continue
        return
        end
*
*       Carries out back substitution to get X
*
        subroutine BackSb(A, B, X, N, M)
        integer I, J, N, M
        real A(N, N), B(N), X(N), Sum
```

```
      do 200 I = M, 1, -1
          Sum = 0.0
          do 100 J = I+1, M
              Sum = Sum + A(I,J) * X(J)
100       continue
          X(I) = ( B(I) - Sum ) / A(I,I)
200   continue
      return
      end
```

Using the preceding set of equations, the output from this program will be

```
Enter size and coefficient matrix
Enter constant vector
Solution is
    -1.0000      1.0000      2.0000
```

and you can easily verify that this is a solution. Notice the way in which the sizes of the matrices were declared in the main program. Using a parameter value to define the sizes of a set of matrices with the same dimensions simplifies the task of changing their sizes during maintenance. If the program is written in this way, the sizes of all of the matrices can be changed by changing the **parameter** statement. This avoids errors due to changing sizes in some places and not others.

Changing a matrix into triangular form can also be used as part of another useful matrix operation: evaluating the determinant of a matrix. Calculating the determinant of a matrix is an expensive operation—by the usual method it takes about $n!$ operations. However, the elementary row operations that we used to transform a matrix into upper triangular form do not change the determinant of the matrix being transformed. The determinant of any matrix corresponding to an elementary row operation is 1. Therefore, the determinant of the upper triangular matrix is the same as the determinant of the original matrix. Evaluating the determinant of a triangular matrix is easy—it is the product of the elements on the main diagonal. (To see why, evaluate the determinant using elements in the first column and their corresponding minors.) The complexity of transforming the matrix into triangular form is $O(n^3)$, so this is clearly a much better way of evaluating determinants.

Now we know that there are sets of equations for which there is not a unique solution and yet our program does not even consider this possibility. What could go wrong with the preceding calculations if the set of equations had no solution? It turns out that this problem shows itself when there is a zero value on the diagonal of the matrix. If this happens, then the algorithm will not work because we divide by this value. Now a zero on the diagonal does not, by itself, mean that the set of equations has no solution. However, we must clearly take this possibility into account in writing the program. The order in which simultaneous equations are written clearly does not affect their solution. Therefore, we are always free to change the order of the rows of the matrix A, provided that we change the order of the elements of B in an identical way. Thus, if we encounter a zero on the diagonal, we can, if we wish, swap the row that has the zero on the diagonal with another row further down the matrix that does not have a zero in that column.

Now if there is no row without a zero, then the system of equations does not have a unique solution and we cannot continue with the triangularization. It is easy to see why this must be—if a zero appears on the diagonal and nothing can be done to remove it, then the determinant of the matrix A must be zero. Hence A is singular and no solution exists.

Although we are free to permute rows of the matrix within a matrix equation, it is important to remember that if we rearrange rows, the sign of the determinant changes. Therefore we must keep a count of the number of row swaps done if we want to use this technique to calculate determinants.

It also improves the algorithm to select not only a nonzero element to lie on the diagonal, but also the element with the largest absolute value. The addition of this step to the algorithm is called **pivoting**. Strictly speaking, the best thing to do is to select the largest element in the submatrix for which the present diagonal element forms the top left-hand corner, and rearrange rows and columns so that the largest element is moved to the diagonal. We will only consider the column below the diagonal element in selecting the largest element, and therefore will only rearrange rows. This is called **partial pivoting**.

Pivoting requires adding a new step to the algorithm. As we move to each new diagonal position, we must search down the column for the element with the largest absolute value. This element and its row are then swapped with the row presently on the diagonal, and the corresponding elements of B are also swapped. This requires an additional loop for the search and one for the swapping.

The next program illustrates the addition of the partial pivoting operation to the triangularization.

```
*       Program 12.5
*       Triangulates using partial pivoting
*
        program Pivot
        real Coeffs(3,3), Vec(3)
        integer I, J
        logical Triang
        print *, 'Enter equations coefficients in order'
        read *, ((Coeffs(I,J), J=1,3),Vec(I),I=1,3)
        print '(4F10.4)',((Coeffs(I,J), J=1,3),Vec(I),I=1,3)
        print *
        if (Triang(Coeffs, Vec, 3, 3)) then
            print '(4F10.4)',((Coeffs(I,J), J=1,3),Vec(I),I=1,3)
        else
            print *, ' Matrix is singular'
        endif
        stop
        end
```

```
*
*         Triangulates if possible. It returns
*         false if the matrix is singular.
*
          logical function Triang(A, B, N, M)
          integer I, J, K, N, M, Row
          real A(N,N), B(N), Factor, Max, Temp
          Triang = .true.
          do 500 I = 1, M - 1
             Max = abs(A(I,I))
             Row = I
             do 100 J = I + 1, M
                if (abs(A(J,I)) .gt. Max) then
                   Max = abs(A(J,I))
                   Row = J
                endif
100          continue
             if (Max .lt. 10E-6) then
                Triang = .false.
             else
                do 200 K = I,M
                   Temp = A(I,K)
                   A(I,K) = A(Row,K)
                   A(Row,K) = Temp
200             continue
                Temp = B(I)
                B(I) = B(Row)
                B(Row) = Temp
                do 400 J = I + 1, M
                   Factor = A(J,I) / A(I,I)
                   do 300 K = I, M
                      A(J,K) = A(J,K) - Factor * A(I,K)
300                continue
                   B(J) = B(J) - Factor * B(I)
400             continue
             endif
500       continue
          return
          end
```

(Notice that the input is entered in a different way from the previous program — see which you prefer.) The triangularization computation has been written as a logical function because the possibility now exists for the triangularization to fail. If it does fail, there is no point in trying to back substitute, so the main program must be informed if the matrix is singular. We next show the output from the program.

```
       3.0000          2.0000          1.0000           1.0000
      -1.0000          5.0000          2.0000          10.0000
       6.0000          1.0000          2.0000          -1.0000


       6.0000          1.0000          2.0000          -1.0000
       0.0000          5.1667          2.3333           9.8333
       0.0000          0.0000         -0.6774          -1.3548
```

Another problem that can arise with sets of simultaneous equations is called **ill-conditioning**. Equations are ill-conditioned when very small changes in the coefficients cause very large changes in the solutions. For example, the equations

$$x_1 + 5x_2 = 17$$
$$1.5x_1 + 7.501x_2 = 25.503$$

have solutions $x_1 = 2$, $x_2 = 3$. However, the equations

$$x_1 + 5x_2 = 17$$
$$1.5x_1 + 7.501x_2 = 25.500$$

have solutions $x_1 = 17$, $x_2 = 0$. If the coefficients used in the simultaneous equations come from experimental measurements, then this behavior is a serious problem. Ill-conditioned equations are not susceptible to solution by simple numerical techniques.

In general, detecting when equations are ill-conditioned is not straightforward. We can consider the expressions

$$\cos \theta_{ij} = \frac{\sum_{k=1}^{n} a_{ik} a_{jk}}{\sqrt{\sum_{k=1}^{n} a_{ik}^2} \sqrt{\sum_{k=1}^{n} a_{jk}^2}}$$

If any of these is near 1, then the set of equations is probably ill-conditioned.

Ill-conditioning generally arises because the lines represented by the equations are very nearly parallel in some appropriately dimensioned space. Thus, very small changes in coefficients cause small changes in the slope of lines, but large changes in their intersection points. The preceding formula measures the cosines of the angles that the lines make with each other.

12.5 Inverting a Matrix

Finding the inverse of a matrix is another important problem in linear algebra. In fact, we saw in the previous section that finding the inverse of a matrix was enough to solve the corresponding set of simultaneous equations. Gaussian elimination can be used to find inverses of matrices.

There are other methods of finding inverses, such as **Cramer's rule**, which are not computationally efficient. Calculating the determinant of an n by n matrix takes $n!$ operations. Cramer's rule requires calculating $n + 1$ determinants and therefore is computationally ridiculous.

Gaussian elimination can be used to find the inverse of a matrix in the following way: take the matrix A and write it down beside an n by n identity matrix, giving an n by $2n$ matrix. Now transform the left-hand half, the original matrix A, into upper triangular form, and repeat the same operations on the rows of the right-hand half. When the left-hand half has been reduced to upper triangular form, then the right-hand half will, of course, no longer be the identity matrix.

The second phase involves removing all of the values above the diagonal in the matrix A by subtracting multiples of appropriate rows. For example, all elements in the last column can be reduced to zero by subtracting multiples of the last row. As before, the same operations are repeated on the right-hand side of the matrix.

Now the left-hand half consists only of values on the diagonal. Each row of the matrix is divided by the appropriate value to reduce the diagonal element to 1 and the right-hand half of the row is divided by the same amount. When the left-hand half of the matrix (which was A) has been reduced to an identity matrix, then the right-hand half is the inverse of A. It is easy to see why this is so. The composition of the operations that were done to the left-hand half can be expressed as a matrix R. R is a product of elementary row operations. Now we know that

$$RA = I$$

so R is an inverse of A. But the right-hand side is

$$RI = R$$

so that the right-hand side is actually the matrix R.

```
*       Program 12.6
*       Finds the inverse of a matrix using a
*       variation of Gaussian elimination
*
        program InvTri
        real A(10,20), B(10,10), C(10,10)
        integer I, J, N
        print *, 'Enter matrix size'
        read *, N
        print *, 'Enter the matrix'
```

```
           do 100 I = 1, N
              read *, (A(I,J), J = 1,N)
*             save the input matrix for later multiplying
              do 200 J = 1, N
                 B(I,J) = A(I,J)
  200         continue
              print '(10F6.2)', (A(I, J), J=1, N)
  100      continue
           print *
*          destroy left-hand part to put inverse in right-hand part
           call Invert(A, N)
           print *, 'Inverse'
           do 600 I = 1, N
              print '(10F6.2)', (A(I,J), J = N+1, 2*N)
  600      continue
           print *
*          parameters: original, inverse (part of A), size, result
           call TriMul(B, A, N, C)
           print *, 'Product of inverse and original'
           do 500 I = 1, N
              print '(10F6.2)', (C(I,J), J=1, N)
  500      continue
           stop
           end
*
*          Checks inverse by multiplying
*
           subroutine TriMul(B, A, N, C)
           real B(10, 10)
           real A(10, 20)
           integer N
           real C(10, 10)
           integer I, J, K
           do 400 I = 1,N
              do 300 J = 1,N
                 Sum = 0.0
                 do 200 K = 1,N
                    Sum = Sum + B(I,K) * A(K, N + J)
  200            continue
                 C(I,J) = Sum
  300         continue
  400      continue
           return
           end
```

```
*
*         Transforms left half of A to identity
*
          subroutine Invert(A, N)
          real A(10, 20)
          integer I, J
          do 300 I = 1,N
             do 200 J = 1,N
                if (I.eq.J) then
                   A(I, N + J) = 1.0
                else
                   A(I, N + J) = 0.0
                endif
    200      continue
    300   continue
          call UpperT(A, N, 2*N)
          call LowerT(A, N, 2*N)
          do 500 I = 1,N
             do 400 J = N + 1, 2*N
                A(I,J) = A(I,J) / A(I,I)
    400      continue
    500   continue
          return
          end
*
*         Transforms left half to upper triangle
*
          subroutine UpperT(A, N, M)
          integer I, J, K, N, M
          real A(10,20), Factor
          do 300 I = 1, N
             do 200 J = I + 1, N
                Factor = A(J,I) / A(I,I)
                do 100 K = I, M
                   A(J,K) = A(J,K) - Factor * A(I,K)
    100         continue
    200      continue
    300   continue
          return
          end
*
*         Transforms left half to lower triangle
*
          subroutine LowerT(A, N, M)
          integer I, J, K, N, M
          real A(10,20), Factor
```

```
        do 300 I = N, 1, -1
          do 200 J = I - 1, 1, -1
            Factor = A(J,I) / A(I,I)
            do 100 K = I, M
              A(J,K) = A(J,K) - Factor * A(I,K)
100       continue
200     continue
300 continue
    return
    end
```

The output of this program shows the input matrix (the same one used in the previous example), the inverse, and the product of the input and the computed inverse.

```
Enter matrix size
Enter the matrix
    3.00            2.00            1.00
   -1.00            5.00            2.00
    6.00            1.00            2.00

Inverse
    0.38           -0.14           -0.05
    0.67            0.00           -0.33
   -1.48            0.43            0.81

Product of inverse and original
    1.00            0.00            0.00
    0.00            1.00            0.00
    0.00            0.00            1.00
```

12.6   LU Decomposition

Any matrix A can be expressed as the product of two matrices, L, a lower triangular matrix, and U, an upper triangular matrix. The matrix U has 1s on its diagonal. If we write these matrices (using a 3 by 3 example) they look like this:

$$\begin{bmatrix} l_{11} & 0 & 0 \\ l_{21} & l_{22} & 0 \\ l_{31} & l_{32} & l_{33} \end{bmatrix} \begin{bmatrix} 1 & u_{12} & u_{13} \\ 0 & 1 & u_{23} \\ 0 & 0 & 1 \end{bmatrix} = \begin{bmatrix} a_{11} & a_{12} & a_{13} \\ a_{21} & a_{22} & a_{23} \\ a_{31} & a_{32} & a_{33} \end{bmatrix}$$

You can see that there are some equations that the value l_{ij} and u_{ij} must satisfy. For example, we know that $l_{11} = a_{11}$, that $l_{11} \times u_{12} = a_{12}$ and so on. In general, the values l_{ij} and u_{ij} are given by the recurrences

Numerical Linear Algebra

$$l_{ij} = a_{ij} - \sum_{k=1}^{j-1} l_{ik} u_{kj}$$

$$u_{ij} = \frac{a_{ij} - \sum_{k=1}^{i-1} l_{ik} u_{kj}}{l_{ii}}$$

where the various terms can only be calculated in a particular order. However, in general, we can calculate the coefficients of the L and U matrices, given the matrix A. The matrices L and U are useful in calculations. For example, if we want to solve the equation

$$A X = B$$

then it can be rewritten as

$$L U X = B$$

The product UX is a vector, say Y, so that we now have two equations

$$L Y = B$$
$$U X = Y$$

Since both of the matrices L and U are triangular, these equations can both be solved by substitution only. We solve the first equation for Y and then use Y to solve the second equation for X.

To write an equation solver using this method, we need three pieces: one to construct the matrices L and U, one to solve the first equation for Y using forward substitution (because L is lower triangular), and one to solve for X using the back substitution we have seen. This next subroutine creates L and U, given A.

```
*       Program 12.7
*       Decomposes a matrix into an upper triangular
*       and a lower triangular matrix.
*
        program Manip
        real A(10,20), B(10,10), C(10,10)
        integer I, J, N
        print *, 'Enter the size and matrix '
        read *, N, ((A(I,J), J=1, N), I=1, N)
        call LUDec(N, A, B, C)
        print '(A, /)', 'Original matrix'
        print '(3G12.4)', ((A(I,J), J=1,N), I=1,N)
        print '(A, /)', 'Lower triangular matrix'
        print '(3G12.4)', ((B(I,J), J=1,N), I=1,N)
```

```fortran
      print '(A, /)', 'Upper triangular matrix'
      print '(3G12.4)', ((C(I,J), J=1,N), I=1,N)
      stop
      end
*
*     Creates L and U from A
*
      subroutine LUDec(N, A, L, U)
      real A(10,10), L(10,10), U(10,10)
      integer I, J, K, N
      real SumL, SumU
      do 400 I = 1,N
        do 300 J = 1,N
          if (I.lt.J) then
            L(I,J) = 0.0
          elseif (I.gt.J) then
            U(I,J) = 0.0
          else
            U(I,J) = 1.0
          endif
300     continue
400   continue
      do 900 I = 1,N
        do 800 J = I,N
          SumL = 0.0
          do 600 K = 1,I-1
            SumL = SumL + L(J,K) * U(K,I)
600       continue
          L(J,I) = A(J,I) - SumL
          SumU = 0.0
          do 700 K = 1,I-1
            SumU = SumU + L(I,K) * U(K,J)
700       continue
          U(I,J) = (A(I,J) - SumU) / L(I,I)
800     continue
900   continue
      return
      end
```

We next show the output from a sample run of the program.

Enter the size and matrix
Original matrix

3.000	2.000	1.000
-1.000	5.000	2.000
6.000	1.000	2.000

Lower triangular matrix

3.000	0.0000E+00	0.0000E+00
-1.000	5.667	0.0000E+00
6.000	-3.000	1.235

Upper triangular matrix

1.000	0.6667	0.3333
0.0000E+00	1.000	0.4118
0.0000E+00	0.0000E+00	1.000

Once we have L and U, we can go ahead and solve the two equations. To find a solution for

$$L Y = B$$

we need a subprogram that will solve a set of equations using forward substitution. We solve first for y_1, then for y_2, and so on.

```
*       Program 12.8
*       Carries out forward substitution
*
        program Manip
        real P(10,10), Q(10), R(10)
        integer I, J
        print *, 'Enter matrix and vector'
        read *, ((P(I, J), J=1, 3), I=1, 3), (Q(I), I=1, 3)
        call ForSb(3, P, Q, R)
        print *, 'Matrix'
        print '(3G12.4)', ((P(I, J), J=1, 3), I=1, 3)
        print *, 'Constant vector'
        print '(/, 3G12.4)', (Q(I), I=1, 3)
        print *, 'Solution'
        print '(/, 3G12.4)', (R(I), I=1, 3)
        stop
        end
```

```
*
*       Does the forward substitution
*
        subroutine ForSb(N, A, B, X)
        integer I, J, N
        real A(10, 10), B(10), X(10), Sum
        do 200 I = 1,N
           Sum = 0.0
           do 100 J = 1,I-1
              Sum = Sum + A(I,J) * X(J)
100        continue
           X(I) = (B(I) - Sum) / A(I,I)
200     continue
        return
        end
```

In this sample run we solve a simple system.

```
Enter matrix and vector
Matrix
        1.000           0.0000E+00      0.0000E+00
        1.000           1.000           0.0000E+00
        1.000           1.000           1.000
Constant vector

        1.000           3.000           6.000
Solution

        1.000           2.000           3.000
```

We now know the value of the vector Y, so we can solve the second equation

$$UX = Y$$

Because U is in upper triangular form, we can use the code we have already written for the Gaussian elimination program to solve it. Finally, we put the whole thing together to produce a new simultaneous equation solver.

```
*       Program 12.9
*       Uses LU decomposition, backward and forward
*       substitution to solve linear equations
*
        program Solve
        real A(10,10), L(10,10), U(10,10), X(10), Y(10), B(10)
        integer I, J, N
```

```
      print *, 'Enter size and coefficient matrix'
      read *, N, ((A(I,J), J = 1, N), I=1, N)
      print *, 'Enter constant vector'
      read *, (B(I), I = 1,N)
      call LUDec(N, A, L, U)
      call ForSb(N, L, B, Y)
      call BackSb(N, U, Y, X)
      print *, 'Solution is... '
      print '(10F7.4)', (X(I), I = 1,N)
      stop
      end
*
*     Carries out LU decomposition
*
      subroutine LUDec(N, A, L, U)
      real A(10,10), L(10,10), U(10,10)
      integer I, J, K, N
      real SumL, SumU
      do 400 I = 1,N
         do 300 J = 1,N
            if (I.lt.J) then
               L(I,J) = 0.0
            elseif (I.gt.J) then
               U(I,J) = 0.0
            else
               U(I,J) = 1.0
            endif
300      continue
400   continue
      do 900 I = 1,N
         do 800 J = I,N
            SumL = 0.0
            do 600 K = 1,I-1
               SumL = SumL + L(J,K) * U(K,I)
600         continue
            L(J,I) = A(J,I) - SumL
            SumU = 0.0
            do 700 K = 1,I-1
               SumU = SumU + L(I,K) * U(K,J)
700         continue
            U(I,J) = (A(I,J) - SumU) / L(I,I)
800      continue
900   continue
      return
      end
```

```
*
*       Carries out backsubstitution
*
        subroutine BackSb(N, A, B, X)
        integer I, J, N
        real A(10, 10), B(10), X(10), Sum
        do 200 I = N, 1, -1
           Sum = 0.0
           do 100 J = I+1,N
              Sum = Sum + A(I,J) * X(J)
100        continue
           X(I) = (B(I) - Sum) / A(I,I)
200     continue
        return
        end
*
*       Carries out forward substitution
*
        subroutine ForSb(N, A, B, X)
        integer I, J, N
        real A(10, 10), B(10), X(10), Sum
        do 200 I = 1,N
           Sum = 0.0
           do 100 J = 1,I-1
              Sum = Sum + A(I,J) * X(J)
100        continue
           X(I) = (B(I) - Sum) / A(I,I)
200     continue
        return
        end
```

When this program is run with the example we've been using for the last few programs, the following output is produced.

```
Enter size and coefficient matrix
Enter constant vectors
Solution is...
-1.0000 1.0000 2.0000
```

12.7 Eigenvalues and Eigenvectors

We have already seen that multiplying a vector by a matrix produces another vector. If we consider all possible vectors in a particular vector space, then a matrix

A can be regarded as mapping each vector into a new vector and thus mapping the space into itself.

It is interesting (and, it turns out, practical) to ask whether there are any vectors in the vector space that are unaffected by the multiplication by A. These vectors are, in some sense, "stable" under the action of the matrix. In many physical situations, vectors with this behavior represent stable or equilibrium states of some system.

In general, we are interested in vectors X such that multiplying X by a matrix A produces a new vector that has the same direction as X, although its magnitude may have changed. If we imagine vectors as arrows from the origin, then these vectors can be stretched by the action of A but must still have their original orientation.

We can write a matrix equation that describes this property more formally. A vector X that is a solution to the matrix equation

$$A X = \lambda X$$

for some λ has the property we want. Such a vector is called an **eigenvector** and the amount λ by which it is stretched is called an **eigenvalue**. We can rewrite the matrix equation as

$$(A - \lambda I) X = 0$$

This is a set of n equations in n unknowns (the elements of X) and will only have a nontrivial solution if $A - \lambda I$ is singular. This corresponds to having a determinant of zero.

$$\det (A - \lambda I) = 0$$

Writing out the evaluation of the determinant gives an equation of degree n with one unknown, λ. Thus there are n different values for λ, that is, n eigenvalues (not necessarily distinct).

The preceding equations can be used to find eigenvalues and eigenvectors using algebra, but they are not particularly suitable for use in computations. There is, however, an algorithm that finds the eigenvector corresponding to the largest eigenvalue in most practical situations.

Notice that if X is an eigenvector, then any nonzero multiple of X is also an eigenvector (try multiplying it by A). Therefore, we can only talk about n eigenvectors by agreeing on which multiples we will choose. We might agree to normalize the eigenvectors by dividing each element by the last element so that the last element of an eigenvector is always 1.

Suppose that we choose a nonzero vector at random and multiply it by A. Then we multiply the result by A again, and go on doing so. If, at any stage, we find that two successive vectors calculated in this way are the same (after normalization), then we have found an eigenvector. There is no guarantee that this will actually happen, but in practice it usually does (and within only a few iterations). This is called the **power method**. More formally, the algorithm is:

given X_i multiply it by A to get X_{i+1}'
normalize X_{i+1}' to give X_{i+1}
compare X_{i+1} and X_i
if they are equal then stop
    (X_{i+1} is an eigenvector;
    the normalizing constant is an eigenvalue)
repeat for some limited number of iterations

The normalization step stops the actual magnitudes of the vectors from growing large. The algorithm usually performs well if the initial vector chosen consists of zeros except for a 1 in the row corresponding to the row of A with the largest element. It is also a good idea to use this element as the one that determines the normalization. The next program uses the power method to find the eigenvector corresponding to the largest eigenvalue.

```
*       Program 12.10
*       Finds eigenvectors using the power method
*
        program Eigen
        real A(10,10), X1(10), X2(10), Eigenv
        integer I, J, N, Count, Row, FndMax
        logical Found, EqVec
        print *, 'Enter size of matrix'
        read *, N
        print *, 'Enter rows of matrix'
        do 200 I=1, N
            read *, (A(I,J), J = 1,N)
200     continue
        Row = Fndmax(A, N)
        do 300 I = 1,N
            X1(I) = 0.0
300     continue
        X1(Row) = 1.0
        Count = 1
        Found = .false.
```

```
          while (Count.le.25.and..not.Found) do
              call Mult(A, X1, X2, N)
              Eigenv = X2(Row)
              do 400 I = 1,N
                  X2(I) = X2(I) / Eigenv
400           continue
              Found = EqVec (X1, X2, N)
              do 500 I = 1,N
                  X1(I) = X2(I)
500           continue
              Count = Count + 1
          end while
          if (Found) then
              print '(A,F7.4)', ' Eigenvalue is ', Eigenv
              print '(/ A /)', 'Eigenvector is'
              print '(10F7.4)', (X1(I), I = 1,N)
          else
              print *, ' Power method did not converge'
          endif
          stop
          end
*
*         Finds the maximum value in the array
*
          integer function Fndmax(A, N)
          real A(10,10), Max
          integer I, J, N
          Max = A(1,1)
          Row = 1
          do 200 I = 1,N
              do 100 J = 1,N
                  if (A(I,J).gt.Max) then
                      Max = A(I,J)
                      Row = I
                  endif
100           continue
200       continue
          Fndmax = Row
          return
          end
*
*         Calculates vector that results from multiplying X by A
*
          subroutine Mult(A, X, Y, N)
          integer I, J, N
          real A(10,10), X(N), Y(N), Sum
```

```
      do 200 I = 1,N
         Sum = 0.0
         do 100 J = 1,N
            Sum = Sum + A(I,J) * X(J)
100      continue
         Y(I) = Sum
200   continue
      return
      end
*
*     Checks if two vectors are equal
*
      logical function EqVec(X, Y, N)
      integer I, N
      real X(N), Y(N)
      EqVec = .true.
      do 100 I = 1,N
         if (X(I).ne.Y(I)) then
            EqVec = .false.
         endif
100   continue
      return
      end
```

When this program is run using the same matrix we have been using, the following output is produced for an eigenvalue and eigenvector.

```
Enter size of matrix
Enter rows of matrix
Eigenvalue is   6.7622
Eigenvector is
0.6677  0.7560  1.0000
```

12.8 Plotting Functions of Two Variables

We have already seen how we can use simple output devices to graph a function of a single variable. Although such graphs aren't accurate enough to allow measurements to be taken from them, they are very useful because they show the shape of the function. Often this is important either because it suggests the class of function to which the particular curve belongs or because it suggests the type of physical relationship that exists in the system being represented.

The same arguments apply to functions in two dimensions, but they are harder to present in such a way that their shape can be seen. For example, suppose that we have a function $f(x,y)$ of two variables, and we want to have our program display some output that represents the shape of this function. Normally we would use a projection into two dimensions of the surface of the function above the x–y plane. This is very hard to do within the limitations of many computer output devices.

The easiest way of presenting 3-dimensional information is simply to use a contour plot. Imagine that you are above the x–y plane looking down. Then you can see every point in the plane. If we mark each point with an indication of its height relative to the x–y plane, we have captured information about the shape of the function in terms of its height. The easiest way to indicate the height is to mark each point with a number indicating the value of the function at that point.

There are two problems that still have to be resolved. The first is that there are infinitely many points in the plane and we can't represent the function value at all of them. We solve this by selecting a finite subset of points, usually evenly spaced, and only showing the function value at those points. The second problem is that the function value is real and so must, in general, be written using many digits. We solve this by scaling the value of the function into (say) single-digit integers and using the integers as the representation of function magnitude. For example, if the function can take on values in the range [0, 100], then we might scale them so that only the first digit of the function value was used (giving possible representations 0, 1, ..., 9). This works only because we assume that the function is continuous. If it is, then we can use the fact that the function has the value a at one point and the value b at another to infer that the values in between are close to the mean of a and b. This assumption is more accurate as the points become closer together.

We have lost a great deal of information about the function by using this kind of representation. It is surprising how good a picture of the function's shape it is still capable of giving.

The next program displays a function of two variables in the plane.

```
*     Program 12.11
*     Purpose: show contours of surface given by a function
      program Grid
      real A(0:20,0:20), ZMin, ZMax, G
      integer IA(0:20,0:20)
      external G
      call Eval(A,15,0.,2.,0.,2.,G)
      call MinMax(A,15,Zmin,ZMax)
      call Scale(A,IA,15,ZMin,ZMax)
      call Output(IA,15)
      stop
      end
```

```
*
*         This is the function being tabulated
*
          real function G(X,Y)
          real X,Y
          G=exp(X-Y)*sin(5*X)*cos(2*Y)
          return
          end
*
*         Evaluates the function at each point on the
*         grid
*
          subroutine Eval(Z,N,XMin,XMax,YMin,YMax,F)
          integer N
          real Z(0:N,0:N),XMin,XMax,YMin,YMax,F
          integer I,J
          do 100 I=0,N
              do 200 J=0,N
                  Z(I,J) = F(XMin+I*(XMax-XMin)/20,
     &                       YMin+J*(YMax-YMin)/20)
  200         continue
  100     continue
          return
          end
*
*         Find the largest and smallest value of the function
*         in the grid
*
          subroutine MinMax(Z,N,ZMin,ZMax)
          integer N
          real Z(0:N,0:N),ZMin,ZMax
          integer I,J
          ZMin = Z(0,0)
          ZMax = Z(0,0)
          do 300 I=0,N
              do 400 J=0,N
                  if (Z(I,J).lt.ZMin) then
                      ZMin = Z(I,J)
                  else if (Z(I,J).gt.ZMax) then
                      ZMax = Z(I,J)
                  endif
  400         continue
  300     continue
          return
          end
```

```
*
*         Scale the function value into the range 0 to 9
*
          subroutine Scale(Z,IZ,N,ZMin,ZMax)
          integer N,IZ(0:N,0:N)
          real Z(0:N,0:N),ZMin,ZMax
          integer I,J
          do 600 I=0,N
             do 500 J=0,N
                IZ(I,J) = 10*(Z(I,J)-ZMin)/(ZMax-ZMin)
500          continue
600       continue
          return
          end
*
*         Print the scaled values in a grid pattern
*
          subroutine Output(IZ,N)
          integer N,IZ(0:N,0:N)
          integer I,J
          do 700 I=0,N
             print '(1X,26I2)',(IZ(I,J),J=0,N)
700       continue
          return
          end
```

Its output is

```
3 3 3 3 3 3 3 3 3 3 3 3 3 3
4 4 4 4 4 3 3 3 3 3 3 3 3 3
5 5 4 4 4 4 3 3 3 3 3 3 3 3
5 5 5 5 4 4 3 3 3 3 3 3 3 3
5 5 5 5 4 4 3 3 3 3 3 3 3 3
5 5 4 4 4 4 3 3 3 3 3 3 3 3
4 4 4 4 3 3 3 3 3 3 3 3 3 3
2 2 3 3 3 3 3 3 3 4 4 4 4 4
1 1 1 2 2 3 3 3 4 4 4 4 4 4
0 0 1 1 2 2 3 3 3 4 4 4 4 4
0 0 0 1 2 2 3 3 3 4 4 4 4 4
0 1 1 1 2 2 3 3 3 4 4 4 4 4
2 2 2 2 3 3 3 3 3 4 4 4 4 4
4 4 4 4 4 4 3 3 3 3 3 3 3 3
7 7 6 6 5 5 4 4 3 3 3 2 2 2
9 9 8 7 6 5 5 4 3 3 2 2 2 2 2
```

12.9 Linear Regression

An experiment often consists of a set of measurements of some physical process. There is usually an input to the experiment whose value we can determine and one or more outputs, which are the values in which we are interested. We call the input that is under our control the independent variable and the outputs of the experiment dependent variables (because their values depend in some way on the input).

One very common situation is to take some set of measurements (x_i, y_i), where the x's are the independent variables and the y's the dependent ones, to which we want to fit some particular function. For example, if we measure the force of gravitational attraction at various distances from a massive body, then we expect that the force will decrease according to an inverse square law.

We can summarize this general problem as follows: given some set (x_i, y_i), we wish to find a function $f(x)$ such that $f(x_i)$ approximates y_i at each of the measured data points.

If each of the measured values y_i was exactly accurate, then we could fit a polynomial to the data points. However, if there are errors in the values of the y_i's, then the "best" fitted curve to the data points may not pass through any of the y_i's exactly. Rather, it will follow the general sense of the data points. Such a function is likely to capture the underlying reality better than a function that follows the exact form of the data.

If we know something about the likely errors in the y_i's, then we can do better. Suppose that we have an error estimate e_i for each y_i, in the sense that there is some known probability that the actual value lies somewhere between $y_i + e_i$ and $y_i - e_i$. Then, when it comes to fitting a function to the data, we would like to give less attention to a data value that has a large error estimate than one that has a small error estimate. Thus we want to find a function that passes "close" to all of the data points, but "closest" to those points with a small error.

One way to do this is to construct a measure of closeness and try to adjust the fitted curve so that the closeness measure is minimized. We could, for example, consider the absolute vertical distance between the function and each data point and minimize the sum of these distances. This is what humans do when they attempt to fit a function by "eyeballing" the data. However, there is a computational problem because the absolute value function is not differentiable at zero, and the simplest way to find the minimum of some measure is to differentiate it and set the result to zero. The simplest (easiest to handle) measure of "closeness" is to minimize the sum of the squares of the differences between the data points and the fitting function, weighting each point according to its error estimate. Thus, we want to find the function $f(x)$ that minimizes the following expression:

$$\sum_{i=1}^{n} \frac{(y_i - f(x_i))^2}{c_i^2}$$

We must know something about the general form of the function $f(x)$ before we can do this. In the discussion that follows we will assume that the function $f(x)$ is a straight line. Fitting a straight line to a set of points is called **linear regression**. The method generalizes easily to any polynomial.

If $f(x)$ is a straight line, then its equation must be

$$f(x) = ax + b$$

for some coefficients a and b, which we want to find. Now we can find the minimum of the quantity

$$\sum_{i=1}^{n} \frac{(y_i - ax_i - b)^2}{e_i^2}$$

by differentiating with respect to a and b and setting the results to zero. We get:

$$-2 \sum \left[\frac{y_i - ax_i - b}{e_i} \right] \frac{x_i}{e_i} = 0$$

and

$$-2 \sum \left[\frac{y_i - ax_i - b}{e_i} \right] \frac{1}{e_i} = 0$$

Using some simple algebra, we see that a and b must satisfy

$$\sum_{i=0}^{n} \frac{y_i x_i}{e_i} - a \sum_{i=0}^{n} \frac{x_i^2}{e_i^2} - b \sum_{i=0}^{n} \frac{x_i}{e_i^2} = 0$$

$$\sum_{i=0}^{n} \frac{y_i}{e_i} - a \sum_{i=0}^{n} \frac{x_i}{e_i^2} - b \sum_{i=0}^{n} \frac{1}{e_i^2} = 0$$

We have two simultaneous equations in two unknowns (a and b), which we can solve either directly or by using the numerical techniques we have seen. The next program calculates the coefficients of the simultaneous equations and hence calculates the coefficients of the straight line approximating the data points.

```
*       Program 12.12
*       Does linear regression
*
        program Regres
        real X, Y, E
        real A(2, 2), B(2), Z(2)
        integer N, I, J
        do 200 I = 1,2
            do 100 J = 1,2
                A(I, J) = 0.0
100         continue
            B(I) = 0.0
200     continue
        print *, 'Enter # of points'
        read (5, *) N
        do 500 I = 1,N
            print *, 'Enter X and Y values and error estimate'
            read (5, *) X, Y, E
            A(1, 1) = A(1, 1) + X * X / (E * E)
            A(1, 2) = A(1, 2) + X / (E * E)
            A(2, 1) = A(1, 2)
            A(2, 2) = A(2, 2) + 1.0 / (E * E)
            B(1) = B(1) + X * Y / (E * E)
            B(2) = B(2) + Y / (E * E)
500     continue
        call Triang(A, B, 2)
        call Backsub(A, B, Z, 2)
        write (6, '(2F12.4)') (Z(I), I = 1,2)
        stop
        end
```

```
      subroutine Triang(A, B, N)
      integer I, J, K, N
      real A(2, 2), B(2), Factor
      do 300 I = 1, N - 1
         do 200 J = I + 1, N
            Factor = A(J,I) / A(I,I)
            do 100 K = I, N
               A(J,K) = A(J,K) - Factor * A(I,K)
100         continue
            B(J) = B(J) - Factor * B(I)
200      continue
300   continue
      return
      end

      subroutine BackSub(A, B, X, N)
      integer I, J, N
      real A(2, 2), B(2), X(2), Sum
      do 200 I = N, 1, -1
         Sum = 0.0
         do 100 J = I+1,N
            Sum = Sum + A(I,J) * X(J)
100      continue
         X(I) = ( B(I) - Sum ) / A(I,I)
200   continue
      return
      end
```

Suppose that we have taken a set of four measurements of a function that we suspect to be $y = 3x + 2$. The set of measurements that we get is:

X	Y	E
0.0	2.15	0.02
1.0	5.1	0.1
2.0	7.97	0.08
3.0	10.83	0.2

When these figures are used by the program, it calculates that the coefficients of the straight line that best fits this data are $2.91 x + 2.15$.

Programming Example

Problem Statement

Extend the Gaussian elimination algorithm described in this chapter to carry out full pivoting.

Inputs

Matrix representation of an equation.

Outputs

Solution to the equation.

Discussion

The entire matrix below and to the right of the current diagonal element must be searched for the element with the largest absolute value. This element must then be moved to the diagonal position by suitable row and column permutation. Permuting columns means changing the order of the x values. The program will have to remember the changes. One way to do this is to keep a vector of the permuted positions of the values. This is done using a vector whose ith entry remembers where the ith x value has been moved to. The same table also remembers how the columns of the coefficient matrix have been permuted.

Program

```
*       Program 12.13
*       Carries out Gaussian elimination with full
*       pivoting on a coefficient matrix
*
*           A - coefficient matrix
*           B - constant matrix
*           X - unknown vector
*           N - dynamic size of matrices and vectors
*
        program Pivot
        integer Size
        parameter (Size = 15)
        real A(Size, Size), B(Size), X(Size)
        integer I, J, N
```

```
      print *, 'Enter size, coefficients, and constant vector'
      read *, N, (( A(I, J), J=1,N), I=1,N), (B(I), I=1,N)
      print '(5X, I4)', N
      print '(2F10.2)', ((A(I, J), J=1,N), I=1,N)
      print '(2F10.2)', ( B(I), I=1,N)
      call Linear(A, B, X, N, Size)
      print '(A, (2F10.2))', 'Values of X are', (X(I), I=1,N)
      stop
      end
*
*     Does the pivot, elimination and back substitution
*
*     A, B - coefficients
*     X - unknown vector
*     N - size of matrices and vectors
*     PR - pivot row
*     PC - pivot column
*     PMap - vector of permutations of columns
*     PX - permuted x values
*
      subroutine Linear(A, B, X, N, S)
      integer N, S
      real B(N), A(S,S), X(N)
      integer I, J, K, PR, PC
      integer PMap(100)
      real PX(100), M
*     Prepare for full pivoting
      do 100 I=1,N
          PMap(I) = I
  100 continue
*     Eliminate
      do 800 I=1,N-1
*     Full Pivoting
          PR = I
          PC = I
          do 300 J=I,N
              do 200 K=I,N
                  if(abs(A(J,K)) .gt. abs(A(PR,PC))) then
                      PR = J
                      PC = K
                  endif
  200         continue
  300     continue
```

```
              do 400 J=I,N
                 M = A(I,J)
                 A(I,J) = A(PR,J)
                 A(PR,J) = M
        400   continue
              M = B(I)
              B(I) = B(PR)
              B(PR) = M
              do 500 J=1,N
                 M = A(J,I)
                 A(J,I) = A(J,PC)
                 A(J,PC) = M
        500   continue
              M = PMap(I)
              PMap(I) = PMap(PC)
              PMap(PC) = M
*             End partial Pivoting
              do 700 J=I+1,N
                 M = A(J,I)/A(I,I)
                 A(J,I) = 0.
                 do 600 K=I+1,N
                    A(J,K) = A(J,K) - M * A(I,K)
        600      continue
                 B(J) = B(J) - M * B(I)
        700   continue
        800 continue
*       Substitute
        do 1000 I=N,1,-1
           M = B(I)
           do 900 J=I+1,N
              M = M - A(I,J) * PX(J)
        900 continue
           PX(I) = M/A(I,I)
       1000 continue
*      Undo the pivoting
       do 1100 I=1,N
          X(Pmap(I)) = PX(I)
       1100 continue
       return
       end
```

Numerical Linear Algebra 381

Testing

```
Enter size, coefficients, and constant vector
        2
        1.00        2.00
        2.00        6.00
        1.00        6.00
Values of X are     -3.00       2.00
```

Discussion

This program declares the arrays in the main program and subprogram a little differently from any of the examples in the chapter. We would like to write the subprogram in full generality, so that it could be used from any main program. Almost everything in the subprogram allows this: the dimensions of the arrays are passed in as parameters. Because the array A is two-dimensional, its actual declared size has to be passed in, as well as the part that is being used. The problem is with the arrays PMap and PX. We would like these arrays to be of dimension S, but cannot as they are not passed in as parameters. Remember that arrays passed as parameters are actually implemented by passing in the address at which the array is located in the calling program's storage area. However, arrays that are local to the subprogram are allocated in the storage space of the subprogram, and therefore must have defined sizes. We are reduced to declaring the arrays PMap and PX to be large. The subprogram works generally, as long as the matrix size is smaller than 100 by 100.

Design, Testing, and Debugging

- Always pay attention to places where a 2-dimensional array is being passed as a parameter. More errors occur here than almost anywhere else in programs. Be careful to check that the array is being treated the same in both the invoking program and subprogram and that the effective limits of the array are not being violated.
- We have seen in the problems in this chapter that there are many situations in which an anomaly can exist. Try to think about these in advance and plan for your program to handle them. Will the matrix have an inverse? Will the technique converge?
- We have also seen that there are numerical pitfalls behind most of the algorithms that we have presented. It is not safe to write numerical programs when you really care about the accuracy of the output without investigating the mathematics behind these algorithms. A whole field of computer science called numerical analysis is devoted to studying algorithms of this kind.

- We have now seen programs that are moderately large. We hope that you're starting to understand the value of a good decomposition of the algorithm and a matching decomposition of the program.

Style and Presentation

- Two-dimensional arrays require much more careful thought about the formatting of output than we have seen so far. It's absolutely crucial for humans to see matrices in their tabular form with the values laid out in rows and columns. This is often difficult to do while still preserving the generality of programs.
- When you write numerical programs, do not promise to do more than the program can do. Be careful of output messages that indicate that you have found an exact value when you have found an approximation. Try to have your program give warnings about possible problems with the output that it has produced. Provide extra data, such as the number of iterations taken, that can be used to assess the reliability of the program output.

Chapter Summary

- An array may have more than one index. Thus it can be thought of as extending in more than one dimension.
- A subprogram can declare arrays whose size depends on the value of some parameter. The parameter must appear in the declarations before the array it describes.
- Input and output of an entire array as a unit can be confusing for multi-dimensional arrays. Values are matched with arrays positions in column major order, that is, so that the first subscript varies fastest.
- Two-dimensional arrays can be used to represent matrices. Many important algorithms of linear algebra can be implemented as programs. Examples include: upper triangularization, Gaussian elimination, solving simultaneous linear equations, matrix inversion, LU decomposition, and determining eigenvectors.
- Linear regression is an important technique in processing experimental data. It fits a straight line to a set of points which may contain experimental variations. It can be transformed into a problem of solving simultaneous equations.

Define These Concepts and Terms

Linear equations
Gaussian elimination
Back substitution
Pivoting
Partial pivoting
Ill-conditioning

Cramer's rule
Eigenvector
Eigenvalue
Power method
Linear regression

Exercises

1. Write a subprogram that accepts a 2-dimensional matrix of arbitrary size and transposes it. The transpose of a matrix has its rows and columns reversed—reflected in the main diagonal.
2. Write a subprogram that accepts a 2-dimensional matrix of arbitrary size and exchanges rows and columns so that the largest element is in the top left-hand corner.
3. Using the triangularization method discussed in this chapter, write a program to calculate determinants efficiently. What is the complexity of the overall program? Verify this by taking timing measurements of your program on different sizes of matrix, if your system has a way to do this.
4. Using the power algorithm, investigate the number of iterations required to find an eigenvector for various matrices. Can you identify classes of matrices for which the algorithm is particularly good? Particularly bad? (You could try matrices with patterns, such as upper or lower triangular matrices, or symmetric matrices.)
5. Generalize Program 12.11 by rewriting it as a function that can be used to print a contour plot of any reasonable function passed as a parameter.
6. Calculate the required simultaneous equations for fitting quadratic and cubic polynomials to a set of experimental data. When you see the general pattern of the coefficients, write a program that reads experimental data, calculates coefficients, and then uses Gaussian elimination to solve for the coefficients of the fitting function.
7. Write a function that finds the length of a vector (the square root of the sum of the squares of the components).
8. Write functions that add two matrices and subtract one matrix from another pointwise.
9. Write a function to determine if an array is idempotent, that is, if $A^2 = A$.
10. Write a function that returns a **logical** value indicating whether a matrix is in upper triangular form. Write another to check if a matrix is in lower triangular form. Does it make sense to combine these two into one function, with

another parameter indicating whether to check for the upper or lower triangular property?
11. Write a program that checks if a matrix has two rows that are related by one row being a multiple of another.
12. Write a program to count the frequency of pairs of letters in a file. Print a table with columns and rows for each letter, in which the entry in row i, column j shows the number of times that this pair appeared in the file. Run the program on some English files and some Fortran files. What are the most common pairs in each? Modify your program to print percentages instead of counts.
13. Extend the program from Exercise 12 so that rather than printing a table it will print a list of the letter pairs, together with their frequencies, sorted from most frequent to least frequent.
14. Write a subprogram that accepts as parameters two two-dimensional arrays of the same shape, their sizes in each dimension, and a threshold value. The first of the arrays contains values; the second array is to be filled in by the subprogram. Every entry in the first array that is greater than or equal to the threshold will cause a 1 to be placed in the corresponding position in the second array. The other entries of the second array will be zero.
15. A matrix is said to be sparse if many of its entries are zero. Write a function that will determine the percentage of non-zero entries in an integer matrix.

Engineering Problem 13

Closing a Traverse in Surveying

We now consider a standard problem in surveying—closing a traverse. A traverse is made up of points on the territory being surveyed, known as stations, and lines between them, called traverse lines. A traverse is therefore a polygon surrounding some territory. Each edge of the polygon is a traverse line. Each traverse line must afford a line of sight. The horizontal lengths of the traverse lines are measured using a steel band tape or an electronic distance-measuring device. The horizontal angles between the traverse lines are measured using a transit or theodolite.

Because errors in measurement arise in all surveying, a surveyor always tries to check his or her work. A closed traverse is one in which the traverse lines return to the original starting point. The amount of error in this closure indicates something about the error in the whole traverse.

The horizontal angles are usually measured as internal angles, that is, the angle between the traverse lines inside the polygon they create. A set of measurements for a traverse therefore consists of lengths for each traverse and an internal angle for each station. The error in the polygon can be determined using trigonometric checks and redistributed to give a polygon that is more probably correct than the measured one. Here are some of the checks that can be made:

1. For an n-sided polygon, the sum of the internal angles is given by $180(n-2)$. This check is usually made at the time the measurements are taken and a correction is made if necessary. A suitable correction is to adjust each angle by the error divided by n.
2. A direction is assigned to one of the traverse lines. This is called the azimuth and is the position of the traverse line measured clockwise in degrees from some predefined direction. It is therefore a vector quantity. The azimuth is usually calculated either by tying the survey to an existing survey or by using an astronomical observation.

   Once the direction of one of the traverse lines has been assigned, the azimuths of the other lines can be determined from the internal angles.

Now if we take the lengths and azimuths of a set of traverse lines that are supposed to represent a closed polygon, we will usually find that, as we follow the lines around the polygon, we end up at a point A' rather than A. The line AA' represents the *error of closure* and its length measures the amount of error. The precision of a survey is usually expressed as the relative error of closure, called the *precision ratio*.

$$\text{Precision Ratio} = \frac{\text{Length of } AA'}{\text{Total Length of Traverses}}$$

We wish to find the error of closure, given the lengths and azimuths of the traverse lines. We first compute, for each line, its north-south component, called *latitude*, and its east-west component, called *departure*.

$$\text{Latitude } AB = AB \cos \text{Azimuth}$$
$$\text{Departure } AB = AB \sin \text{Azimuth}$$

North latitudes and east departures are positive, while south latitudes and west departures are negative. Now for a true closed polygon, the sum of the latitudes and the departures will each equal zero over all sides. However, for a real traverse, the sums will have some small value that represents the gap at the end of the traverse. Thus the sums of the latitudes and departures give the latitude and departure of the line AA'. The length of AA' is given by

$$\sqrt{\sum \text{Latitudes}^2 + \sum \text{Departures}^2}$$

We can use this to calculate the precision ratio.

3. If the precision ratio calculated is acceptable, then the traverse can be rearranged, or balanced, to remove the error of closure. We will use the *compass rule* to make adjustments. This rule assumes that the error in a traverse is proportional to its length. Longer lines receive a larger correction. This rule states that:

$$\text{Correct Latitude of } AB = \text{Old Latitude of } AB -$$
$$\sum \text{Latitudes} \left(\frac{AB}{\text{Length of all Traverses}} \right)$$

$$\text{Correct Departure of } AB = \text{Old Departure of } AB -$$
$$\sum \text{Departures} \left(\frac{AB}{\text{Length of all Traverses}} \right)$$

When all latitudes and departures have been corrected in this way, a check should be made that the new values do sum to zero. The lengths and azimuths of the new traverse lines can be calculated from these latitudes and departures as

$$\text{Length} = \sqrt{\text{New Latitude}^2 + \text{New Departure}^2}$$

$$\text{Azimuth} = \arctan \frac{\text{New Departure}}{\text{New Latitude}}$$

Care must be taken to get the arctan in the right quadrant (for example, $\arctan(-2/1)$ is in the fourth quadrant while $\arctan(2/-1)$ is in the second quadrant — see the atan2 function in Appendix D).

This computation completes the balancing of the traverse. The corrected lengths and azimuths define a closed polygon to the precision of the computation.

It is common to compute coordinates for the stations to make it easier to plot them. This is done in the following way. The most southwesterly station is given coordinates (1000, 1000). The coordinates of the other stations are calculated by proceeding clockwise around the polygon using the formulas:

$$X_{i+1} = X_i + \text{Departure}(i, i+1)$$
$$Y_{i+1} = Y_i + \text{Latitude}(i, i+1)$$

Programming Problems

1. Write a program to close and balance a closed traverse, given the measured lengths and azimuths of the traverse lines. Also compute the coordinates of the stations. Angular measurement is given to a precision of 0.1 minutes and linear measurements to 0.001 m. Output data should be presented to the same precision. Also calculate the precision ratio before correction. Here is a table of measured values. Your program should read only the values from the first two columns. The other columns allow you to check the initial part of your calculation. The point A has coordinates (1000, 1000).

	Length	Azimuth	Lat	Dep
AB	328.321	10°17.3'	323.042	58.639
BC	491.267	86°12.7'	32.458	490.194
CD	566.877	159°23.4'	-530.596	199.544
DE	386.927	228°14.6'	-257.681	-288.640
EA	631.171	313°15.9'	432.588	-459.613

Engineering Problem 14

Text Compression

Computer networks allow computers all over the country to talk to each other. These computers are used to send mail electronically between users, make airline reservations, transfer money, and many other things. Applications such as these require large amounts of data to be moved along the communication lines between computers. Because each communication line costs money to install and maintain, computer network owners are interested in reducing the amount of data that must be transferred. This has the added benefit that communication is faster, since data do not have to wait in queues for communication lines as often.

One way to reduce the amount of data moving is to use *text compression* on the data to be sent. This means recoding the information so that the total volume that has to be sent gets smaller. For example, the standard encoding using ASCII or EBCDIC uses 8 bits for each character. However, we know that, in English, some letters are much more common than others. If we can come up with an encoding scheme that uses a short bit string for common letters (such as "e") and longer bit strings for uncommon letters (such as "z"), then the *average* number of bits that need to be sent can be substantially reduced.

One such encoding scheme is called Huffman coding. In a later problem we will explore how the Huffman coding can be created for particular kinds of text to be transmitted (for example, the character frequency in English text might be different from that in program text). For this problem, we assume the coding illustrated next.

' '   is 11111
'a'   is 1101
'b'   is 010010
'c'   is 01000
'd'   is 11100
'e'   is 011
'f'   is 10010
'g'   is 111101

'h' is 0011
'i' is 1000
'j' is 1110101110
'k' is 11101010
'l' is 10011
'm' is 00101
'n' is 1011
'o' is 1100
'p' is 111100
'q' is 1110101101
'r' is 1010
's' is 0101
't' is 000
'u' is 00100
'v' is 1110100
'w' is 010011
'x' is 1110101111
'y' is 111011
'z' is 1110101100

We want to develop a program that will use this table to create an encoded version of some input text and calculate the compression ratio achieved.

Programming Problems

1. Write a program to read the preceding Huffman data from a file. The program should then read text from another file and produce the characters representing the binary encoding of the compressed text. It should also calculate the compression ratio based on the number of characters printed and the number of characters in the input text (assuming an 8-bit byte). Thus, if the input file of text contained only the word

    hi

    then the output would be

    00111000

    and the compression ratio would be 0.5 (since the original took two characters, while the encoded version would fit into one).

Engineering Problem 15

Calculating the Huffman Encoding

In the previous problem we discussed how the Huffman encoding could be used for text compression. In this problem, we develop the Huffman encoding, based on the relative frequencies of letters in the kind of text for which the encoding will be used.

We begin with a table of the relative frequencies of letters. As an example, suppose that we deal only with the characters "a" to "e" and that we know their relative frequencies in the text we plan to compress. Thus, we have a table like this:

a   0.35
b   0.26
c   0.15
d   0.13
e   0.11

We start with the two entries with smallest relative frequency and combine them to form a single entry, adding their relative frequencies. The combined symbol is inserted in the table in order of its new relative frequency. We get the new table

a   0.35
b   0.26
de   0.24
c   0.15

Putting "d" and "e" together also means that their eventual encoding will be the same, except that it will differ in the last bit, "d" ending in a 1 and "e" ending in a 0. We now repeat the merging operation on the new table, combining the bottom two entries. This gives the new table

cde   0.39
a   0.35
b   0.26

The representation for "c" will end in a 0 and the representations for "d" and "e" will have a 1 in their second last position. The next step gives a table

ab   0.61
cde   0.39

so that the representations of "a" and "b" will end in 1 and 0, respectively. Finally we merge the final two entries in the table. The representations for each of the characters will be

a   11
b   10
c   00
d   011
e   010

The algorithm is

- Merge the two symbols with the lowest remaining frequencies.
- Insert the entry for the combined symbol back into the table so that it remains in descending order of frequency.
- Add a 1 to the front of the representation of the symbols in the highest element that was merged. Add a zero to the front of the representation of the symbols in the lowest element that was merged.

This continues until the list is reduced to one line.

Programming Problems

1. Write a program that will read a table of relative frequencies from a file. The program then computes the Huffman encodings for each of the characters in the table and prints a table of characters and encodings. Try to make the program general enough that any set of characters can be used.
2. (Optional) Use the frequency-counting programs given earlier to produce frequency counts for different kinds of text. Are there differences in the distribution of characters in ordinary text, programs, and special kinds of text?

As possible test data we provide a table of the relative frequencies of letters in ordinary text.

'e'	0.1204
't'	0.0945
'a'	0.0756
'o'	0.0747
'n'	0.0707
'r'	0.0677
'i'	0.0627
's'	0.0607
'h'	0.0528
' '	0.0450
'd'	0.0378
'l'	0.0339
'f'	0.0289
'c'	0.0279
'm'	0.0249
'u'	0.0249
'g'	0.0199
'p'	0.0199
'y'	0.0199
'w'	0.0149
'b'	0.0139
'v'	0.0092
'k'	0.0042
'x'	0.0017
'j'	0.0013
'q'	0.0012
'z'	0.0008

This frequency table should produce the Huffman encoding given in the earlier problem.

Models, Simulation, and Games

13.1 Models and the World
13.2 Static Simulation
13.3 Randomness
13.4 Generating Pseudo-Random Numbers
13.5 Monte Carlo Methods
13.6 Dynamic Simulations
13.7 Playing Games

13

13.1 Models and the World

Many real world systems are too complex to be easily understood, particularly when they are encountered for the first time. One of the main ways in which we handle this as humans is to build (mental) **models** of the world we encounter and then reason about the models. A model is an abstraction of the real world system in which we are interested that conceals some of the real world complexities.

Models are particularly important in engineering because they provide a way to understand processes in the physical world and also to develop new processes. Models can be of two kinds: abstract models that we construct to reason about a system or explore its properties, and **prototypes**, which are small scale models of the real world that we develop to prove that some real world system is viable.

We will restrict our attention to the first kind of model, although much of what we say applies to prototypes as well. Modeling itself is a skill that takes some practice, although we won't talk much about it explicitly. At least part of the difficulty of programming, however, comes from the necessity to build a computer program model of the problem that we are interested in.

Once we have constructed a model, there are two ways in which we can investigate it. The first way is to use mathematical techniques. For example, many engineering problems have been modeled using continuous functions and are analyzed using calculus. When we set up a framework of girders, we can use well established techniques to calculate the forces acting on them and hence decide whether or not they will stay standing. However, the analysis assumes certain things about the system being analyzed, in this case that the material making up the girders has some ideal properties and isn't filled with cracks. Our model girders also ignore aspects of the real world system that are "irrelevant," such as the color of the girders.

The second main way of investigating models is to use simulation. Simulation means implementing a model of the system as part of a computer program and then manipulating the program itself and investigating what happens. Because we can build models of many systems as computer programs and because the programs allow the models to be manipulated quickly, we can use simulation to investigate a wide class of problems. It is usually much easier to alter the parameters of a program than to alter the real world environment.

Before we talk more about these techniques, it is important to realize that the results obtained from either mathematical analysis or simulation apply to the model from which we have been working. They do *not* necessarily apply to the real world system from which the model was built. This is because the model is an abstraction of the real world system and hence omits many aspects of the real world system that could be important. The applicability of results derived from the model to the real world system depends crucially on exactly how the abstraction was done. As an example, a nineteenth-century traffic engineer predicted that by the year 1920 the streets of London would be six feet deep in horse manure. The conclusion he reached by mathematical analysis was valid for his model, but his model did not take into account possible changes in motive power. On a slightly more serious level, Malthus predicted the demise of the human race from overpopulation—a conclusion that has not yet eventuated because his model did not include information about new sources of food and the effects of war in keeping populations down. So be warned; models can be powerful tools for understanding new situations, but conclusions from models must always be tested in the real world system before any confidence can be given to them (this is another reason for building prototypes).

There are two main types of simulation: static and dynamic. **Static simulations** are those in which time does not play a significant role. **Dynamic simulations** usually involve time as a major factor, and several important techniques have been developed for such simulations. We will look at examples of both kinds of simulation in the remainder of this chapter.

13.2 Static Simulation

As an example of a static simulation, we will look at a sampling procedure. The general idea behind this simulation can be applied to a number of different fields.

Suppose that we are interested in finding out how many fish are present in a lake. Someone suggests the following procedure: go out into the lake and catch some number n of fish. Mark these fish and throw them back. Then go out again and catch a sample s of fish and see how many are marked. Suppose that k of the s are marked. Then it seems reasonable to conclude that the ratio of the marked fish in the second sample to the total size of the sample might approximate the ratio of the number of fish marked in the first sample to the total number of fish in the lake. Thus, if N is the total number of fish in the lake, then we might expect the ratios

$$\frac{n}{N} = \frac{k}{s}$$

to be equal. We can then solve this equation for N.

We need to decide how big to make both of the samples (how big to make n and s) so that we get a reasonable approximation. It would be possible, but difficult, to find a lake with known population and try the method; it is a great deal easier to simulate the lake and the procedure for sets of values of n and s.

We do this by using a large 2-dimensional array to represent the lake. At the beginning we place N fish in the lake at random positions. Then we simulate the first phase by marking n fish at random. In the second phase, we select s fish from the array at random and see how many are marked. From this information we can calculate the value n from the preceding equation and compare it to the known value of N. We now show the program.

```
*       Program 13.1
*       Simulates a fish-counting method
*       to assess its accuracy
*
        program Sample
        integer Lake(500,500), Side
        integer N, Ndash, S, K, I, J, Count
        Side = 500
        call SetSd(time())
        do 200  I = 1,Side
           do 100  J = 1,Side
              Lake(I, J) = 0
100        continue
200     continue
        print *, ' Enter # of fish in lake'
        read *, N
        Count = 0
        while (Count.le.N) do
           I = random() * Side + 1
           J = random() * Side + 1
           if (Lake(I, J).eq.0) then
              Lake(I, J) = 1
              Count = Count + 1
           endif
        end while
        print *, ' Enter first sample size '
        read *, Ndash
        Count = 0
        while (Count.le.Ndash) do
           I = random() * Side + 1
           J = random() * Side + 1
           if (Lake(I, J).eq.1) then
              Lake(I, J) = 2
              Count = Count + 1
           endif
        end while
        print *, ' Enter second sample size '
        read *, S
        Count = 0
        K = 0
```

```
        while (Count.le.S) do
            I = random() * Side + 1
            J = random() * Side + 1
            if (Lake(I, J).ne.0) then
                Count = Count + 1
                if (Lake(I, J).eq.2) then
                    K = K + 1
                endif
            endif
        end while
        print '(A,F10.2)', 'Predicted number ', Ndash * S / K
        print '(A,I6)', ' Actual number ', N
        stop
        end
```

To carry out the various random operations, we must have some way of having our programs behave nondeterministically. This is done using a function that produces pseudo-random numbers. We will discuss this more fully in Section 13.4. For the time being, note how we generate such numbers in this program. Unfortunately, generating **random numbers** is not a standard part of the definition of Fortran and so details will vary from compiler to compiler. You will have to consult a manual for your compiler to discover how to do it on your system. We will assume that we are provided with a subroutine called SetSd to initialize the random number sequence, a function called Time returning the current time, and a function called Random returning a random value between 0 and 1.

Because the random number function returns a value between 0 and 1 (including 0 but not 1), we apply a transformation to produce a value between 1 and 1000. This is done by multiplying the value obtained from the random number routine by 1000, adding 1, and taking the integer part. Convince yourself that this does indeed produce values in the range 1 to 1000 inclusive.

13.3 Randomness

In the last section we needed to provide some randomness for our simulations to be useful. In this section we discuss more precisely what we mean by some process being "random."

Suppose that we are standing at the door of a bank and we want to consider the way in which people arrive. Intuitively we would consider their arrivals to be random since, except for the obvious exceptions of people who meet outside the bank, there is no collusion between customers. The time at which one customer arrives should not affect the time at which any other customer arrives. This notion of customers not affecting each other's behavior is at the heart of what we mean by randomness.

More formally we say that some events are random if we cannot use any information about past happenings to predict specifically what is going to happen in the future. Outside the bank, the fact that a customer has just arrived should have absolutely no predictive value concerning the next customer. If we can deduce anything about specific future behavior of the system from past behavior, then the system is not random. Failing to understand this is the reason behind some large gambling losses. A gambler may start to imagine that because there has been a long run of "heads," then "tails" is due; but he would be failing to understand that coin flipping is a random procedure. (Of course, in principle it isn't random either, since it is possible to calculate, from the force applied to the coin, how it will land.)

Notice that this doesn't exclude finding a distribution function describing how random events happen over the long term. We might be able to write down a function describing the probability of a customer arriving at a particular time and be able to say that, on average, customers arrive at a specific rate. The essence of the randomness is that we cannot make absolute statements, although we may be able to make probabilistic ones.

When we want to include truly random behavior in programs, we are at something of a loss because there aren't very many truly random processes around. In fact, the only way in which to get truly random numbers is to examine physical events that are currently understood to be random. For example, we could point a scintilliscope at a lump of radioactive material and use the decay of atoms as random events to drive our simulations.

Most programs settle for pseudo-random numbers, rather than truly random ones. These numbers exhibit the property that they appear, for short periods of time, to be random. The fact that they aren't is another weakness of the simulation method. In the next section we discuss some of the ways in which pseudo-random numbers are generated.

13.4 Generating Pseudo-Random Numbers

The most common simple method for generating pseudo-random numbers is to use a recurrence called a **linear congruential generator**. This recurrence has the form:

$$x_{i+1} = (ax_i + c) \bmod m$$

where a, c, and m are suitably chosen constants. The constant m is usually chosen to be the size of the maximum positive integer that the machine can represent, a is chosen to be positive, and c is chosen to be non-negative. Thus the x_i values are all positive integers between 0 and $m - 1$. These are subsequently scaled by division by m to produce values in the range 0 (included) to 1 (excluded).

If the constants a and c are carefully chosen, this recurrence contains all of the integers between 0 and $m - 1$ exactly once before it repeats. Such a sequence is said to be "full." Since m is typically 2^{31}, this sequence is much longer than the number of random values required by any simulation program, so that the repetition of the sequence is not of practical significance.

A good sequence will also have the property that successive terms will differ widely (that is, there are no obvious correlations between terms that are close in the sequence).

Because every value appears in the sequence before any repetitions, the distribution of values produced is uniform. If we considered the values between 0 and 1 that were produced in the entire sequence and drew a histogram of the values between 0 and 0.1, between 0.1 and 0.2, and so on, the heights of each column would be the same.

To allow a program to execute several times with different random numbers, it is necessary to vary the point in the sequence at which the program starts. This is done by providing the initial value x_0, called the *seed*. The seed can be any element of the sequence. In some systems the program must provide the seed to the random number generator. Of course, if the seed is, say, a constant in the program, then the program will behave deterministically because it will always get the same sequence of pseudo-random values. This can be useful when the program is being debugged. However, if we want the program to behave differently each time it is executed, then we must arrange for it to use a different seed each time.

There are several ways in which the first value for the seed can be determined. The simplest is to ask the user to enter a number that will serve as the seed. Unfortunately, humans tend to pick numbers from a small set so this doesn't provide a wonderful solution. (Ask your friends to pick any number and see how they answer—most will probably pick a number less than 10.) The second method, commonly used by simpler microcomputers, is to set the seed by measuring how long the user takes to respond to some request for input. This is the reason that some video games ask apparently redundant questions as they begin. The third and best solution is to use some function of the time at which the program begins. Many systems have a time-of-day clock whose value can be used as an initial seed. If the date is included as well, the value used to generate the seed will always be different and doesn't depend on the user's behavior at all.

We now give examples of SetSd and Random routines. If your system doesn't have a way to generate random numbers, you may want to use them. Some of the features used in these subprograms are unusual and aren't talked about in the text. If you want to know more about them you should refer to Appendix C. Here is the SetSd subroutine:

```
*       Initializes a seed in a common block for use
*       by subsequent random number routines
*
        subroutine SetSd (X)
        integer X, Y, M
        common /Rand/ Y
        parameter (M = 233280)
        Y = mod(X, M)
        return
        end
```

Its sole purpose is to provide an initial point from which to begin the recurrence. The second important piece of code is the function Random, which does not have any parameters and returns a random value that is uniformly distributed in the interval (0,1).

```
*       Returns a random number in [0, 1)
*
        real function Random ()
        integer Y, A, C, M
        common /Rand/ Y
        parameter (A = 9301, C = 49297, M = 233280)
        Y = A * Y + C
*       zero high order bit if necessary
        if (Y.lt.0) then
            Y = Y + 2147483647 + 1
        endif
        Y = mod(Y, M)
        Random = real(Y) / real(M)
        return
        end
```

The values for *a*, *c*, and *m* were taken from a table in *Numerical Recipes*, by W.H. Press, B.P. Flannery, S.A. Teukolsky and W.T. Vetterling, and published by Cambridge University Press, 1986. You may be interested in reading their discussion of generating random numbers. Another possible set of values that they give is $a = 4096$, $c = 150889$, and $m = 714205$.

Many random number systems do not allow you to set the initial seed and then forget about it. Instead they allow the previous value of the seed to be one of the parameters of the random number routine. This has both good and bad features. It makes it possible to have several different sets of random numbers in use at the same time, which can be useful in complicated simulations. On the other hand, the programmer has to manage the seed correctly, passing it to the random number routine as needed and not altering it otherwise. This next function works by being explicitly given the value of the seed.

```
*       Generates a random number passing seed in and out
*
        subroutine Random (N, U)
        integer N
        real U
```

```
        N = N * 843314861 + 453816693
        if (N.lt.0) then
            N = N + 2147483647 + 1
        endif
        U = N * 0.4656612e-9
        return
        end
```

This subroutine uses the input value of N to produce a new value that is output in N, and also scales the integer N to an output real value U, in the range 0 (included) to 1 (excluded).

13.5 Monte Carlo Methods

One important class of static simulations is called Monte Carlo methods (after the casinos of the Principality of Monaco). As an example we will illustrate several forms of **Monte Carlo integration**.

Suppose we are asked to calculate the integral of some complicated function between two points, *a* and *b*. We have already seen numerical ways to do this. For some functions, however, these numerical techniques are too expensive. It may also happen that we only require an approximation for the integral. In both these situations, Monte Carlo integration can be used.

The basic idea is to sample the function at a set of N random points in the interval from *a* to *b*. The integral is then approximated by calculating the sum of the function values at these points, multiplied by their average distance apart. For example, if we are finding the integral of a function over an interval of size 1 and we sample at ten points, then the sum of the function values is multiplied by 0.1.

In a program to implement this method, we must generate random values in the interval between *a* and *b* and then sum the function values at these points. We must then calculate the average distance between points and multiply by this to get the approximation to the area. Here is the program:

```
*       Program 13.2
*       Does a Monte Carlo integration of a function
*
        program Monte
        real A, B
        real X, Sum, Width
        integer N, I
        call SetSd(time())
        print *, ' Enter limits for integration
        read *, A, B
        print *, ' Enter # of points to sample'
        read *, N
```

Models, Simulation, and Games

```
      Sum = 0.0
      do 500 I = 1, N
         X = (B - A) * Random() + A
         Sum = Sum + F(X)
500   continue
      Width = (B - A)
      Sum = Width * (Sum / N)
      print '(A, F10.4)', 'Integral is ', Sum
      stop
      end
      real function F(X)
      real X
      F = sin(X) * exp(X)
      return
      end
```

The output from this program for the interval between 0 and 1 with 100 sample points is

```
Enter limits for integration
Enter # of points to sample
Integral is      0.8757
```

Try experimenting with it to see how it compares with the other numerical methods we have discussed. If your system has a means of finding out how much time a particular program used, then compare the running times of the Monte Carlo and Simpson's rule programs on the same functions.

Another approach to Monte Carlo simulation is to consider the rectangle whose opposite corners are the points $(a, minf(x))$ and $(b, maxf(x))$ where $minf(x)$ is the smallest value of the function in the interval $[a, b]$ and $maxf(x)$ is the largest. The function itself divides this rectangle into two regions: one is the region under the function (which is the integral we want to calculate), and the other is the area above the function. We can also calculate the area of the rectangle itself, since we know its sides. Of course, we need to take into account the location of the axis in this. We'll assume that the function is positive over the interval of interest, and hence that *minfx* is always zero.

In this form of Monte Carlo integration, points in the rectangle are randomly generated by producing random coordinate pairs. The x coordinates are forced to be in the range a to b and the y coordinates in the range $minf(x)$ to $maxf(x)$. Each coordinate pair is then tested to determine whether it is above the function or below it. The number of pairs of coordinates in each part is recorded.

To calculate the value of the integral, the following reasoning is used: the ratio of the number of points above the function to the number of points below the function should approximate the ratio of the area above the function to the area below it. This is a reasonable assumption if the points chosen were uniformly distributed in the plane. As we know the total area of the rectangle, we can find the area below the function by multiplying by the fraction of points below it.

For example, suppose that there were $H = 300$ points above the function and $L = 150$ points below it. If the area of the rectangle is 6, then the area below the function is approximately

$$\frac{150}{450} \times 6 = 2$$

The following program carries out this kind of integration.

```
*       Program 13.3
*       Uses Monte Carlo integration
*
        program MonteC
        integer Below, Above, Total, I
        real Area, Totar, A, B, X, Y, Maxf
        call SetSd(time())
        Below = 0
        Above = 0
        print *, ' Enter integral limits '
        read *, A, B
        print *, ' Enter # of sample points'
        read *, Total
        print *, 'Enter upper bound on function in interval'
        read *, Maxf
        Totar = abs(B - A) * Maxf
        do 500 I = 1, Total
            X = Random() * abs(B - A) + A
            Y = Random() * Maxf
            if (Y.le.F(X)) then
                Below = Below + 1
            else
                Above = Above + 1
            endif
500     continue
        Area = real(Below) / Total * Totar
        print '(A, F10.4)', ' Approximate area is ', Area
        stop
        end
        real function F(X)
        real X
        F = sin(X) * exp(X)
        return
        end
```

Notice that we have not actually used the number of points above the function in this program at all. That is because we are only interested in the fraction of points below the function and we already know the total number of points. We used a certain number of sample points as a terminating condition. Another way to terminate would be to calculate at intervals the ratio of points above and below the function. We could terminate when this ratio showed signs of converging, that is, when successive values of the ratio were very close together.

Monte Carlo integration can also be used to integrate in higher-dimensional spaces. This means that we can get approximations for the volumes of solids in 3-dimensional space. We use the same technique as before. If we know an enclosing cube for the objects whose volume we want to calculate, then we can select random points in that volume and see how many fall within the objects. This gives an approximation for the volume occupied by the objects as a fraction of the enclosing volume.

Suppose we want to find the volume enclosed by two spheres, both of diameter 2 with centers at (0,0,0) and (1,1,1). These spheres overlap and so analytic calculation of the volume they enclose is nontrivial. However, a Monte Carlo integration can easily calculate an approximate answer.

Assuming a Euclidean coordinate system in three dimensions, we can write equations for the spheres as follows:

$$x^2 + y^2 + z^2 = 2$$
$$(x - 1)^2 + (y - 1)^2 + (z - 1)^2 = 2$$

A point with coordinates (x, y, z) is inside one of these spheres if the value produced by substituting its coordinates into these equations makes the left-hand side less than or equal to 2. We use this to determine if random points are within the sphere or not. The enclosing cube is one such that

$$-2 \leq x, y, z \leq 3$$

The next program calculates this integral.

```
*       Program 13.4
*       Calculates volumes using Monte Carlo integration
*
        program MonteC
        integer Inside, Outsid, Total
        real Volume, Totvol
        real X, Y, Z
        integer I
        call SetSd(time())
        Inside = 0
        Outsid = 0
        print *, ' Enter # of sample points'
        read *, Total
```

```
              Totvol = 5 * 5 * 5
              do 300 I = 1, Total
                   X = Random() * 5
                   Y = Random() * 5
                   Z = Random() * 5
                   if (Sph1(X, Y, Z).le.4.0.or.Sph2(X, Y, Z)
           &           .le.4.0) then
                        Inside = Inside + 1
                   else
                        Outsid = Outsid + 1
                   endif
        300   continue
              Volume = real(Inside) / Total * Totvol
              print '(A,F10.4)', ' Approximate volume is ', Volume
              stop
              end

              real function Sph1(X, Y, Z)
              real X, Y, Z
              Sph1 = X * X + Y * Y + Z * Z
              return
              end

              real function Sph2(X, Y, Z)
              real X, Y, Z
              Sph2 = (X-1) * (X-1) + (Y-1) * (Y-1) + (Z-1) * (Z-1)
              return
              end
```

13.6 Dynamic Simulations

We now turn to consider dynamic simulations—those in which time plays a major role. Many of the problems for which dynamic simulation is used are based on a small set of models, called queuing models. In fact, the mathematical analysis of these models forms a branch of mathematics called **queuing theory**.

Consider the following situation: a teller at a bank performs various services for customers who arrive at random times. At times customers who arrive will have to wait because the teller is busy with someone else. Those customers who are waiting form a queue.

This general situation of a server and a queue of requests for service occurs in many different engineering disciplines—wherever some facility provides a service to "customers." The most obvious example would be the replacement by an automatic teller of the service provided by a human teller at a bank.

The sorts of questions that are of interest in queuing systems are things like: how long does the queue get, how long does an average customer have to wait, how are these things affected by speeding up the server, and so on. Mathematical techniques can be applied for simple cases, but for more complex situations, simulation can be used.

If we are going to write simulation programs to model queuing systems, then we need to be able to deal with random arrivals. Usually we have some information about the average *rate* of arrival and we have to convert this into a *set* of random arrivals. The easiest way to do this is to convert the rate of arrival into a probability of arrival in some small time interval. For example, if the average arrival rate is one arrival every 10 seconds, then we can treat this as a probability of 0.1 that an arrival will occur in any second. Similarly, we could treat it as a probability of 0.01 of an arrival in 0.1 seconds and so on. Usually the granularity we choose depends on the rate itself and how much accuracy we require. Notice that this method allows for more than one arrival in a 10-second period, but this is all right as long as the average arrival rate stays fixed.

We could use the same idea to measure the amount of work that each customer brings with her. For example, each customer in a bank ties up a teller for a different length of time and the length of time that one customer takes is independent of the time that the next customer will take. If we know that a customer requires service for an average of 10 seconds, we can model this by deciding that the current customer will finish being served with probability 0.1 in any second. However, for simplicity we will assume that each customer requires the same amount of service.

Our simulation program will thus model arrivals with a specified probability. We maintain a count of how many customers are waiting in the queue. This count is incremented whenever a customer arrives and decremented every time a customer finishes being served. We must be careful to correctly handle the case of a customer who doesn't have to wait.

Let us suppose then that we have a single server, that requests for service appear on average every 10 seconds, and that each request will tie up the server for 9 seconds.

```
*       Program 13.5
*       A dynamic simulation of a server and its queue
*
        program Queue
        integer QLen, TotTim, Time, Ttogo
        real TotQL
        call SetSd(1024)
        print *, ' Enter total simulation time'
        read *, Tottim
        print '(I8)', Tottim
```

```
            QLen = 0
            Ttogo = 9
            TotQL = 0.0
            do 300 Time = 1, Tottim
                if (Random() .lt. 0.1) then
                    Qlen = Qlen + 1
                endif
                Ttogo = Ttogo - 1
                if (Ttogo.le.0.and.QLen.gt.0) then
                    QLen = QLen - 1
                    Ttogo = 9
                endif
                TotQL = TotQL + QLen
300         continue
            print '(A,F10.4)',' Average queue length', TotQL/Tottim
            stop
            end

*       Initializes a seed in a common block for use
*       by subsequent random number routines
*
        subroutine SetSd(X)
        integer X, Y, M
        common /Rand/ Y
        parameter (M = 233280)
        Y = mod(X, M)
        return
        end

*       Returns a random number in [0, 1)
*
        real function Random()
        integer Y, A, C, M
        common /Rand/ Y
        parameter (A = 9301, C = 49297, M = 233280)
        Y = A * Y + C
*       zero high order bit if necessary
        if (Y.lt.0) then
            Y = Y + 2147483647 + 1
        endif
        Y = mod(Y, M)
        Random = real(Y) / real(M)
        return
        end
```

The output from this program follows.

```
Enter total simulation time
  2000
Average queue length     1.4235
```

Notice how we have used the random number generator to generate a probability. The numbers produced by the random number generator are uniformly spread through the interval from 0 to 1. Therefore, a random number is smaller than 0.1 about 10 percent of the time. We have used a test for this to cause an arrival to occur when the generated random number is smaller than 0.1.

One of the reasons for running the simulation is to discover how the system behaves. To aid in this, we must collect appropriate statistics, measuring properties of the system that are of interest. In this program we have calculated the average length of the queue.

13.7 Playing Games

Although engineers are not often paid to play games, they are sometimes called on to design them. Some games have practical reasons for existing, other than simply entertainment. Some examples are simulators that are used to train pilots, astronauts, and air traffic controllers. These exist for the reasons that most simulation tools exist—because using real systems is too expensive or dangerous. Other examples are business and war game programs that are used to provide people with a feel for the way in which complex systems behave (some would say a dangerously unsophisticated feel). Games are also a useful and enjoyable environment in which to practice some of the programming techniques we have discussed. In particular, games are a good place to talk about the human interface to software. Programs should always be designed with the users in mind. This makes good economic sense because it's much more expensive to waste 5 minutes for each user while they figure out how to use the program than to spend an extra half hour making the user interface simple. Although this applies to all programs, it is especially relevant for games.

As an example, we'll consider the children's game of Snakes and Ladders. This is simple enough that we can give an easy implementation, but it is surprisingly more complex to build an interesting interface for it.

The game is played on a board with squares numbered 1 to 100 inclusive. A player starts with a token placed on square number 1. On each turn a single die is rolled, and the token is moved the indicated number of squares. Some squares are at the foot of a ladder; if the token lands on such a square it immediately advances to the top of the ladder. Some squares are at the tail of a snake; if the token lands on such a square it immediately moves to the head of the snake, which will be a square with a lower number. If the token is closer to square 100 than the number of squares indicated by the roll of the die, it is blocked and cannot move.

Suppose we wanted to determine the length of an average game of Snakes and Ladders, that is, the average number of moves taken by a single player. Since players do not interfere with each other, save in the alternation of turns, we can ignore other players. The simulation of the game will require a random die roller and a mechanism for representing the snakes and the ladders. A simple way to handle this latter aspect is to associate a slide target with each square. For most squares, the slide target is the square itself; this corresponds to having no slide associated with that square. But for some squares, the slide target is a larger number (these slides represent the ladders), and for some squares, the slide target is a smaller number (these slides represent the snakes).

To determine the average number of moves, we will write a program that reads a random number seed and a number of games to play, and then plays the specified number of games, keeping track of the number of moves made.

```
*       Program 13.6
*       Simulates the game of snakes and ladders
*
        program Snake1
        common /Rand/IR
        integer IR
        integer I, Games, Count, Total
        print *, ' Enter random seed and number of games:'
        read *, IR, Games
        Total = 0
        do 100 I = 1, Games
            call Snake(Count)
            Total = Total + Count
100     continue
        print *,' Total moves:',Total,' Average:',Total/Games
        stop
        end
*
*       Actually plays the game
*
        subroutine Snake(Count)
        integer Count
        common /Rand/IR
        integer Pos, Move, I, Slide (100)
        real X
*       most squares have no slide
        do 100 I=1,100
            Slide(I) = I
100     continue
*       ladders
        Slide( 3) = 37
        Slide( 6) = 16
        Slide(14) = 32
        Slide(27) = 56
        Slide(41) = 85
```

```
            Slide(69)  =  87
            Slide(79)  =  98
            Slide(89)  =  91
 *       snakes
            Slide(15)  =   9
            Slide(42)  =  17
            Slide(49)  =  12
            Slide(58)  =  45
            Slide(61)  =  22
            Slide(75)  =  47
            Slide(88)  =  36
            Slide(94)  =  64
            Slide(97)  =  65
            Pos  =  1
            Count  =  0
            while (Pos .ne. 100) do
                call Random(IR, X)
                Move = X * 6 + 1
                if (Move + Pos .le. 100) then
                    Pos = Slide (Move + Pos)
                endif
                Count = Count + 1
            end while
            return
            end
            subroutine Random(N, U)
            integer N
            real U
            N = N * 843314861 + 453816693
            if (N.lt.0) then
                N = N + 2147483647 + 1
            endif
            U = N * 0.4656612e-9
            return
            end
```

The random number generator used here takes an integer as an input parameter, generates the next integer in the series, and also generates a real value that is the random integer scaled to be in the range [0, 1), that is, greater than or equal to zero and less than one. Notice that by multiplying this value by 6, we get a number in the range [0, 6); then by adding 1, we get a number in the range [1, 7), and by truncating this to an integer, we get an integer in the range [1, 6]. This random integer is used to represent the value of the die. This scaling technique can be used to produce real numbers or integers in any desired range.

We have run this program with various random number seeds for over 100,000 simulated games and determined the average number of moves to be 45.

A variant of this game involves always letting the token move even though it's close to square 100; the moves in this range involve moving up to 100 and then turning back. Thus if the token is at 98 and the die comes up 5, the token moves forward 2 squares to 100 and then backward 3 squares to 97. It's straightforward to modify the program to do this.

```
*         Program 13.7
*         Simulates the game of snakes and ladders
*         over a number of games
*
          program Snake2
          common /Rand/IR
          integer IR
          integer I, Games, Count, Total
          print *, ' Enter random seed and number of games:'
          read *, IR, Games
          Total = 0
          do 100 I = 1, Games
              call Snake(Count)
              Total = Total + Count
  100     continue
          print *,' Total moves:',Total,' Average:',Total/Games
          stop
          end
*
*         Plays the game
*
          subroutine Snake(Count)
          integer Count
          common /Rand/IR
          integer Pos, Move, I, Slide (100)
          real X
*         most squares have no slide
          do 100 I=1,100
              Slide(I) = I
  100     continue
*         ladders
          Slide( 3) = 37
          Slide( 6) = 16
          Slide(14) = 32
          Slide(27) = 56
          Slide(41) = 85
          Slide(69) = 87
          Slide(79) = 98
          Slide(89) = 91
```

```
*     snakes
      Slide(15)  =   9
      Slide(42)  =  17
      Slide(49)  =  12
      Slide(58)  =  45
      Slide(61)  =  22
      Slide(75)  =  47
      Slide(88)  =  36
      Slide(94)  =  64
      Slide(97)  =  65
      Pos = 1
      Count = 0
      while (Pos .ne. 100) do
         call Random(IR, X)
         Move = X * 6 + 1
         if (Move + Pos .le. 100) then
             Pos = Slide(Move + Pos)
         else
             Pos = Slide(200 - (Pos + Move))
         endif
         Count = Count + 1
      end while
      return
      end

      subroutine Random(N, U)
      integer N
      real U
      N = N * 843314861 + 453816693
      if (N.lt.0) then
          N = N + 2147483647 + 1
      endif
      U = N * 0.4656612e-9
      return
      end
```

We have run this one with various seeds for a large number of games and it produces an average number of moves of 60.

Now, both these programs are quite straightforward, and the modification to produce the second from the first was accomplished by adding a single **else** clause. To say this another way, the basic algorithm representing the game is simple to implement and to modify. But now consider enhancing the program in a number of ways.

First, we could very simply modify it so that a single human player could play against the program. This isn't a very exciting thing to do, but it illustrates the general point we want to make. Now there will be two tokens to keep track of, and there should be a display of information to the human player on each move.

```
*       Program 13.8
*       Simulates playing snakes and ladders
*       against the user
*
        program Snake3
        common /Rand/IR
        integer IR
        print *, ' Enter random seed:'
        read *, IR
        call Snake
        stop
        end
*
*       Plays the game
*
        subroutine Snake
        common /Board/Slide
        integer PosMe, PosYou, I, Slide (100)
*       most squares have no slide
        do 100 I=1,100
            Slide(I) = I
  100   continue
*       ladders
        Slide( 3) = 37
        Slide( 6) = 16
        Slide(14) = 32
        Slide(27) = 56
        Slide(41) = 85
        Slide(69) = 87
        Slide(79) = 98
        Slide(89) = 91
*       snakes
        Slide(15) =  9
        Slide(42) = 17
        Slide(49) = 12
        Slide(58) = 45
        Slide(61) = 22
        Slide(75) = 47
        Slide(88) = 36
        Slide(94) = 64
        Slide(97) = 65
```

```
        PosMe = 1
        PosYou = 1
        while (PosMe .ne. 100 .and. PosYou .ne. 100) do
            call Turn('I  ', PosMe)
            if (PosMe .ne. 100) then
                call Turn('You', PosYou)
            endif
        end while
        if (PosMe .eq. 100) then
            print *, 'I won'
        else
            print *, 'You won'
        endif
        return
        end
*
*
*       This subroutine plays a move for the user or the computer
*
        subroutine Turn(Who, Where)
        character*3 Who
        integer Where
        common /Rand/IR
        integer IR
        common /Board/Slide
        integer Slide(100)
        integer Move
        real X
        call Random(IR, X)
        Move = X * 6 + 1
        if (Move + Where .le. 100) then
            Where = Move + Where
            print *, Who, ' rolled', Move
            if (Slide(Where) .lt. Where) then
                print *, ' A snake!'
            else if (Slide(Where) .gt. Where) then
                print *, ' A ladder!'
            endif
            Where = Slide(Where)
            print *, ' New position is', Where
        else
            print *, Who, ' rolled', Move, ' and cannot move'
        endif
        return
        end
```

```
subroutine Random(N, U)
integer N
real U
N = N * 843314861 + 453816693
if (N.lt.0) then
    N = N + 2147483647 + 1
endif
U = N * 0.4656612e-9
return
end
```

Here's a sample of output from this program. You'll see that it communicates everything that needs to be known, but the display of information—although simple to program—is not attractive.

```
Enter random seed:
I  rolled        5
A ladder!
New position is        16
You rolled        2
A ladder!
New position is        37
I  rolled        4
New position is        20
You rolled        4
A ladder!
New position is        85
I  rolled        1
New position is        21
You rolled        5
New position is        90
I  rolled        4
New position is        25
You rolled        4
A snake!
New position is        64
I  rolled        4
New position is        29
You rolled        6
New position is        70
I  rolled        5
New position is        34
You rolled        3
New position is        73
I  rolled        5
New position is        39
```

Models, Simulation, and Games

```
 You rolled         5
  New position is       78
 I  rolled         3
 A snake!
  New position is       17
 You rolled         6
  New position is       84
 I  rolled         4
  New position is       21
 You rolled         2
  New position is       86
 I  rolled         5
  New position is       26
 You rolled         1
  New position is       87
 I  rolled         3
  New position is       29
 You rolled         5
  New position is       92
 I  rolled         3
  New position is       32
 You rolled         6
  New position is       98
 I  rolled         2
  New position is       34
 You rolled         4 and cannot move
 I  rolled         6
  New position is       40
 You rolled         2
  New position is       100
 You won
```

To improve the output, think about displaying the board grid on the screen. The tokens, snakes, and ladders would have to be represented. To make the game visually appealing, it would be good to actually see the tokens move after the die is rolled, and to slide up ladders or down snakes. While this all sounds simple, if you think about it (see the exercises), you'll find that it's not.

What is true of this simple application is true in general. The basic algorithms are straightforward for many of the simulations for which computers can be truly valuable. At least, they are considerably simpler than the programming required to support a smooth and powerful user interaction.

We'll end this chapter with one more example of a game. This time it's a simple card game, a type of solitaire. The basic idea of the game is that the player holds the deck of cards and deals one card each onto 9 piles, face up. If there is a pair showing, two more cards are played on top of the pair. The player wins if all the cards are played.

This program illustrates a couple of interesting things. The shuffling of the cards is accomplished by generating 52 random numbers and associating one with each card. The random numbers are then sorted into order, but the association between the numbers and the cards is maintained; that is, when the numbers are finally ordered, they are still paired with the cards, but now the cards appear to be in a random order.

The program also uses a logical variable to assist in determining if a pair of cards of the same denomination is visible on the piles.

```
*       Program 13.9
*       Plays solitaire
*
        program Sol
        character*2 Card(52)
        integer Seed, I,N,Remain,Win,LoseBy
        read *, Seed, N
        call Fresh(Card)
        Win = 0
        LoseBy = 0
        do 100 I=1,N
            call Shuf(Card, Seed)
            call SimSol(Card,Remain)
            if (Remain .eq. 1) then
                Win = Win + 1
            else
                LoseBy = LoseBy + Remain
            endif
100     continue
        print *, ' Games played:', N
        print *, '         won:   ', Win
        print *, '         lost:  ', N-Win
        if (N .gt. Win) then
            print *,' Average left:',(1.*Loseby)/(N-Win)
        endif
        stop
        end
*
*       This subroutine shuffles the deck of cards randomly
*
        subroutine Shuf(Deck, Seed)
        character*2 Deck(52)
        integer Seed
        integer Order(52), I, J, T
        character*2 S
        real X
```

Models, Simulation, and Games

```
      do 100 I=1,52
          call Random (Seed, X)
          Order (I) = Seed
100   continue
      do 200 I=52,2,-1
          do 300 J=1,I-1
              if (Order(J).gt.Order(J+1)) then
                  T = Order(J)
                  Order(J) = Order(J+1)
                  Order(J+1) = T
                  S = Deck(J)
                  Deck(J) = Deck(J+1)
                  Deck(J+1) = S
              endif
300       continue
200   continue
      return
      end
*
*     This subroutine produces a sorted deck
*
      subroutine Fresh(Deck)
      character*2 Deck(52)
      character*4 Suit
      character*13 Which
      integer I,J
      Suit = 'CDHS'
      Which = 'A234567890JQK'
      do 100 I=1,4
          do 200 J=1,13
              Deck((I-1)*13+J) = Suit(I:I)//Which(J:J)
200       continue
100   continue
      return
      end
*
*     This subroutine plays solitaire
*
      subroutine SimSol(Deck, Left)
      character*2 Deck(52)
      integer Left
      character*2 Pile(9)
      integer Next, I, J
      logical Pair
```

```
      do 100 I=1,9
         Pile(I)  =  Deck(I)
100   continue
      Next  =  10
      Pair  =  .true.
      while (Pair .and. Next .lt. 52) do
         Pair  =  .false.
         I  =  1
         while (I .lt. 9 .and. .not. Pair) do
            J  =  I + 1
            while ((J .le. 9) .and. .not. Pair) do
               if (Pile(I)(2:2) .eq. Pile(J)(2:2)) then
                  Pile(I)  =  Deck(Next)
                  Pile(J)  =  Deck(Next+1)
                  Next  =  Next + 2
                  Pair  =  .true.
               else
                  J  =  J + 1
               endif
            end while
            I  =  I + 1
         end while
      end while
      Left  =  52 - Next + 1
      return
      end

      subroutine Random(N, U)
      integer N
      real U
      N  =  N * 843314861 + 453816693
      if (N.lt.0) then
         N  =  N + 2147483647 + 1
      endif
      U  =  N * 0.4656612e-9
      return
      end
```

This program was run with a particular random number seed and produced this set of output lines.

```
Games played:         100
       won:            29
      lost:            71
Average left:          30.9718170
```

Programming Example

Problem Statement

Write a program to play blackjack. The player is dealt one card that the dealer does not see. The player is then dealt further cards, face up, until he or she decides to stop. If the player's total exceeds 21 (including the hidden card), then the player busts and the game is over. If not, then the dealer deals cards to himself, face up, for as long as he wishes. When the dealer sticks, the player's hidden card is turned face up. The winner is the person closest to, but not exceeding, 21. The dealer works on the basis of knowledge of all but one of the player's cards.

Inputs

The player enters a stake for each game. The player also decides when to stop.

Outputs

The game keeps the player informed about total amount of money, cards dealt by both sides, and decisions of the dealer.

Discussion

We will not implement the more esoteric rules of blackjack.

Program

```
*       Program 13.10
*       Plays a version of blackjack
*
        program Black
        integer Total, Stake, Seed
        logical Win
*       This sets the initial seed
        print *, 'Enter a number between 1 and 1000'
        read *, Seed
        call SetSd(Seed)
*       Starting stake is 24 beers
        Total = 24
        print '(A, I4, A)', 'You have ', Total, ' Beers'
```

```
      while (Total.gt.0) do
         print *, 'Place your bets'
         read *, Stake
         if (Stake.le.Total) then
            if (Win()) then
               Total = Total + Stake
            else
               Total = Total - Stake
            endif
         else
            print *, 'You don''t have that much'
         endif
         print '(A, I5, A)', 'You have', Total, ' Beers'
      end while
      stop
      end
*
*     This function plays one game and determines if the player won
*
      logical function Win()
      integer Usrtot, Dlrtot, Init
      logical UPlay, DPlay, UBust
      character*1 C
      Usrtot = 0
      UPlay = .true.
      UBust = .false.
      print *, 'Player plays'
      print *, 'First card:'
      call Move(Usrtot)
*     Init holds value of hidden card
      Init = Usrtot
      while (UPlay) do
         call Move(Usrtot)
         print '(A, I5)', 'Your total is now', Usrtot
         if (Usrtot.gt.21) then
            print *, '****Player busts****'
            UBust = .true.
            UPLay = .false.
         else
            print *, 'Want to continue?'
            read '(A1)', C
            if (C.eq.'n') then
               UPLay = .false.
            endif
         endif
      end while
```

```
*       If player didn't bust, dealer plays
        Dlrtot = 0
        if (.not.UBust) then
            print *, 'Dealer plays'
            print *
*           Dealer plays til 16 and to exceed visible player sum
            while (Dlrtot.le.15.or.Dlrtot.lt.Usrtot-Init) do
                call Move(Dlrtot)
                print '(A, I5)', 'Dealer total is', Dlrtot
            end while
        end if
*       dealer wins if he got more than player but not more than 21
*       or if player busts
        if (Dlrtot.ge.Usrtot.and.Dlrtot.le.21.or.UBust) then
            print *, '****Dealer wins****'
            Win = .false.
        else
            print *, '****You win****'
            Win = .true.
        endif
        return
        end

*
*       This subroutine makes a move for a player
*
        subroutine Move(N)
        integer N
        integer Card, Deal
*       Aceflg signals if card was ace. Can count as 1 or 11
        logical Aceflg
        Card = Deal(Aceflg)
        N = N + Card
*       If counting ace as 10 will reach 21 then do so
        if (N.eq.11.and.Aceflg) then
            N = 21
        endif
        return
        end
```

```
*
*       This function deals a card
*
        integer function Deal(Flg)
        logical Flg
        integer Suit, Card
        Flg = .false.
        Suit = random() * 4 + 1
        Card = random() * 13 + 1
        if (Suit.eq.1) then
            print '(I3, A)', Card, ' of Spades'
        elseif (Suit.eq.2) then
            print '(I3, A)', Card, ' of Hearts'
        elseif (Suit.eq.3) then
            print '(I3, A)', Card, ' of Diamonds'
        else
            print '(I3, A)', Card, ' of Clubs'
        endif
        if (Card.eq.1) then
            Flg = .true.
        endif
        if (Card.gt.10) then
            Card = 10
        endif
        Deal = Card
        return
        end
*
*       This subroutine initializes a seed in a common
*       block for use by subsequent random number routines
*
        subroutine SetSd(X)
        integer X, Y, M
        common /Rand/ Y
        parameter (M = 233280)
        Y = mod(X, M)
        return
        end
```

```
*       Returns a random number in [0, 1)
*
        real function Random()
        integer Y, A, C, M
        common /Rand/ Y
        parameter (A = 9301, C = 49297, M = 233280)
        Y = A * Y + C
*       zero high order bit if necessary
        if (Y.lt.0) then
            Y = Y + 2147483647 + 1
        endif
        Y = mod(Y, M)
        Random = real(Y) / real(M)
        return
        end
```

Testing

```
Enter a number between 1 and 1000
You have    24 Beers
Place your bets
Player plays
First card:
 1 of Spades
10 of Clubs
Your total is now     11
Want to continue?
 5 of Diamonds
Your total is now     16
Want to continue?
Dealer plays

 8 of Spades
Dealer total is      8
10 of Clubs
Dealer total is     18
****Dealer wins****
You have     0 Beers
```

Design, Testing, and Debugging

- Simulation models and programs to implement them should always be validated. Try to run your simulation program in a situation that's already well understood and see how well it compares to what you expect. If it doesn't agree in the known application, then no one will trust it in a new and unknown situation.
- Simulations are particularly susceptible to edge effects—caused by the time it takes for an artificially started situation to settle into a more typical one. Simulations should always be run for longer than you think they need to be to limit these effects.

Style and Presentation

- Good graphical output is important in simulation and in games. Investigate your machine and see what kinds of tools you have available for graphical display. Use these tools where they are appropriate.

Chapter Summary

- Simulations are programs that behave as some system in the real world would behave. Static simulations do not have a time component. Dynamic simulations do.
- Simulations rely on the generation of random numbers. It is very hard to find processes that are truly random. Most computer programs actually generate pseudo-random numbers that are produced using a recurrence.
- The most popular recurrence technique for generating pseudo-random numbers is called linear congruential generation. This uses a recurrence that multiplies the previous value by a large constant, adds another large constant, and reduces the result modulo another large number.
- Monte Carlo integration is an algorithm that estimates the area under a curve by counting the number of randomly generated points that lie under the curve as a fraction of the total number of points. It can be easily extended to higher-dimensional spaces.
- Games are really a special form of simulation, and often have a serious purpose. The quality of the user interface may be the most important factor in the design of a game.

Define These Concepts and Terms

Models
Prototypes
Static simulation
Dynamic simulation

Random numbers
Linear congruential generator
Monte Carlo integration
Queuing theory

Exercises

1. Modify the fish-catching program to use your system's random number generator. Run the program to see how good this sampling method is.
2. Try writing your own linear congruential generator. Evaluate its effectiveness by producing a long sequence of pseudo-random numbers and calculating their mean. What other tests could you apply to try and detect patterns in the pseudo-random numbers?
3. Use Monte Carlo integration to calculate the total volume enclosed by a hollow sphere with thickness 6 inches and inner radius 25 inches, overlapped by a solid sphere with radius 7 inches and centered at some arbitrary point on the inner surface of the hollow sphere.
4. Implement the terminating condition suggested in Section 13.5 using convergence of the estimated area and compare it with the version in that section.
5. Run the dynamic simulation program in Section 13.6, varying the amount of service required by each arrival from 1 to 9 units. Calculate the average queue length in each case and plot them.
6. Run the first two versions of the snakes and ladders program to verify the reported results. Assuming the number 45 for the first version is correct, can you argue convincingly that the simple change to the second version should add 1/3 to the number of moves on average?
7. Modify the final version of the snakes and ladders program to display the board after each move. That is, don't try to show the tokens moving, just their final positions.
8. Write a subprogram to simulate rolling two dice. Use this program to simulate 10,000 rolls and tabulate the results. Can you explain them?
9. Modify the result of Exercise 7 to also show the tokens as they move and slide.
10. Write a program to play some other type of solitaire.
11. Could the snakes and ladders style of simulation be extended to encompass other board games that you know? Could it be extended to checkers in any sensible way? Backgammon? Chess?
12. Write a program to simulate tossing a coin. In 10,000 simulated tosses, what is the longest run of equal results (either heads or tails)? Try several different random number seeds. Does the value vary much as the seeds are changed?

13. Check that the random number generator described in the chapter performs as expected. Compute the mean and standard deviation of a long sequence of random numbers. Graph the output as a frequency histogram.
14. Write a program that includes the queue simulation of Program 13.5 combined with the plotting of Program 9.22 (or a modified version if you have written one). Run the simulation with a range of expected service times (that is, departure probabilities) and plot the relationship of average queue length to service time.
15. Modify the blackjack program to call a subprogram to decide whether to take another card. The present version can be recast using a subprogram that interactively queries the user. Write another subroutine that implements a simple strategy of taking another card if the score is below a threshold. Let the threshold vary over what you consider to be a reasonable range. Run 100 experiments (that is, with 100 different random seeds) for each threshold value. If you were to adopt this simple strategy in practice, what threshold value do your experiments indicate you should use?

Engineering Problem 16

Detecting Transmission Errors

When messages are transmitted through networks, errors can occur. These arise because of faults in equipment, atmospheric conditions, or other environmental factors. The effect is that the characters that are transmitted are not necessarily the same as those that are received. To allow the receiver of the message to determine whether errors have occurred, some additional information must be transmitted.

We could, for example, think of transmitting each message twice, and then the receiver could compare them; if they are the same, the transmission has occurred correctly; if they are different, the pair of messages could be retransmitted. This is a very high price to pay; when the communication system is working without errors, it is also working at half its theoretical capacity.

The usual approach is to send some redundant information to allow checking for transmission errors, but not as much as resending the entire message.

In this problem we investigate the effectiveness of a particular scheme for checking transmissions for errors. For every k characters transmitted, an additional checking character will be transmitted. The check character is produced as follows. As we transmit the first character of a sequence of k characters, we copy that character into the check character. For each subsequent character transmitted, we interpret the check character and the character to be transmitted as 8-bit integers and multiply them, producing (in general) a 16-bit result; the middle 8 bits are retained as the check character. For example, if the current check character had the binary representation 10000100 (decimal value 132) and the current character to transmit had the binary representation 00011011 (decimal value 27), the product would have the binary representation 0000110111101100 (decimal value 3564) and the middle eight bits are 11011110 (decimal value 222), which would be retained as the new check character. After k characters have been transmitted, the check character is transmitted.

We will not actually transmit the characters. Instead, we will construct random sequences of characters that we will consider to be messages. A check character will be calculated for the messages. We will then introduce random errors into the messages and recalculate the check character. If the new check character differs from the old one, the method we are using has detected the error.

Programming Problems

1. Write a subprogram that will randomly change one bit in a given character.
2. Write a function that will compute a check character on a sequence of k characters, using the method described previously.
3. Write a subprogram that will randomly modify a sequence of characters, given a value that is the probability of introducing an error in a single character.
4. Write a subprogram that will generate a random sequence of characters.
5. Combine all these pieces into a program that will read a probability (the probability that a transmitted character will contain an error), a value for k, and a number of trials. It should then construct the specified number of random strings of length k, introduce errors at random as determined by the given probability, and count how many errors are introduced and how many are detected.
6. Use your program to investigate the effectiveness of this method of checking transmissions over a range of probabilities and a range of k values.

Engineering Problem 17

Placing Chips onto Circuit Boards

Large pieces of electronics such as computers are made up of silicon chips packaged in plastic or ceramic carriers. These chips have pins that are used to connect them to other chips, and are placed on circuit boards, which then sit in racks sharing power supplies.

A typical large computer has many thousands of chips placed in as many as twenty boards. The number of pins per chip can vary from around twenty up to several hundred.

Chips that are placed on the same circuit board can be connected together by metal paths that are etched onto the surface of the board. Chips that need to be connected but are on different boards have to be connected by a much longer wire that leaves one board and travels to the other. This greatly increases the expense of the connection because the delay caused by the longer wire can create a bottleneck for the system and also because the wiring itself is expensive and hard to do.

This creates a problem for the system designer who may know which chips are required and how they are logically connected, but must come up with an implementation that divides the chips up among boards. This must be done in such a way that the number of off-board connections is minimized.

The trivial solution that eliminates off-board connections is to place all of the chips onto a single board. This is impractical because the amount of space on a board is limited. The minimization problem therefore has two constraints: keeping the number of off-board connections small, and keeping the number of chips within the capacity of each board.

We can represent the logical connections between the chips using a graph. The vertices, or nodes, of the graph represent the chips and the edges represent the connections required between them. We can label each edge with a value (its weight) that represents the cost of placing the chips connected by that edge on different boards. The graph can be represented by a cost matrix. If there are N chips, then the cost matrix is an N by N matrix in which entry (i, j) is 0 if there is no connection between chip i and chip j; it has the value c if there is a connection between i and j and the cost of separating i and j is c.

If we are given an adjacency matrix for a particular system and we know how many chips can be placed on a single board (say M), then we can randomly allocate chips to boards as an initial configuration. We can compute the cost of such a configuration using the adjacency matrix. For each pair (i, j) we add the cost of their connecting edge if our allocation places them on distinct boards. If they are on the same board, we assume that the cost of connecting them is zero.

One way to construct a reasonable configuration is to begin with a randomly chosen configuration and use a *greedy algorithm*. Such an algorithm makes a random perturbation in the initial configuration (such as moving a chip from one board to another) and calculates the new cost. If the new cost is smaller, then the new configuration is accepted and becomes the starting configuration for the next move. If the new cost is larger than the old one, then the perturbation is rejected and a new perturbation to the old configuration is tried. When some number of perturbations in a row have not produced any cost improvement, the algorithm terminates.

If we think of the cost space as a function of the configuration, then we can see that this algorithm is really making moves in the configuration space (the search space) and assessing their effect on the cost function. If the cost function decreases, then a move is made. Therefore it always moves "downhill" in the search space. If the cost function has local minima, then this algorithm will become trapped in them and so will not always find the global smallest cost. (In the next problem we will consider ways to avoid this difficulty.)

Programming Problems

1. Write a procedure to read a cost matrix and the number of chips allowed per board and generate a random starting configuration.
2. Include the procedure in a program that uses the greedy algorithm to find a good configuration.
3. Run the program using different seeds to see how the best answer found depends on the initial configuration. What is the ratio of best answer to worst? Do answers improve if you change the terminating condition to allow longer stretches without improvement before terminating?
4. Suggest easy ways to improve this algorithm.

Use $M = 2$. We next show a sample input cost matrix.

```
0  1  3  5  3  1  2
1  0  2  3  5  8  7
3  2  0  2  1  5  3
5  3  2  0  2  3  3
3  5  1  2  0  5  6
1  8  5  3  5  0  4
2  7  3  3  6  4  0
```

Generate other cost matrices of your own and test your program with them. Can you find types of cost matrices that seem particularly difficult to partition?

Engineering Problem 18

Simulated Annealing

This problem is based on the chip layout minimization problem (Problem 17). We want to improve the greedy algorithm to avoid getting trapped in local minima. We would like to be able to jump out of local minima so that we have a better chance of finding the global minimum.

One way to allow this is to accept perturbations that increase the cost with some nonzero probability. This means that even if we get caught in a local minimum there is a possibility of getting out of it by a sequence of moves that (temporarily) increase the cost. The difficulty is to do this in a controlled way.

One important approach to doing this is called simulated annealing. The algorithm works by analogy with physical annealing. When a solid is heated it melts. If it is cooled again very quickly (quenched), then the resulting solid has many imperfections caused by the fact that the atoms didn't have time to arrange themselves into regular (and hence low energy) arrangements. On the other hand, if the liquid is cooled very slowly, particularly around the temperature at which it will solidify, then a regular structure with low energy tends to develop.

Simulated annealing uses the same approach to minimization problems. We begin with an initial configuration chosen randomly as before. This configuration is perturbed, and changes that reduce the cost are always accepted. But now some changes that increase the cost are also accepted. When the (simulated) temperature is high, perturbations that increase the cost by a large amount will be accepted. As the temperature is reduced, smaller and smaller positive cost changes are accepted, until eventually only negative cost changes are permitted. This corresponds to the freezing point. The rationale is that if the temperature is reduced very slowly around the freezing point, the cost function will tend to settle into a global minimum. Local minima found early in the search will probably be discarded by later cost-increasing perturbations.

The algorithm works as follows: an initial allocation of chips to boards is created randomly as before. A small perturbation is generated and the change in the cost function (ΔC) is calculated. If ΔC is negative, then the change is accepted. If ΔC is positive, then it is accepted with probability

$$P = e^{-\Delta C/T}$$

where T is the simulated temperature.

The temperature starts off with a value such that almost all perturbations are accepted. It is slowly reduced according to some annealing schedule, often requiring a fixed number of perturbations to be tried at each temperature.

Programming Problems

1. Write a program to add the simulated annealing technique to the minimization program written previously.
2. Investigate a suitable initial temperature for the program based on a given adjacency matrix.
3. Try several different annealing schedules. Are some types better than others?
4. Investigate how to determine when "freezing" is occurring.

Engineering Problem 19

Neutron Scattering

Many nuclear processes (including bombs) produce neutrons. In this problem we assess the amount of shielding needed to provide protection against neutrons.

Consider a wall of thickness L made of a shielding substance. Neutrons have no charge and thus are unaffected by the electrical fields of the shield as they pass through it. They are, however, deflected when they suffer collisions with the nuclei of the atoms in the wall. These collisions can be modeled as elastic collisions with fixed objects. As a result of a collision, a neutron is scattered onto a new path in a random direction.

We wish to simulate the passage of neutrons by having neutrons strike the inside face of the wall at random angles, travel through the wall until they collide with a nucleus, follow a new path until they collide with another nucleus, and so on. Eventually a neutron either escapes the wall (to the outside or back to the inside) or is absorbed. We will assume that neutrons are absorbed in their tenth collision (somewhat unrealistically).

We need to be able to calculate how long, on average, a neutron will travel before being involved in a collision. This is done by considering the *mean free path*, which is defined to be

$$MFP = \frac{1}{n\sigma}$$

where n is the number of molecules per unit volume in the target substance and σ is the *microscopic collision cross section*, which depends on the radii of the target and entering objects. We can calculate the number of molecules per unit volume from the density and the atomic weight. For lead, n is 3.401×10^{28} molecules/m³ and for water is 3.444×10^{28} molecules/m³.

The value of σ can be calculated as

$$\sigma = \pi(r_t + r_n)^2$$

where r_t is the radius of the molecules of the target and r_n is the radius of a neutron. The radius of a neutron is about 10^{-14} m, whereas the radius of a water molecule is about 10^{-10} m and the radius of a lead atom is about 5×10^{-14} m.

Programming Problems

1. Calculate the mean free path of neutrons in water and lead.
2. Write a program to simulate the approach of neutrons to a wall of given thickness and mean free path. You should assume that neutrons approach the wall from the inside at all possible angles with equal probability. Also assume that the collision diverts a neutron to a new path whose angle with the old path is uniformly chosen from the range $[0, 2\pi)$. Keep track of the penetration depth of each neutron and print the fraction of neutrons (a) absorbed, (b) passing through, and (c) returning to the interior. Run the program for enough neutrons that the fractions are stable.
3. Investigate how the fraction of neutrons that pass through the wall depends on the wall's thickness. Graph the function that results. How thick do walls of lead and water have to be to contain 99 percent of all neutrons?

Computing with Other Numeric Types

14.1 Double Precision Variables
14.2 Complex Numbers

14

14.1 Double Precision Variables

We have already seen that round-off can be a serious problem in real world computations. In situations where the number of significant figures that can be retained using **real** variables is too small to be useful, Fortran has a type called *double precision*. As its name implies, **double precision** variables allow approximately twice as many significant figures to be retained during computations. We say that the computation is being done with extra precision. Variables of this type occupy twice as much memory as real variables, allowing extra accuracy.

Although **double precision** variables allow about twice the accuracy, it takes longer to manipulate them on most machines. Adding two **double precision** variables will probably take twice as long as adding two real variables and multiplication will take even longer. So **double precision** is not always used—there is a trade-off between accuracy and speed.

Variables of type **double precision** are declared like this

   double precision X, Y, Z(10)

and referred to in programs by their names just like variables of other types.

When expressions contain both **double precision** and **real** variables, the evaluation behaves according to the same rules as for expressions containing mixtures of real and integer variables. If the expression contains real values, then they are converted automatically to **double precision** before any calculations take place. These calculations are done as **double precision** calculations and, if necessary, are converted back to **real** values before being assigned. For instance, if we have these declarations

   real X, Y
   double precision Z, T

then the expression

   X = Y * Z / T

is calculated by converting the value of Y to the equivalent **double precision** representation, performing the multiplication to give a **double precision** value, and then performing the division giving another **double precision** value. Finally, this value is converted to type real (by truncating any significant figures beyond those that could be represented by the real representation) and assigned to X.

Constants in double precision expressions are converted in a similar way, as needed. If required, a constant can be specified as **double precision** using the exponent form, but written with a D instead of an E.

   5.1278588214683D-19

The format code that is used for input and output is of the form

   D*width*.*fractionWidth*

where the values can be large enough to allow for up to 14 significant figures. The same rules apply as for E format codes—space must be left for the exponent, signs, and leading zeros.

In all other respects, **double precision** variables behave as **real** variables. In particular, a function can be of type **double precision**. Many of the examples in earlier chapters can be trivially rewritten to use **double precision**. We suggest some appropriate ones in the exercises at the end of this chapter.

14.2   Complex Numbers

You are probably familiar with the idea of a **complex number**. A complex number has two parts: a real part and an imaginary part. Complex numbers were originally defined in the context of finding roots of equations. For example, a quadratic may have two (real) roots, only one, or possibly none. However, every quadratic equation has exactly two roots if imaginary roots are allowed. It turns out that complex numbers are very useful in electrical engineering, and this section is probably more important for you if you are going to be an electrical engineer than if you are specializing in some other area.

When a complex number is written in ordinary mathematics, it is usually done in terms of its real and imaginary values as $2.3 + i5.6$, where the value 2.3 is the real part and the 5.6 the imaginary part. Of course, $i^2 = -1$. Unfortunately, the symbol i is used in electrical engineering for current and so electrical engineers write complex numbers as $2.3 + j5.6$ and we will accordingly adhere to this notation.

Fortran provides a built-in type to allow complex numbers to be represented. It is called type **complex**. A variable is declared to be of type **complex** by a declaration of the form

   complex l

A **complex** variable is represented internally by keeping its real and imaginary parts as real values.

When we wish to write a complex constant such as the one in the preceding example, then we specify the real and imaginary part in parentheses, like this

   l = (2.3, 5.6)

The rules for operations on complex numbers are slightly more complicated than for real numbers, and Fortran understands them. Thus the definition of addition and multiplication for complex numbers is

$$(a + jb) + (c + jd) = (a + b) + j(c + d)$$
$$(a + jb) \times (c + jd) = (ac - bd) + j(bc + ad)$$

If we multiply two **complex** variables, then the result agrees with these definitions.

Complex variables can be multiplied, divided, added, and subtracted. However, it is not legal to raise a **complex** value to a **complex** or **real** power (using the exponentiation operator). A **complex** value may be raised to an integral power. Expressions involving complex numbers may also include **real** and **integer** values. As you would expect, expressions are converted to type **complex** before operations are carried out. Thus the expression

```
C = C + 3.0
```

where C is a **complex** variable, is exactly equivalent to

```
C = C + (3.0, 0.0)
```

For addition and subtraction, therefore, a real value will only affect the real part of the **complex** variable.

Input and output of **complex** values is handled using E or F format codes, as we have already seen them used. Each **complex** value, however, requires *two* format codes.

Complex values cannot be compared using any of the relational operators such as .le., .gt. and so on.

There are two functions that allow explicit conversion between **complex** values and pairs of **real** values. The function

```
Z = cmplx(X, Y)
```

takes as operands two **real** values (in this case X and Y) and "packs" them together to form the real and imaginary parts of a complex number. The functions

```
X = real(Z)
Y = aimag(Z)
```

take a **complex** value and return its real and imaginary parts as **real** values. (If you are wondering why the name aimag is used, see Appendix C under the discussion of the **implicit** statement.)

Functions can be of type **complex** (so that they actually do return more than one value, although we think of it as a single one).

Programming Example

Problem Statement

Write a program to evaluate the exponential function using double precision to get as much accuracy as possible.

Inputs

A value at which the exponential function is to be calculated.

Outputs

The exponential at the input value.

Program

```
*       Program 14.1
*       Evaluates the exponential function using the
*       MacLaurin series expansion. It terminates when
*       the term being added is small. Double precision
*       is used for accuracy
*
        program  Dexp2
        double precision X, Sum, Term
        integer I
        Sum = 1.0D0
        write (*, 100)
100     format ('Enter X value')
        read (*, 200) X
200     format(D14.6)
        Term = X
        I = 1
        while (abs(Term).gt.1.0d-9 .and. I.le.100) do
            I = I + 1
            Sum = Sum + Term
            Term = Term * X / I
        end while
        write (*, 300) X, Sum
300     format('Value of exp at ', D12.5,' is ', D14.7)
        stop
        end
```

Testing

```
Enter X value
Value of exp at   0.10000D+01 is   0.2718282D+01
```

Design, Testing, and Debugging

- Using double precision always raises difficult issues in the trade-off between increased accuracy and increased execution time. Think about these trade-offs before writing programs that use double precision.
- Be aware of the behavior of expressions involving real (single precision) and double precision variables and constants. Make sure that you do not lose accuracy through thoughtlessly mixing the two types.

Style and Presentation

- When printing complex values, try to indicate what is going on by delineating the real and complex parts.

Fortran Statement Summary

Declaration Statements

The **double precision** and **complex** declarations declare variables of the appropriate type, just like other declarations.

> **complex** *variableList*
> **double precision** *variableList*

> **complex** I, Current
> **double precision** X, Y

Chapter Summary

- The **double precision** type uses about twice as much memory to hold each floating point value and therefore represents about twice as many significant figures. This type should be used when accuracy is critical or when round-off error is a problem. However, operations on **double precision** variables take longer on most machines.
- **Double precision** constants are written as floating point constants except that the symbol "D" is used is place of "E", and more significant figures can be used. The D format code controls the input and output of **double precision** values.
- The type **complex** represents complex numbers. It is implemented using pairs of floating point values. However, ordinary arithmetic operators will operate correctly on **complex** variables.

Define These Concepts and Terms

Precision

Complex number

Exercises

1. Write a program to find the roots of a quadratic under all circumstances. The output should print the roots in as readable a manner as possible.
2. Write a function that computes a double precision value for cos x.
3. The *modulus* of a complex number represents its magnitude in some sense. In particular, if a complex number is drawn on an *Argand diagram*, with a horizontal real axis and vertical imaginary axis, then the modulus represents the length of the vector from the origin to the complex number. For a number $x + jy$, the modulus is given by $x^2 + y^2$. The *argument* of a complex number is the angle it makes with the positive real axis in an Argand diagram. It is therefore a value between 0 and 2π. Write a program to read complex numbers and calculate their moduli and arguments.
4. Compute π as accurately as you can in double precision.
5. Modify Program 8.14 to use double precision. Compute the integral of the function as accurately as it is sensible to do, considering running time.

Engineering Problem 20

Fractal Geometry

This problem is a departure from the others in that it is not, strictly speaking, an application of computing technology to a real world engineering problem. It is, though, based on a concept that is increasingly being used in building models—especially graphical models—of natural and artificial processes.

Fractal sets are useful in many settings. One of the most commonly seen and easiest to understand is their use in creating computer-generated images. One of the best-known fractal sets is the Mandelbrot set, named after the mathematician Benoit Mandelbrot. It is a set of points in the complex plane.

The set is defined by a simple complex function of a complex value. The function is applied at a point in the plane, yielding a new value. The function is then applied to this new value, yielding yet another value. At each stage the absolute value of the complex result of the function is computed, to be used in the terminating condition of an indefinite iteration. (The absolute value of a complex number is the distance from the origin to the point represented by the number.)

The generating function has the property that arguments with large absolute value (larger than 2) will quickly grow to have much larger absolute value. Think of this as meaning that the function will map values that are some distance away from the origin to points that are further away from the origin with each iteration. These points are outside the Mandelbrot set. By contrast, points inside the set do *not* move arbitrarily far away from the origin as the function is repeatedly applied.

The generating function is repeatedly applied until either a result is found that has a large absolute value, or some maximum number of applications has been made. The interpretation of this latter outcome is that the point is inside the Mandelbrot set. In fact, the determination is only a probable one near the boundaries of the set, but our confidence increases as the number of function applications increases. To say this another way, the number of applications in some sense corresponds to how far outside the set a point is.

Most of the representations of the Mandelbrot set that are shown not only indicate the points that are inside the set, but also show the number of function applications that were required to determine that the point had achieved a large absolute value. There is beautiful structure and symmetry observable near the boundary between the points that are in the set and those that are not in the set when different values are uniformly mapped to different colors. In this problem, we will first work with textual, as opposed to graphical, output. Even though it is extremely limited, it can show some interesting aspects of the set.

Programming Problems

1. Use the function $f(z) = z^2 + c$ to investigate the extent of the Mandelbrot set. Write a program that will repeatedly apply this function at points in a square area in the complex plane. Define a square matrix that has 30 rows and 30 columns, corresponding to a grid of points at which the computation will be performed, to hold the number of applications of the function at each point. The program should read in coordinates of a corner of the square in the plane, the length of the side of the square, and the maximum number of function applications to attempt at each point. Use regularly spaced values in the specified grid. Start by setting z to the complex value (0,0), and c to the grid point. The result of applying the function is fed back into the function for another application. Apply the function repeatedly until the result gets large (absolute value larger than 2) or the limit of the number of possible applications is reached (100). Store the number of function applications at each point in the corresponding array entry.
2. Print the array one row per line. If you use two columns for each output value, and a limit on the number of function applications of 100, those points inside the set will attain 100 applications and the 100 will not fit in the two columns, being printed as two asterisks instead. This will give an interesting visual representation of the set.
3. The entire Mandelbrot set can be represented as in Figure P20.1 by using a square with lower left corner at $(-2, -1.25)$ and a side of length 2.5. Another interesting area is the bulge on the upper right of the complete set, which can be seen in more detail by "zooming" in to the square with lower left corner at $(-0.375, 0.5)$ and a side of length 0.5, as is shown in Figure P20.2.
4. Use the graphics capabilities of your system to show the results of your computation on a graphics screen. You will doubtless want to increase the size of your matrix so as to sample more points in the same interval. You will probably need to increase the limit of the number of applications of the function, perhaps to 1000, because as you have more resolution for the display, you will want a finer calculation at the edges of the set.

Figure P20.1
Mandelbrot Set

```
Enter coordinates (R,I) of lower left corner
          -2.0000000              -1.2500000
Enter length of side and iteration limit
           2.5000000            100
       1 1 1 1 1 1 2 2 2 2 2 3 3 3 3 3 3 3 3 3 3 2 2 2 2 2
       1 1 1 1 1 2 2 2 2 3 3 3 3 3 3 3 3 4 4 4 3 3 2 2 2 2
       1 1 1 1 2 2 2 3 3 3 3 3 3 3 3 4 4 4 5 9 6 5 4 4 3 2 2
       1 1 1 1 2 2 3 3 3 3 3 3 3 3 4 4 4 5 5 81010 4 4 3 3 3
       1 1 1 2 2 3 3 3 3 3 3 3 3 4 4 4 4 5 5 7 925 8 5 5 4 3 3
       1 1 1 2 3 3 3 3 3 3 3 4 4 4 4 5 5 611****17 7 5 5 4 4 3
       1 1 2 3 3 3 3 3 3 3 4 4 4 4 5 6 7 711****38 8 6 6 5 4 3
       1 1 2 3 3 3 3 3 3 4 4 4 5 6 917105123********21 8 9 6 4
       1 2 3 3 3 3 3 3 4 4 5 5 5 6 6 9******************38** 7 5
       1 2 3 3 3 3 3 4 5 5 5 5 6 61617*******************13 7 5
       1 3 3 3 3 4 5 6 8 7 8 7 7 811*************************11 5
       1 3 3 4 4 4 5 5 712151313101023**********************13 6
       1 4 4 4 4 5 5 6 712********13*************************63 5
       1 4 4 4 5 6 71511**************************************** 9 5
       1 5 6 7 7 81215*****************************************20 6 5
       1 5 6 7 7 81215*****************************************20 6 5
       1 4 4 4 5 6 71511**************************************** 9 5
       1 4 4 4 4 5 5 6 712********13*************************63 5
       1 3 3 4 4 4 5 5 712151313101023**********************13 6
       1 3 3 3 3 4 4 5 6 8 7 8 7 7 811*************************11 5
       1 2 3 3 3 3 3 4 5 5 5 5 6 61617*******************13 7 5
       1 2 3 3 3 3 3 3 4 4 5 5 5 6 6 9******************38** 7 5
       1 1 2 3 3 3 3 3 3 4 4 4 5 6 917105123********21 8 9 6 4
       1 1 2 3 3 3 3 3 3 3 4 4 4 4 5 6 7 711****38 8 6 6 5 4 3
       1 1 1 2 3 3 3 3 3 3 3 4 4 4 4 5 5 611****17 7 5 5 4 4 3
       1 1 1 2 2 3 3 3 3 3 3 3 3 4 4 4 4 5 5 7 925 8 5 5 4 3 3
       1 1 1 1 2 2 3 3 3 3 3 3 3 3 4 4 4 5 5 81010 4 4 3 3 3
       1 1 1 1 2 2 2 3 3 3 3 3 3 3 3 4 4 5 9 6 5 4 4 3 2 2
       1 1 1 1 1 2 2 2 2 3 3 3 3 3 3 3 3 4 4 4 3 3 2 2 2 2
       1 1 1 1 1 1 2 2 2 2 2 3 3 3 3 3 3 3 3 3 3 2 2 2 2 2
```

Engineering Problem 20

Figure P20.2
Mandelbrot Detail

```
Enter coordinates (R,I) of lower left corner
          -0.3750000              0.5000000
Enter length of side and iteration limit
          0.5000000              100
        5 5 5 5 5 5 6 6 6 7 8 9111811 9 917111311 8 6 5 5 4 4 4 4
        5 5 5 5 5 5 6 6 7 7 8 91046161111152213 9 7 6 6 5 5 4 4 4
        5 5 5 5 5 6 6 7 7 8 8 910123115232511 9 7 7 6 6 5 5 5 4 4
        5 5 5 5 5 6 6 7 7 8 8 9101114201310 9 8 7 7 6 6 5 5 5 4 4
        5 5 5 5 6 6 7 7 7 8 8 9101243201311 9 8 7 7 6 6 6 5 5 5 5
        5 5 5 5 6 6 7 7 7 8 8 91115141925131210 8 8 7 7 6 6 5 5 5 5
        5 5 5 6 6 7 7 8 9 9 9116037193439163311 9 8 8 7 6 6 5 5 5 5
        5 5 6 6 6 7 810131010111419****7566191310 9 8 8 7 6 6 5 5 5
        5 6 6 6 7 91012211312141622******60151211101115 8 6 6 5 5 5
        6 6 6 71231252059161521223752****24172512133214 8 7 6 6 5 5
        6 6 7 7 810121461222430************422315162014 9 7 6 6 5 5
        6 7 7 7 8 9111345********************834233211812 7 7 6 6 5
        7 7 7 8 8 910132361********************343011 9 7 7 7 6 6
        7 7 7 8 8 9113719********************40191311 9 8 7 7 6 6
        7 7 7 8 8 9262131********************84202733 9 8 7 7 7 6
        7 7 8 8 910121449**********************1911 9 8 8 7 7 7
        7 8 8 9 91011148878**********************211310 9 8 8 7 7
        8 9 9 9101011131734********************9652121110 9 9 8 8 8
      10101010101112131521**********************231413111110 9 9 911
      2349121214141314161818******************251715141213151111101113
      44461515357332**5730513041********41292770****20186716121443
      **44551977*****************************************39**19168247
      ****74******************************************************23****
      34***********************************************************
      *************************************************************
      *************************************************************
      *************************************************************
      *************************************************************
      *************************************************************
      *************************************************************
```

Fractal Geometry

Appendices

Appendix A Coding Conventions
Appendix B Fortran Summary
Appendix C Other Fortran Features
Appendix D Intrinsic Functions
Appendix E Accessing External Files
Appendix F Building Your Own While Statement
Appendix G Selected Solutions

Appendix A
Coding Conventions

As we pointed out several times in the body of the book, there are ways of writing programs that make them easier for humans to read and understand. We'll describe the conventions we've used. You might want to adopt all or some of them yourself. Whether or not you use our conventions, you should use *some* consistent rules so your programs are easy for you and others to read.

Formatting Programs

Fortran imposes some constraints on program layout. A line has four parts: a statement number part in the first five columns, which may be blank; a continuation column, column 6, which may be blank; a statement body part in columns 7 to 72; a sequence number in columns 73 to 80 (which may be left blank). Each statement goes on a separate line. Logical lines too long to fit on a physical line are broken at convenient points and as many continuation lines as necessary are used. A continuation line has a nonblank character in column 6. The continuation lines are indented one level from the initial line.

Blank lines can be used optionally to set off subroutines and functions, or any logical grouping of statements.

Blank characters can be used optionally around operators.

Structured statements are laid out with all subordinate clauses on separate lines, indented one level from the initial line of the statement. The terminating clause should appear at the same indentation as the initial line (for example, **end** and **continue**).

Naming

Mnemonic names should be used for subroutines, functions, and variables. Names may comprise several words or word fragments. These should be written with the initial letter of each word or fragment in uppercase and all other letters in lowercase (for example, I, NewX, Count, OneTry). This can require considerable creativity, since Fortran limits names to six characters.

Comments

Comments can be used to indicate use of identifiers (on declarations); indicate the author, date, and so on of a program; indicate the purpose of a subprogram; or clarify the intent of difficult expressions.

Appendix B
Fortran summary

This appendix gathers together the syntax specification of the Fortran subset used in this book. We use the following notational conventions.

[x] means x is optional
{x} means x can appear zero or more times

Thus if we write

x [y] z

we can match the two strings

x z
x y z

and if we write

x {y} z

we can match the infinite set of strings

x z
x y x
x y y z
x y y y z
...

We now give the definition of our language subset.

a Program is:
program *id*
{Declaration}
{Statement}
end

{Subprogram}

Now, just to make this clear, we'll say explicitly what this part of the definition means. A program is the keyword **program** followed by an identifier (this occurs so frequently that it improves readability to shorten it to *id*), followed by any number of occurrences (zero or more) of Declaration (which has not yet been defined), followed by any number of occurrences (zero or more) of Statement (which has not yet been defined), followed by the keyword **end**, followed by any number of occurrences (zero or more) of Subprogram (which has not yet been defined). We will use boldface for keywords (like **program**), italics for things in Fortran that represent one of a number of specific choices (like *id*), and capitalized words or phrases for things that are defined in this definition.

We'll now present the entire definition without any more explanatory text. This definition actually allows some forms that are not legal; making the definition precise enough to eliminate these would also make it considerably longer and, we judge, less usable.

a Program is:
program *id*
{Declaration}
{Statement}
end

{Subprogram}

a Declaration is one of:
   a. TypeSpecifier Name {, Name}
   b. **parameter** (*id* = Expression {, *id* = Expression})
   c. **external** *id* {, *id*}
   d. **common** /*id*/ *id* {, *id*}

a TypeSpecifier is one of:
   a. **integer**
   b. **real**
   c. **character***integer
   d. **logical**

a Name is:
id [(*integer* [: *integer*] {, *integer* [: *integer*]})]

a Statement is one of:
   a. Variable = Expression
   b. **print** Control [Expression {, Expression}]
   c. **call** Variable
   d. **return**
   e. **read** Control Variable {, Variable}
   f. **do** *integer* Variable = Expression,Expression [,Expression]
       {Statement}
    integer **continue**
   g. **while** (Expression) **do**
       {Statement}
    end while
   h. **if** (Expression) **then**
       {Statement}
    {**elseif** (Expression)
       {Statement}}
    [**else**
       {Statement}]
    end if
   i. **write** Control [(Expression {, Expression})]
   j. *integer* **format** (FormatCode {, FormatCode})

a Variable is:
   id [(Expression {, Expression})] [([Expression]:[Expression])]

a Control is one of:
   a. Format [,]
   b. (Device, Format [, **end**=*integer*])

a Format is one of:
   a. *
   b. '(FormatCode {, FormatCode})'
   c. *integer*

a FormatCode is one of:
   a. *string*
   b. [*integer*] I*integer*
   c. [*integer*] A*integer*
   d. [*integer*] F*integer.integer*
   e. [*integer*] E*integer.integer*
   f. *integer*X
   g. T*integer*
   h. TR*integer*
   i. TL*integer*
   j. [*integer*] D*integer.integer*
   k. [*integer*] G*integer.integer*

l. / {/}
m. [*integer*] (FormatCode {, FormatCode})

a Device is one of:
  a. *
  b. *integer*

a Subprogram is one of:
  a. Subroutine
  b. Function

a Subroutine is:
  subroutine *id* [(*id* {, *id*})]
  {Declaration}
  {Statement}
  end

a Function is:
  TypeSpecifier **function** *id* [(*id* {, *id*})]
  {Declaration}
  {Statement}
  end

an Expression is:
  Operand {Operator Operand}

an Operand is one of:
  a. Variable
  b. *integer*
  c. *real*
  d. *string*
  e. .true.
  f. .false.
  g. (Expression)
  h. − Expression
  i. .not. Expression

an Operator is one of:
  a. +
  b. −
  c. *
  d. /
  e. **
  f. .and.
  g. .or.
  h. //

    i. .lt.
    j. .le.
    k. .gt.
    l. .ge.
    m. .eq.
    n. .ne.

This definition allows many expressions that are illegal, such as

3 .and. 'Hi there'

We will not formally define type compatibility rules that preclude these constructs. Another source of ambiguity is precedence. Operators are arranged in several groups as follows:

    ** (exponentiation)
    * (multiplication), / (division)
    + (addition), − (subtraction)
    // (string catenation)
    .le., .lt., .ge., .gt., .eq., .ne.
    .not.
    .and.
    .or.

All groups are applied in the order listed. Operators within a group are applied left to right. Parentheses can be used when the default precedence is unacceptable.

    An *integer* is a sequence of digits.

    A *string* begins with a single quotation mark and ends with a single quotation mark and can contain any sequence of characters. If a single quotation mark is to be represented in the string, it is shown as two successive single quotation marks.

    A *real* number comprises a fraction part followed by an exponent part. The fraction part is a sequence of digits, optionally either preceded by, containing, or followed by a decimal point. The exponent part is the letter "e" followed by an optional plus or minus sign, followed by one or two digits.

Appendix C
Other Fortran Features

In this appendix we discuss briefly the other statements that exist in Fortran. These statements are not usually used by novice programmers and so we include them here primarily for recognition purposes. You may, however, come across them in programs that you read, so it is useful to understand their purpose. Many are included to allow programs written in earlier dialects of Fortran such as Fortran IV (1966) to be compiled by Fortran77 compilers. Fortran77 provides different ways of doing some of these things so that some statements described here will appear redundant.

Data Statement

The **data** statement is a way of initializing variables at compile time. The compiler produces values for the specified variables and loads them with the program. It is an alternative to a set of assignment statements at the beginning of the executable part of the program. Because the initialization is done at compile time, it is done exactly once. The form of the **data** statement is

    data *variableList* /*valueList*/

For example, the **data** statement

    data A, B, I / 2.0, 3.0, 5/

initializes the values of A, B, and I positionally so that A gets the value 2.0, B gets 3.0, and I gets 5. In fact, the variables and their values can be mixed (as long as the values are surrounded by slashes) so that this form of the **data** statement

    data A /2.0/, B /3.0/, I /5/

is exactly equivalent to the first. The association of values and variables is done positionally, with the only restriction that the name of the variable must appear before its value.

In situations where there are many values to be initialized, it is convenient to provide a repetition count for values. This is often useful when an array is being initialized to zero. In this form, the repetition count is included before the value in the following way:

    data A / 10 * 0 /

where A is presumably an array with at least 10 elements.

Implied do loops can be used where appropriate to describe the variables to be initialized. For example, the statement

    data (A(I), I=1,10,2) /5 * 1.0/,
    &   (A(I), I=2,10,2) /5 * 2.0/

initializes the odd-numbered elements of A to the value 1.0 and the even-numbered ones to the value 2.0.

The same kind of syntax can be used for 2-dimensional arrays. If X is declared to be a 3 by 4 array such as

    real X(3,4)

then the **data** statement

    data X / 12 * 1.0 /

initializes the whole array to 1.0. Of course, if all the values were not the same it would be important to know in which order the locations would be filled. The usual rule is followed: the array is filled in the order $X(1,1)$, $X(2,1)$, $X(3,1)$, $X(1,2)$, $X(2,2)$, and so on. If the other order is desired then a nested implied loop can be used such as

    data ((X(I,J), J=1,4), I=1,3)
    &   /1.0, 6*5.0, 7.6, 4.9, 4*0.0/

Although **data** statements can appear at the beginning of any subprogram they always only have effect once, at the very first execution of the subprogram. For this reason, we suggest that they only be used in the main program. This avoids any expectation that the initial values set up in a subprogram will somehow regenerate themselves between invocations.

The **data** statement is placed after declarations but before the executable part of any program unit in which it is used.

Common Blocks

We have seen how the variables of each subprogram are completely independent of each other except where values are passed from one subprogram to another at invocation or return. For programs in which a large amount of data is shared, the overhead of writing parameter lists and of actually passing data at invocations can become large. The **common** statement is a way of avoiding this by declaring a globally visible set of variables that can be used in many places.

The form of the **common** statement is

> **common** /name/ variableList

The variables mentioned in the list are placed into a common block. Any other subprogram that contains a similar **common** statement using the same name is deemed to refer to the same set of storage locations. Thus the values in these locations are shared between different subprograms.

The matching of corresponding values is done positionally. The names appearing in the list of variables in different places can be different. They just refer to the same actual locations. Here is an example of the same common block referenced in two different subprograms.

> **subroutine** X
> **integer** I, J
> **common** /Example/ I, J
>
> ...
>
> **subroutine** Y
> **integer** K, L
> **common** /Example/ K, L
>
> ...

The value of K is the same value as that of I at all times because these names are simply aliases for the same memory location.

The number of values described in each location does not have to be the same. If one list is shorter than the other, then those locations that are described in the short list are matched with the corresponding locations in the longer list. Those that are not matched are not accessible in the subprogram with the shorter list. For example, if the code had been

> **subroutine** X
> **integer** I, J
> **common** /Example/ I, J
>
> ...

```
subroutine Y
integer K
common /Example/ K
...
```

then I and K are the same location but the value of J is not accessible in the second subprogram.

If a common block is going to be referenced in many subprograms, then it is a good idea to place the most commonly referenced variables first. The reason for this is that you don't have to name all the variables in a common block in a particular subprogram; if you only want to refer to the first few variables in the block, you declare it as though it only contained those variables. If it is absolutely necessary to refer to a variable that appears after one to which you don't want to refer, the following scheme can be used:

```
subroutine X
real A(10, 10), B(10), C(10)
common /Partial/ A, B, C
...

subroutine Y
real X(10), Y(10), Dummy(10, 10)
common /Partial/ Dummy, X, Y
...
```

This signals that in the second subprogram only the values of X and Y are being legitimately referenced. The choice of the name Dummy signals that this variable is a placeholder.

If there is only one common block in the program then its label or name can be omitted, together with the slashes that surround it. A common block without a name is called *blank common*, whereas one with a name is called *labeled common*. As many common blocks as necessary can be used in a program. However, no variable can appear in more than one common block. Since a common block represents locations that are shared, it is clearly impossible for a variable to be shared in two different places.

Common block descriptions should appear at the beginning of program units, after declarations but before **data** statements.

A word of warning—the sizes of variables are not usually explicitly mentioned in **common** statements. This can cause obscure errors if the list of variables in a matching **common** statement does not indicate that some or all of the referenced variables are arrays. For example, this pair of **common** statements

```
subroutine X
integer A(5), B
common A, B
...
```

```
    subroutine Y
    integer I, J
    common I, J
    ...
```

equates the location referenced as J in the second subprogram to that named A(2) in the first. The compiler cannot detect this kind of mistake (since equating these two variables might be a reasonable thing to do under some circumstances) so that errors of this kind can be very hard to find.

Block Data Subprograms

A block data subprogram is a way of initializing a large number of variables in a single place in the program. It is usually used to initialize the contents of a common block that can then be referenced in many places. Its effect is the same as a sequence of **data** statements except that it has the syntactic form of a subprogram. Here is a typical example:

```
    block data Init
    common /Main/ A, X, Y
    data A /100 * 0.0/
    data X /10 * 1.0/
    data Y /10 * 2.0/
    end
```

You are unlikely to encounter a block data subprogram until you work with a fairly large program.

Dimension Statement

This statement allows the sizes of arrays to be specified. It is redundant because the sizes can be included with the declaration of the variable itself, and this is a much better way to do it. It has the form

```
    dimension A(10)
```

where A is declared somewhere earlier (usually the preceding line). It is thus silly. The **dimension** statement is placed after the declarations in a program unit.

Equivalence Statement

This statement is used to place two different variables in the same memory location(s). It does have its uses but should be treated with extreme caution. Almost invariably, anything that can be done with an **equivalence** statement ought to be done in some other way. We next show its form, in case you need it.

```
real A, B
equivalence (A, B)
```

This declares what appear to be two variables but then places them in the same memory location. Thus references to A or B both end up referring to the same piece of data. Special care is needed when one of the variables involved is an array. For example, the equivalence illustrated here

```
integer X(3, 4), Y(3)
equivalence (Y, X(1,2))
```

has the effect of equating the vector Y to the second column of the matrix X.

The **equivalence** statement is placed after the declarations in a program unit.

Entry Statement

The **entry** statement provides an alternate place from which a subprogram may be executed. Thus, instead of beginning the execution of an invoked subprogram from its first executable statement, its execution can begin somewhere else. The **entry** statement can have its own argument list, which does not have to match that at the beginning of the subprogram. This statement is a potential disaster and you should never use it. Should you come across a program which uses an **entry** statement, you should rewrite it as a service to humanity.

Statement Functions

Statement functions are a way of providing a shorthand notation for some expression that is used regularly in a program. Statement functions appear between the declaration part of a program and the executable part.

The form of a statement function is like that of an ordinary function, compressed to a single line. It has parameters that can be replaced by arguments when it is invoked. It may also reference variables of the program segment in which it appears.

The next program segment uses a statement function.

```
program Show
integer I, J
Cube(X) = X * X * X
I = 57
J = 34
print *, Cube(I), Cube(J)
stop
end
```

Statement functions may only be referenced in the program segment in which they are declared. They cannot be passed to other subprograms.

Statement functions are usually expanded by the compiler so that they are *in line*. This means that the compiler inserts the instructions to execute the function into the instructions of the program wherever the statement function would have been invoked. This means that there is no overhead involved in the function's invocation.

Save Statement

When an invocation of a subprogram is completed, then the variables that are local to the subprogram "lose" their values, so that on the next invocation they are undefined. This is a useful property because it means that a subprogram behaves in exactly the same way at every invocation, regardless of its past history. In some special situations it is useful to be able to get around this. The **save** statement provides one way to do this.

A **save** statement has the form

> **save** *variableList*

and specifies that all variables listed are not to lose their values between invocations. If no variables are listed, then the **save** statement applies to all variables local to that subprogram.

The **save** statement appears after the declarations in a program unit but before any **data** statements.

This statement can be useful when debugging, to find out how many times a subprogram is entered. For example, the following subprogram keeps track of how many times it has been invoked.

```
subroutine Times (X)
integer X, N
save N
data N /1/
print *,'On invocation number', N, ' value is', X
N = N + 1
return
end
```

The value of N is set to 1 exactly once at compile time and is incremented on each invocation; the intermediate values are retained between invocations.

Implicit Statement

Although we haven't talked about it, Fortran does not require that all variables be declared. Instead it makes assumptions about the types of undeclared variables, based on the letters with which their names begin. Names that begin with letters in the range I through to N are assumed to be of type **integer**, while all other names are assumed to belong to **real** variables. Of course, arrays must be explicitly declared since there is no automatic way to decide how large an array must be.

The **implicit** statement is a way of changing these default rules about which letters are used to represent each type. The **implicit** statement allows us to specify a type followed by the letters that should be assumed to begin the names of variables of that type. Several types may be used in the same **implicit** statement. Here is an example.

>   **implicit logical** (L), **character** (C), **real** (A-B, D-K, M-Z)

The **implicit** statement must appear before any declarations, which forces it to be the first statement in the program.

Since we do not encourage using variables without declaring them, we do not concern ourselves with the **implicit** statement. There is one way in which it can be useful and it is as well to be familiar with it in case you encounter it. Using the statement

>   **implicit logical** (A-Z)

is one way in which most mistakes caused by mistyping a variable name can be caught. This **implicit** statement causes all variables that are not explicitly declared to be assumed to be of type **logical**. If they are in fact mistypings of declared variables, then they are probably used in arithmetic expressions and the like. So the compiler will detect the apparent misuse of a **logical** variable and produce an error message. (Of course, if the mistyping is of the name of a **logical** variable, then this will not help.) We do not particularly recommend this approach but we don't denigrate it either. You will have to make up your own mind.

GoTo Statement

The **go to** statement is a way of transferring control in programs. We have already seen that statements such as **while** and **if** statements have implicit transfers of control (so that at the end of the repeated statements in a **while** loop, control passes to the top of the loop). The **go to** statement allows this kind of transfer to be done in arbitrary ways.

The form of the **go to** statement is

>   **go to** *statementNumber*

and causes the statement labeled with the matching statement number in columns 1 through 5 to be executed next. It is therefore an unconditional transfer of control. We will shortly see some conditional ways of transferring control.

The problem with the **go to** statement is that it allows the program statements to appear in an order that does not reflect the logical progression through the algorithm. This leads to what has been descriptively called "spaghetti" code, where conceptually related groups of statements are textually intermixed and it quickly becomes very hard to see what actually happens when the program is executed.

This problem gets much worse when the size of the program increases. It becomes very difficult to find where a particular **go to** statement transfers because it can be literally anywhere in the program segment. This is one of the reasons why we have been careful to use increasing sequences of numbers in the statement number field. If this is done, the destination of a transfer can be easily found.

We counsel against using the **go to** statement as an alternative to the structured statements that we have presented. The one situation where it can be useful is to provide an elegant "bailout" from a nested set of constructs. If we have a set of nested loops and some abnormal condition can occur in the innermost loop requiring all of the loops to terminate then, rather than having tests in all the loop conditionals for an abnormal situation, we can test for it where it may occur and transfer out of the whole nested set of statements in a single statement.

Logical If Statement

The logical **if** statement can be thought of as a shorthand version of the **if...then** statement in which there is only a single statement. It has the form

    if (*booleanExpression*) *statement*

If the expression evaluates to true, then the single statement is executed; otherwise control passes to the next sequential instruction.

There are two ways in which this can be useful. The first is when there is some calculation that needs to be done only in particular circumstances. This is really just a short form of an ordinary **if** statement. The other way that a logical **if** can be useful is when the statement is a **go to** statement. This allows a conditional transfer of control to some other part of the program. We can thus write

    if (X.gt.3) **go to** 200

rather than

    if (X.gt.3) **then**
       go to 200
    endif

Arithmetic If Statement

The arithmetic **if** statement is a way of providing a three-way transfer of control based on the sign of some expression. This is useful in many mathematical situations where there are three different things to be done and they can be distinguished by the sign of some other formula.

The arithmetic **if** statement has the form

>   **if**(*expression*) *stmtNum1*, *stmtNum2*, *stmtNum3*

Control passes to the statement labeled with statement number 1 if the value of the expression is negative, to the statement labeled with statement number 2 if the value of the expression is zero, and to the statement labeled with statement number 3 if the expression's value is positive. Notice that there is always a transfer to one of the three labeled statements so that the statement textually following the arithmetic **if** cannot be reached by sequential execution. It must therefore always have a statement number attached.

Computed GoTo Statement

This statement also allows a multiway transfer of control based on the value of some expression, just as for the arithmetic **if** statement. It is more general because it allows more than three different paths to be selected.

The form of the computed **go to** statement is

>   **go to** (*stmtNumList*) *expression*

The integer-valued expression is evaluated. If its value is 1, then control passes to the statement labeled with the first statement number in the list. If its value is 2, then control passes to the second statement number, and so on. If the value of the integer expression is such that no statement would be selected (for example, if the expression produces a negative value), then the statement has no effect and execution continues at the statement following the computed **go to**.

Assigned GoTo Statement

The assigned form of the **go to** is very seldom seen. It allows several different paths in the program to come together for a while and then branch again depending on where they came from. It has two parts: an **assign** statement and the **go to** itself. The form of the **assign** statement is

>   **assign** *stmtNum* **to** *variable*

and has the effect of saving the statement number in the integer variable. The variable cannot be used anywhere else in the program for any other purpose. There may be many different **assign** statements.

The second part is the **go to** statement itself. It has the form

>   **go to** *variable* (*stmtNumList*)

and causes a transfer of control to the statement whose label was saved in the integer variable. The set of statement numbers is optional but, if included, is used as a check.

Rewind Statement

Whenever a file is being used for input or output, a record is kept of the current location in the file. Thus a succession of **read** or **write** can be used to move through a file sequentially. The **rewind** statement allows a file to be accessed from its beginning. This can be useful when the data in the file must be accessed more than once. We have already seen how to use an array to store such data to avoid having to access the file twice, but for some practical applications the file is too long to be read once and kept in an array.

The form of the **rewind** statement is

>   **rewind** *unitNumber*

where *unitNumber* is the unit number corresponding to the file concerned (associated with it by an **open** statement).

Usually rewinding is only useful for files that are used for input to the program. If a file is being used for output and is then rewound, all data previously written to it will be effectively destroyed. Therefore, as a general rule, output files should not be rewound.

Inquire Statement

The **inquire** statement allows a program to discover the status of a file or unit number and record this information in variables if required. It comes in two flavors, inquiry about a file and inquiry about a unit number.

It can be written as either

>   **inquire** (*unitNumber*, *optionList*)

or as

>   **inquire** (**file**=*fileName*, *optionList*)

The file name can be any character-valued expression (usually a character constant) whose value is the name of the file concerned. The list of options is slightly complicated:

IOSTAT = integer variable Sets the integer variable to zero if the statement executed correctly and to a positive value if it didn't.

ERR = statement number Transfers to the statement labeled with the statement number if an error occurs. The statement label used is one within the same program or subprogram.

EXIST = logical variable Sets the logical variable to .true. if the file or unit exists and .false. otherwise.

OPENED = logical variable Sets the **logical** variable to .true. if the file or unit has been opened and .false. otherwise.

NAMED = logical variable Sets the **logical** variable to .true. if the file has a name and .false. otherwise. The file or unit must exist.

NAME = character variable Sets the character variable to the name of the file. The file must exist.

SEQUENTIAL = character variable

DIRECT = character variable

FORMATTED = character variable

UNFORMATTED = character variable Assigns to the character variable the character string "YES", "NO", or "UNKNOWN" depending on whether the file or unit has the corresponding property or it cannot be determined.

FORM = character variable The character variable is assigned the character string "FORMATTED" or "UNFORMATTED" depending on the type of access for which the file was opened.

NUMBER = integer variable The **integer** variable is assigned the unit number associated with the file.

RECL = integer variable The **integer** variable is assigned the maximum record size (line size) allowed for the file. The file must be a direct access file for this to apply. The size is given in characters.

NEXTREC = integer variable The **integer** variable is assigned the record number of the record following the one most recently read. If no **read** has yet taken place, this will return the value 1.

BLANK = character variable The character variable should be assigned either the character string "NULL" or "ZERO" depending on whether or not blanks in the input are to be treated as zeros.

Endfile Statement

The **endfile** statement allows a program to write an end-of-file marker into a file. It has the form

    endfile *unitNumber*

Such a marker does not necessarily prevent a program from reading further but does allow files to contain subgroupings of information. The boundaries of these subgroups can be detected by **end** = flags in **read** statements as we have previously seen.

Backspace Statement

The **backspace** statement allows the logical position in a file to move backward. Thus a previously written record can be overwritten by backspacing and rewriting. The statement has the form

> **backspace** *unitNnumber*

Open Statement.

The **open** statement opens a file or device for subsequent input/output operations. It has the form

> **open** (*Options*)

where the list of options is similar to that of the **inquire** statement. In particular, the options **UNIT**=, **IOSTAT**=, **ERR**=, **FILE**=, **RECL**=, and **BLANK**= are identical to those of the **inquire** statement. The following options are also possible:

STATUS= Must be one of "OLD", "NEW", "SCRATCH", or "UNKNOWN". Some operating systems make use of this information about the projected use of the file.

ACCESS= Must be one of "SEQUENTIAL" or "DIRECT". Specifies whether the file will be accessed sequentially, as we have assumed throughout this book, or directly, that is with a record address.

Close Statement

This statement closes a particular unit or file after it has been used. This may have to be done if the operating system sets an upper limit on the number of files that may be active at once. It has the form

> **close** (*Options*)

The options may include **UNIT**=, **IOSTAT**=, and **ERR**=, with the same meanings as before. One further option is

STATUS= which may be assigned "KEEP" or "DELETE", depending on what is to happen to the file when the program completes.

Read Statement

The **read** statement, we have already seen, may contain a unit designator and a format designator. It may also contain an **end** = *statementNumber* option to control what happens if an end of file is encountered during a read operation. It also has one further option, the **ERR**=*statementNumber* option, which specifies a statement label to which control passes if an error is encountered during the read operation. Such an error might be the occurrence of character data when an integer was being read. This provides a clean way of handling the wide variety of possible problems that can occur during input.

Appendix D

Intrinsic Functions

Here is a table of the intrinsic functions that should be provided by all Fortran77 compilers. They are almost all generic, that is, they work with operands of different types and return a value of the appropriate type for those operands. The letters I, R, D, C, S, and L stand for integer, real, double precision, complex, character, and logical, respectively.

Name	Operands	Result	Effect
sqrt	R,D,C	R,D,C	square root of operand
exp	R,D,C	R,D,C	exponential of operand
log	R,D,C	R,D,C	natural logarithm of operand (base e)
log10	R,D	R,D	common logarithm of operand (base 10)
sin	R,D,C	R,D,C	sine of operand (must be radians)
cos	R,D,C	R,D,C	cosine of operand (must be radians)
tan	R,D	R,D	tangent of operand (must be radians)
asin	R,D	R,D	arcsine of operand
acos	R,D	R,D	arcosine of operand
atan	R,D	R,D	arctangent of operand
atan2	R,D	R,D	arctangent of first operand divided by second operand, gets sign right
sinh	R,D	R,D	hyperbolic sine
cosh	R,D	R,D	hyperbolic cosine
tanh	R,D	R,D	hyperbolic tangent

Name	Operands	Result	Effect
int	I,R,D,C	I	converts to integer (truncates)
real	I,R,D,C	R	converts to real (with as much precision as possible)
dble	I,R,D,C	D	converts to double precision
cmplx	I,R,D,C	C	converts to complex. With two operands of the same type uses first as real part and second as imaginary
nint	R,D	I	converts to integer (rounding)
abs	I,R,D,C	I,R,D,C	absolute value of operand
mod	I,R,D	I,R,D	remainder of first operand divided by second operand
sign	I,R,D	I,R,D	returns absolute value of first operand with sign of the second
max	I,R,D	I,R,D	takes at least two operands. Returns maximum.
min	I,R,D	I,R,D	takes at least two operands. Returns minimum.
len	S	I	length of character string
index	S	I	position of second string in first
lge	S	L	true if first string is greater than or equal to second
lgt	S	L	true if first string is greater than second
lle	S	L	true if first string is less than or equal to second
llt	S	L	true if first string is less than second
aimag	C	R	imaginary part of complex number

Appendix E

Accessing External Files

In the body of the book we talk about "standard input" and "standard output," and by those terms we refer to default IO streams used by programs written in Fortran. Some operating systems also have a concept of standard input and output. The standard input of a program in Fortran may or may not be coming from what the operating system considers to be the standard input, and a similar situation can hold for standard output. This can be a bit confusing. In the following paragraphs we describe how to access external files in several systems that are commonly used for engineering instruction. If the system you are using is not mentioned here, you'll have to look up the appropriate information in your system documentation.

MS-DOS

The WATFOR-77 compiler preconnects units 5 and 6 to the keyboard and the screen. If no **open** statement is used, an attempt to read from another unit (say *n*) will try to read from a file whose name is FTddF001. Here *dd* is two digits corresponding to the unit number, where unit 1 is "01" and so on. In the same way, an attempt to write to unit *n* causes the output to go into a file called FTddF001. (Notice that this will cause the overwriting, and hence loss of contents, of that file if it already exists.)

The **open** statement may be used to explicitly connect a unit with an external file. A second method is to preconnect a unit number to an external file using the DOS *set* command.

set *n* = *fileSpecifier*

The fileSpecifier is a DOS file specification, which can be as simple as a file name or as complex as a drive, path, file name, and extension. This statement is executed as a DOS command prior to running the WATFOR-77 system.

Standard input is used for commands and standard output for messages. The standard input and standard output can be redirected.

UNIX

The UNIX *f77* compiler preconnects units 0, 5, and 6 to the standard error, standard input, and standard output. If no **open** statement is used, an attempt to read from another unit (say *n*) will try to read from a file whose name is *fort.n*. In the same way, an attempt to write to unit *n* causes the output to go into a file called *fort.n*. (Notice that this will cause the overwriting, and hence loss of contents, of that file if it already exists.)

The **open** statement explicitly connects a unit with an external file.

The standard input, standard output, and standard error can all be redirected, allowing units 0, 5, and 6 to be associated with other files if required.

VM/CMS

The WATFOR-77 system and the IBM VS FORTRAN compiler both behave as follows. The compiler preconnects units 5 and 6 to the keyboard and the screen. If no **open** statement is used, an attempt to read from another unit (say *n*) will try to read from a file whose name and type are FILE FTddF001. Here *dd* is two digits corresponding to the unit number, where unit 1 is "01", and so on. In the same way, an attempt to write to unit *n* causes the output to go into a file called FILE FTddF001. (Notice that this will cause the overwriting, and hence loss of contents, of that file if it already exists.)

The **open** statement may be used to explicitly connect a unit with an external file. A second method is to preconnect a unit number to an external file using the CMS *filedef* command.

filedef *n* disk *fileSpecifier*

The fileSpecifier is a CMS file specification, which can be a filename and filetype with or without a filemode. This statement is executed as a CMS command prior to running the compiler.

The IBM VS FORTRAN compiler uses uppercase keywords and identifiers.

Appendix F
Building Your Own While Statement

If your system does not allow you to use **while** statements, you can still design your algorithms using these statements, and then use another Fortran mechanism to implement them. This other mechanism is the **go to** statement (see Appendix C). Generally speaking, the **go to** is a dangerous mechanism to use since it can lead to programs that are difficult to understand, and that are accordingly prone to contain errors and difficult to maintain. We virtually never use it in our own programming, and recommend that you avoid it if at all possible.

However, if you had a **while** construct

>   **while** (*BooleanExpression*) **do**
>   *Statements*
>   **end while**

and wanted to implement it using **go to**, you could write:

>   100 **continue**
>   **if** (.not. (*Boolean Expression*)) **go to** 200
>   *Statements*
>   **go to** 100
>   200 **continue**

where the statement labels 100 and 200 are examples, and any otherwise unused labels could be used. There are other ways to rewrite **while** using **go to** but this is simple and will always do the job.

Appendix G

Selected Solutions

Chapter 1

1. (a) Is an algorithm. Won't terminate if number is odd. (b) Is an algorithm. Will terminate. (c) Isn't an algorithm. Pages 157 and 158 are usually back to back.
2. (a) $O(n)$ since we need to examine the whole deck (assuming it's not in any particular order). (b) $O(n)$ since we can place each card on one of four piles as we go through the deck. Then the piles can be placed on top of one another.
4. The best way to test your algorithm is to find a 12-year-old and have him try it.
8. (a) Public phones usually have a simple form of such an algorithm printed on them. The steps should be something like this:(1)lift handset and listen to earpiece (2) if no dial tone present then give up (3) while no more numbers left, dial next number (4) while phone rings and not more than 12 rings, wait (5) if more than 12 rings, hang up and give up (6) if answered, say "Hello" (7) converse (8) place handset back on phone. Notice that we need to consider what should happen if the phone isn't working, if there's no answer and so on.

Chapter 2

1. Make sure that you have the manufacturer's name and not just the name of the distributor. Some possibilities are IBM, Digital, Apple, Zenith, Prime, Hewlett-Packard.
2. Most microcomputers have between 512K and 1Megabyte of storage. Larger machines may have up to 2 Gigabytes (a Gigabyte is 1000 Megabytes).
4. A novel has about 200 pages, each page has about 35 lines, and each line contains about 60 characters, for a total of 420,000 characters. This book contains about 1.6 million characters. Your local library might contain a million books, each of about a million characters for a total of 1000 Gigacharacters.
8. (a) 113 (b) 85 (c) 23 (d) 63

Chapter 3

1. (a) Many text editors cannot do this in a single command (b) Almost all can do this (c) This is also very common (d) And so is this
2. Most large systems would provide languages such as Pascal or PL/1. UNIX systems and some others might have C. Many commercial systems would have Cobol and RPG. Microcomputers often have the language Basic available. Other languages you might encounter are Ada, GPSS, Simscript, SAS, and APL. You might want to look up these languages and see how programs written in them look. Also try to find out why each of them exists (that is, what sort of programming is it good for).
4. The text shows our output.
8. The program prints the integer 5.

Chapter 4

1. The output is

```
-----
XXXX
YYYY
YYYY
ZZZZ
YYYY
YYYY
ZZZZ
-----
```
2. The output is
```
Z
A
A
B
A
C
A
```
4. The output would not change at all since the same pieces of code would be executed in the same order (although they would have different symbolic names).

8. We ran this on three systems. Two of them went into loops printing "A" on one line followed by "B" on the next, infinitely often. The third produced one "A" and one "B" and then the message "attempt to invoke active function/subroutine."

Chapter 5

1. q-113, w-119, e-101, r-114, t-116, y-121, u-117, i-105, o-111, p-112, [-91,]-93

2. The value will be 0 because the difference between the two values will not be represented in the internal encoding of either. Thus their internal representations will be equal and hence their difference will be zero.

4. Adding the value 0.23 ten times gives this sequence 0.46, 0.69, 0.92, 1.2, 1.4, 1.6, 1.8, 2.0, and 2.2. Multiplying 0.23 by 10 gives 2.3.

8. (a) 0 (b) 0 because the real result on the right-hand side is slightly smaller than 1 and assignment to an integer truncates. (c) compile time error because the type of the right-hand side is not compatible with that of the left-hand side.

Chapter 6

1. (a) The program prints these 6 lines:
```
       4          4
       5          5
       5          4
       6          6
       6          5
       6          4
```
(b) The program prints these 10 lines:

```
20
18
16
14
12
10
 8
 6
 4
 2
```

2. The modified program is
```
*    Finds twice the sum of its inputs
     program AddUp2
     real Total, Number
     print *, 'Enter next number'
     read *, Number
     Total = 0
     while (Number .ne. 0) do
        Total = Total + Number
        print *, 'Enter next number'
        read *, Number
     end while
     print *, 'Sum is ', 2 * Total
     stop
     end
```

4. The output of the program is 375.

8. This program tests all possibilities.
```
     program TestL
     logical A, B, First, Second
     A = .true.
     B = .true.
     First = .not. (A.and.B)
     Second = .not.A.or..not.B
     print *, First, Second
     B = .false.
     First = .not. (A.and.B)
     Second = .not.A.or..not.B
     print *, First, Second
     A = .false.
     B = .true.
     First = .not. (A.and.B)
     Second = .not.A.or..not.B
     print *, First, Second
     B = .false.
     First = .not. (A.and.B)
     Second = .not.A.or..not.B
     print *, First, Second
     stop
     end
```
Its output is
```
       F          F
       T          T
       T          T
       T          T
```

Chapter 7

1. This program illustrates both functions.

```
*   Shows functions Plus and Minus
    program Ex71
    real X, Y, Plus, Minus
    read *, X, Y
    print *, Plus(X, Y), Minus(X, Y)
    stop
    end
    real function Plus (A, B)
    real A, B
    Plus = A + B
    return
    end
    real function Minus(A, B)
    real A, B
    Minus = A - B
    return
    end
```

2. This program illustrates both functions.
```
*   Test intrinsic
    program Ex72
    real X, NewExp, NewSin
    read *, X
    print *, NewExp(X), NewSin(X)
    stop
    end
    real function NewExp(Y)
    real Y
    NewExp = exp( - Y * Y / 2.0)
    return
    end
    real function NewSin(Z)
    real Z, Y
    Y = sin(Z)
    NewSin = Y * Y
    return
    end
```

4. The first program checks identity (a).
```
*   Check trig identity
    program Ex74
    real X, Start, Finish, LHS
    logical Same
    parameter (Start = 0.0)
    parameter (Finish = 2 * 3.141592)
    Same = .true.
    do 100 X = Start, Finish, 0.1
        Same = Same.and.
    &        abs(LHS(X)-1.0).lt.1.0e-4
100 continue
    print *, 'Identity is', Same
    stop
    end
    real function LHS(X)
    real X, Y, Z
    Y = sin(X)
    Z = cos(X)
    LHS = Y * Y + Z * Z
    return
    end
```

The second checks identity (b).
```
*   Check trig identity
    program Ex74
    real X, Start, Finish, LHS
    logical Same
    parameter (Start = 0.0)
    parameter (Finish = 2 * 3.141592)
    Same = .true.
    do 100 X = Start, Finish, 0.1
        Same = Same.and.
    &        abs(LHS(X)-RHS(X)).lt.1.0e-4
100 continue
    print *, 'Identity is', Same
    stop
    end
    real function LHS(X)
    real X
    LHS = sin(2 * X)
    return
    end
    real function RHS(X)
    real X
    RHS = 2.0 * sin(X) * cos(X)
    return
    end
```

8. We tried this function on a machine on which we could get 16! which has the value 2 004 189 184. For many microcomputers, it will not work properly for numbers greater than about 11!. Changing the function to real will increase the size of numbers that can be calculated but of course with a loss of precision in the answers.

Chapter 8

1. The rectangle rule is exact on functions that are horizontal in the interval. The trapezoid rule is exact on functions which are straight lines. Simpson's rule is exact on functions which are polynomials of degree two or less.

2. The MacLaurin series for cosine is

$$\cos x = \sum_{i=0}^{\infty} \frac{-1^{i+1} x^{2i}}{(2i)!}$$

The program to calculate it is

```
program Cosine
real X, Sum, Term
integer I
Sum = 1.0
Term = 1.0
print *, 'Enter X value'
read *, X
do 100 I = 1, 20
    Term = -Term*X*X/(2*I)/(2*I-1)
    Sum = Sum + Term
100 continue
print *, 'Approx Value is ', Sum
print *, 'Exact value is ', cos(X)
stop
end
```

4. This program counts the number of terms required to evaluate the series to get six decimal places of accuracy compared to the intrinsic function.

```
program Cosine
real X, Sum, Term, CosX
integer I
Sum = 1.0
Term = 1.0
print *, 'Enter X value'
read *, X
CosX = cos(X)
I = 1
while(abs(Sum-CosX).gt.1.0e-6)do
    Term = -Term*X*X/(2*I)/(2*I-1)
    Sum = Sum + Term
    I = I + 1
end while
print *, 'Approx Value is ', Sum
print *, 'Exact value is ', CosX
print *, '# of iterations used ', I
stop
end
```

8. Compute the value of sine three ways.

```
program Compar
real X
intrinsic cos
print *, 'Enter X value'
read *, X
print *, 'Intrinsic gives ', sin(X)
print *, 'Series gives', Series(X)
print *, 'Integration gives ',
&    Simp(0.0, X, 1000, cos)
stop
end
```

```
function Series(X)
real X, Sum, Term
integer I
Sum = X
Term = X
do 100 I = 1, 20
    Term = - Term*X*X/(2*I+1)/(2*I)
    Sum = Sum + Term
100 continue
Series = Sum
return
end
*   Integrates function that it is passed
real function Simp (A, B, N, F)
real A, B, Width, EvenS, OddS
real Area, EvenX, OddX, F
integer N, I
EvenS = 0.0
OddS = 0.0
Width = (B - A) / N
do 100 I = 0, N/2
    EvenX = A + 2 * I * Width
    OddX = A + (2 * I + 1) * Width
    EvenS = EvenS + F(EvenX)
    OddS = OddS + F(OddX)
100 continue
EvenS = EvenS - F(EvenX)
Simp = Width *
&    (F(A)+F(B)+2*EvenS+4*OddS)/3.0
return
end
```

The output when $X = \pi/2$ is entered is

```
Enter X value
Intrinsic gives           0.9999999
Series gives              1.0000000
Integration gives         1.0009546
```

Chapter 9

1. Here is the program.

```
program Calend
character*2 Days(31)
character*9 Month(12)
integer Year, CurMon, Numday(12)
integer Day
character*2 Chcode
Month(1)  = 'January'
Month(2)  = 'February'
Month(3)  = 'March'
Month(4)  = 'April'
Month(5)  = 'May'
Month(6)  = 'June'
Month(7)  = 'July'
Month(8)  = 'August'
Month(9)  = 'September'
Month(10) = 'October'
Month(11) = 'November'
Month(12) = 'December'
```

```
        Numday(1)  = 31
        Numday(2)  = 28
        Numday(3)  = 31
        Numday(4)  = 30
        Numday(5)  = 31
        Numday(6)  = 30
        Numday(7)  = 31
        Numday(8)  = 31
        Numday(9)  = 30
        Numday(10) = 31
        Numday(11) = 30
        Numday(12) = 31
        print *, 'Enter year and month'
        read *, Year, CurMon
        print *, 'Enter day month begins'
        read *, Day
        print '(T17, I4)', Year
        print '(T15, A)',Month(CurMon)
        print '(1X,7(A1, 6X))',
     &        'M','T','W','Th','F','S','S'
        print '(7(A2,5X))',('   ',I=1,Day),
     &        (Chcode(I),I=1,Numday(CurMon))
        stop
        end
        character*2 function Chcode( J )
        integer J
        character*1 First, Second
        First  = Char(J/10 + Ichar('0'))
        Second = Char(mod(J,10) + Ichar('0'))
        if (First.eq.'0') then
            Chcode = ' ' // Second
        else
            Chcode = First // Second
        endif
        return
        end
```

2. This program builds on the previous one.
```
        program Calend
        character*2 Days(31)
        character*9 Month(12)
        integer Year, CurMon, Numday(12)
        integer Day
        character*2 Chcode
        Month(1)  = 'January'
        Month(2)  = 'February'
        Month(3)  = 'March'
        Month(4)  = 'April'
        Month(5)  = 'May'
        Month(6)  = 'June'
        Month(7)  = 'July'
        Month(8)  = 'August'
        Month(9)  = 'September'
        Month(10) = 'October'
        Month(11) = 'November'
        Month(12) = 'December'
        Numday(1)  = 31
        Numday(2)  = 28
        Numday(3)  = 31
        Numday(4)  = 30
        Numday(5)  = 31
        Numday(6)  = 30
        Numday(7)  = 31
        Numday(8)  = 31
        Numday(9)  = 30
        Numday(10) = 31
        Numday(11) = 30
        Numday(12) = 31
        print *, 'Enter year'
        read *, Year
        Curmon = 1
        print *, 'Enter day January begins'
        read *, Day
        print '(T17, I4)', Year
        do 500 Curmon = 1, 12
            print *
            print '(T15,A)',Month(CurMon)
            print '(1X, 7(A1, 6X))',
     &            'M','T','W','Th','F','S','S'
            print '(7(A2,5X))',('   ',I=1,Day),
     &            (Chcode(I),I=1,Numday(CurMon))
            Day = mod(Day + Numday(Curmon),7)
500     continue
        stop
        end
        character*2 function Chcode( J )
        integer J
        character*1 First, Second
        First  = Char(J/10 + Ichar('0'))
        Second = Char(mod(J,10) + Ichar('0'))
        if (First.eq.'0') then
            Chcode = ' ' // Second
        else
            Chcode = First // Second
        endif
        return
        end
```

4. Here is the revised program.
```
        program ShowMe
        external FOfX
        real FOfX
        call Plot (FOfX, 0., .4, 15)
        stop
        end
        real function FOfX (X)
        real X
        FOfX = sin (X)
        return
        end
*
*       Plots the function values using
*       info about line length and function
*
        subroutine Plot(F,Start,Step,Number)
        real F, Min, Step
        integer Number
        integer Width
```

482 Appendix G

```
              parameter (Width = 35)
              real Here, FHere, FMin, FMax
              integer I, N, Over
       *      find the bounds
              FMax = F(Start)
              FMin = FMax
              do 100 I = 0, Number
                  FHere = F (Start + I * Step)
                  if (FHere .gt. FMax) then
                      FMax = FHere
                  else if (FHere .lt. FMin) then
                      FMin = FHere
                  endif
       100    continue
       *      now plot it
              print 600, FMin, FMax,
         &         ('-', N=1, Width + 2)
       600    format (11X,'Y =',G12.5,' to ',
         &         G12.5,/,1X, 'X =', 9X, 67A1)
              do 200 I = 0, Number
                  Here = Start + I * Step
                  FHere = F (Here)
                  Over = Width * (FHere - FMin)
         &             / (FMax - FMin)
       *          Here's the only change
                  print 610, Here,
         &             ('.', N=1, Over), '*'
       610    format (1X, G12.5, '|', 67A1)
       200    continue
              print 620, ('-', N=1, Width + 2)
       620    format (13X, 67(A1))
              return
              end
   8. Here is the program.
              program Pascal
              integer N, I, J, Fact
              print *, 'Enter number of rows'
              read *, N
              do 100 I = 1, N
                  print *,('        ', J = 1,N-I),
         &             (Fact(I-1, J),'    ',J = 0,I-1)
       100    continue
              stop
              end
              integer function Fact(N, I)
              integer N, I, J, Numer
              real Term
              Numer = N
              Term = 1.0
              do 200 J = I, 1, -1
                  Term = Term * Numer / J
                  Numer = Numer - 1
       200    continue
              Fact = int(Term)
              return
              end
```

Chapter 10

1. This is built from programs in the chapter.
```
           program CharF
           integer Freq(0:255), I
           character*1 KeepC(0:255)
           character*1 C
           do 100 I = 0,255
               Freq(I) = 0
               KeepC(I) = Char(I)
    100    continue
           print *,'Enter chars, one per line'
           while (.true.) do
               read (*, '(A1)', end=300) C
               Freq(Ichar(C)) = Freq(Ichar(C)) + 1
           end while
    300    continue
           call Sort(Freq, KeepC, 256)
           print '(A)','Character        Frequency'
           do 500 I = 0,255
               print '(5X,A3,5X,I5)',
        &         KeepC(I), Freq(I)
    500    continue
           stop
           end
           subroutine Sort (List, List2, Size)
           integer Size, List(Size)
           character*1 List2(Size), Ctemp
           integer I, Last, Minpos, Min, Temp
           do 200 Last = Size, 2, -1
               Min = List(1)
               Minpos = 1
               do 100 I = 1, Last
                   if (List(I).lt.Min) then
                       Min = List(I)
                       Minpos = I
                   endif
    100        continue
               Temp = List(Last)
               List(Last) = List(Minpos)
               List(Minpos) = Temp
               Ctemp = List2(Last)
               List2(Last) = List2(Minpos)
               List2(Minpos) = Ctemp
    200    continue
           return
           end
```
For the frequencies of letters in English and Fortran see the engineering problems about Huffman encoding.

2. Here are the output statements.
```
           do 200 I = 6, 1, -1
               print 100, ('*', J = 1, I)
    100    format(6A1)
    200    continue
```
4. Here is the program.

```
program Integr
integer N, I
real P(0:10), IP(0:11), A, B, C
print *,'Enter degree, coefficients'
read *, N, (P(I), I = 0, N)
print *,'Enter constant of integration'
read *, C
print *,'Enter limits of integration'
read *, A, B
do 100 I = N+1, 1, -1
    IP(I) = P(I-1) / I
100 continue
IP(0) = C
print *, 'Integral is ',
&   Eval(IP,N+1,B) - Eval(IP,N+1,A)
stop
end
real function Eval(P, N, X)
integer N, I
real P(0:N), Sum, X
Sum = 0.0
do 100 I = N, 0, -1
    Sum = Sum * X + P(I)
100 continue
Eval = Sum
return
end
```

8. Here is the program. It is simple revision of the program shown in the chapter.

```
program PSort
integer List(10), N, I
print *, 'Enter size and list'
read *, N, (List(I), I = 1,N)
call Sort(List, N)
print *, 'Sorted list is'
print '(5I10)', (List(I), I = 1,N)
stop
end
subroutine Sort(List, Size)
integer Size, List(Size)
integer I, Last, Final
integer Temp
Last = Size
while (Last .gt.1) do
    Final = 1
    do 100 I = 1, Last-1
        if (List(I).lt.List(I+1)) then
            Temp = List(I)
            List(I) = List(I+1)
            List(I+1) = Temp
            Final = I
        endif
100     continue
    Last = Final
end while
return
end
```

16. Here is the subroutine.

```
program ShowPI
real X(50)
integer I
do 100 I = 1, 50
    X(I) = sin(I / 20.0)
100 continue
call Plot( X, 50, 0.0, 2.5)
stop
end
subroutine Plot(X, N, Left, Right)
integer N, I, J
real X(N), Left, Right, Step
real Height, Max
character*1 Line(50)
Max = X(1)
do 100 I = 2, N
    if (X(I).gt.Max) then
        Max = X(I)
    endif
100 continue
Step = Max / 50.0
do 300 Height = Max, 0.0, -Step
    do 200 J = 1, N
        if (X(J).ge.Height) then
            Line(J) = '*'
        else
            Line(J) = ' '
        endif
200     continue
    print '(50A1)', Line
300 continue
print '(50A1)', ('-', I = 1,N)
print *,Left,(' ',I = 1,N/2-10),Right
return
end
```

Chapter 11

1. I hate computers
2. Here are two versions of the program.

```
program Percen
real Count
integer I
character*80 Line
character*1 C
Count = 0
read *, Line
read *, C
do 100 I = 1,80
    if (Line(I:I).eq.C) then
        Count = Count + 1
    endif
100 continue
print *, 'Percentage of ', C,
&   ' is ', Count * 1.25
stop
end
```

```
    program Percen
    real Count
    integer I
    character*80 Line, Temp
    character*1 C
    Count = 0
    read '(a80)', Line
    read '(a1)', C
    J = index(Line, C)
    while (J.ne.0) do
        Count = Count + 1
        Temp = Line(J+1:)
        Line = Temp
        J = index(Line, C)
    end while
    print *, 'Percentage of ', C,
&        ' is ', Count * 1.25
    stop
    end
```

4. Here is the program.
```
    program Compres
    character*80 Line, Temp
    integer I, J
    print *, 'Enter string'
    read 100, Line
100 format(A80)
    I = 80
    while (Line(I:I).eq.' ') do
        I = I - 1
    end while
    J = index(Line(1:I), ' ')
    while (J.ne.0) do
        Temp = Line(1:J) // Line(J+2:)
        Line = Temp
        I = I - 1
        J = index(Line(1:I), ' ')
    end while
    print 200, Line(1:I)
200 format(A)
    stop
    end
```

8. Here is the program.
```
    program Check
    integer Indx
    print *,Indx('abcdefg',7,'cd',2)
    print *,
&       Indx('abcdefghijklm',13,'efg',3)
    print *,Indx('abcdefgh',8,'z',1)
    stop
    end
    integer function Indx(S1,L1,S2,L2)
    character*(*) S1, S2
    integer L1, L2, I
    logical Found
    Indx = 0
    Found = .false.
    I = 1
    while(I.le.L1-L2+1.and..not.Found)do
        if (S1(I:I+L2-1).eq.S2(1:L2)) then
            Found = .true.
            Indx = I
        endif
        I = I + 1
    end while
    return
    end
```

16. The changes are
* added just before return in Commnd
```
        elseif (Word(2:3) .eq. 'ce') then
            call Center (MaxLin)
```
* added subroutine
*   Center the next output line
```
    subroutine Center (MaxLin)
    integer MaxLin
    character*80 CnLine
    integer I, J, Shift
    read '(a)', CnLine
    I = 80
    while (CnLine(I:I) .eq. ' ') do
        I = I - 1
    end while
    Shift = (MaxLin - I) / 2
    print '(80(a))', ' ', (' ', J=1, Shift),
&        CnLine (1:I)
    return
    end
```

Chapter 12

1. Here is a subroutine to transpose a matrix.
```
    program ShowT
    real A(10,10)
    integer I, J, N
    read *,N,((A(I,J),J=1,N),I=1,N)
    call Trans(A, 10, N)
    print *,((A(I,J),J=1,N),I=1,N)
    stop
    end
    subroutine Trans(A, M, N)
    integer M, N, I, J
    real A(M, M), Temp
    do 200 I = 1, N
        do 100 J = I+1, N
            Temp = A(I, J)
            A(I, J) = A(J, I)
            A(J, I) = Temp
100     continue
200 continue
    return
    end
```

2. This is done in the full pivoting example shown at the end of the chapter.

4. The power method is bad when the eigenvalues of the matrix are about the same magnitude. Therefore matrices which have no eigenvalue much larger than the others will not

work very well. On the other hand, any matrix with one large eigenvector will converge quite quickly. There are some matrices which will converge quite rapidly because of their structure. For example, a circulant matrix, one whose rows are shifts of one another, has an obvious eigenvector that can actually be determined by inspection.

8. Here is the function to add two matrices pointwise.

```
program UseF
real A(10, 10), B(10, 10), C(10 ,10)
integer I, J, N
read *,N,((A(I, J), J = 1,N), I = 1,N)
read *, ((B(I, J), J = 1,N), I = 1,N)
call Add(A, B, C, 10, N)
print *, ((C(I, J), J = 1,N), I = 1,N)
stop
end
subroutine Add(X, Y, Z, M, N)
integer I, J, M, N
real X(M, M), Y(M, M), Z(M, M)
do 200 I = 1, N
    do 100 J = 1, N
        Z(I, J) = X(I, J) + Y(I, J)
100     continue
200 continue
return
end
```

The function to subtract one matrix from another is identical except that the addition operation is replaced by a subtraction.

Chapter 13

1. You have to do this one yourself.
2. This following program allows you to try different combinations of coefficients for the linear congruential generator.

```
program Ran
real Mean, Sum
integer A, C, M, Seed, I
print *, 'Enter A, C and M'
read *, A, C, M
print *, 'Enter seed'
read *, Seed
Sum = 0.0
do 100 I = 1, 1000
    Seed = A * Seed + C
    if (Seed.lt.0) then
        Seed = Seed + 2147483647 + 1
    endif
    Seed = mod(Seed, M)
    Sum = Sum + Seed / real(M)
100 continue
Mean = Sum / 1000
print *, 'Mean is ', Mean
stop
end
```

Other measures that are useful are the standard deviation and the serial correlation. This can be estimated by looking at the distribution of distances between successive random numbers (this can range from $-m$ to m).

4. Here is the program.

```
program MonteC
integer Below,Above,Total,I,Seed
real Area, Totar, A, B, X, Y, Maxf
real Old, New
Below = 0
Above = 0
Seed = 12765439
print *, ' Enter integral limits '
read *, A, B
print *, 'Enter upper bound'
read *, Maxf
Old = 1000
New = 0
Total = 1
Totar = abs(B - A) * Maxf
while (abs(Old - New).gt.1.0e-3
  &  .or.New.eq.0.0) do
    call Random(Seed, X)
    X = X * abs(B - A) + A
    call Random(Seed, Y)
    Y = Y * Maxf
    if (Y.le.F(X)) then
        Below = Below + 1
    else
        Above = Above + 1
    endif
    Total = Total + 1
    Old = New
    New = real(Below) / Total
    print *, New
end while
Area = New * Totar
print '(A, F10.4)',
  &  ' Approximate area is ', Area
stop
end
real function F (X)
real X
F = sin(X) * exp(X)
return
end
subroutine Random (N, U)
integer N
real U
N = N * 843314861 + 453816693
if (N.lt.0) then
    N = N + 2147483647 + 1
endif
U = N * 0.4656612e-9
return
end
```

8. Here is the program.

```
      program Dice
      real Sum(2:12)
      integer I,First,Second,Total,Seed
      print *, ' Enter random number seed'
      read *, Seed
      do 100 I = 2, 12
          Sum(I) = 0.0
100   continue
      do 200 I = 1, 10000
          call Random(Seed, U)
          First = U * 6 + 1
          call Random(Seed, U)
          Second = U * 6 + 1
          Total = First + Second
          Sum(Total) = Sum(Total) + 1
200   continue
      do 300 I = 2, 12
          print *, 'Frequency of total',
     &        I, ' is ', Sum(I)
300   continue
      stop
      end
      subroutine Random(N, U)
      integer N
      real U
      N = N * 843314861 + 453816693
      if (N.lt.0) then
          N = N + 2147483647 + 1
      endif
      U = N * 0.4656612e-9
      return
      end
```

Chapter 14

1. Here is the program.
```
      program Roots
      real A, B, C, Discrm
      complex Dis, Root1, Root2
      print *, ' Enter coefficients'
      read *, A, B, C
      Discrm = B * B - 4.0 * A * C
      Dis = sqrt(cmplx(Discrm,0.0))
      Root1 = (-B + Dis) / (2 * A)
      Root2 = (-B - Dis) / (2 * A)
      print 100, Root1, Root2
100   format('Roots are', 2(F10.3,
     &    '+i', F10.3))
      stop
      end
```

2. Here is the program.
```
      program Cosine
      double precision X, Sum, Term
      integer I
      Sum = 1.0D0
      Term = 1.0D0
      print *, 'Enter X value'
      read *, X
      do 100 I = 1, 20
          Term = - Term*X*X/(2*I)/(2*I-1)
          Sum = Sum + Term
100   continue
      print 200, 'Approx Value is ', Sum
200   format(A, D20.12)
      stop
      end
```

3. The arctan of 1.0 is $\pi/4$. Thus we can calculate π as 4 arctan(1.0).
```
      program GetPi
      double precision Term, Sum
      integer I
      Term = 1.0D0
      Sum = 1.0D0
      do 100 I = 1, 100000
          Term = - Term
          Sum = Sum + Term / (2*I+1)
100   continue
      print *, 4.0 * Sum
      stop
      end
```
Unfortunately this series converges very slowly and it takes many terms to get even four or five decimal places of accuracy.

Glossary

Abstraction: The process of extracting the most significant features from a situation or problem and ignoring all other details. Used to describe the method of design in which pieces of program can invoke other pieces without having to know in detail how the other pieces are built.

Algorithm: An unambiguous set of instructions for doing something. Often refers to a set of instructions that can be easily written in a programming language.

Analog computers: Computers that calculate results using some continuously varying physical phenomenon. Electrical and mechanical devices have been used. Contrast with *digital computers*.

Argument: A value that is passed to a subprogram when it is invoked.

Arithmetic/logic unit: The part of a computer that carries out the actual calculations.

ASCII: Stands for American Standard Code for Information Interchange. The most common encoding of character information. See also *EBCDIC*.

Assembly language: A computer language that symbolically represents the computer's machine language. Used by humans only when absolutely necessary for efficiency reasons.

Back substitution: Method of solving simultaneous linear equations when they have been expressed in triangular form. Computationally very easy.

Binary: The number representation used within almost all computers for both instructions and data. Used because it is easy to build physical devices that can be in two states.

Binary search: Method of searching in an ordered list. Involves sampling the midpoint of the list and deciding whether the element being searched for lies in one half or the other. Takes time $O(\lg n)$ to search a list of length n.

Bisection method: A method of finding the root of an equation. Begins with an interval in which at least one root must lie. This interval is successively divided in half until it becomes small enough to approximate the root. Requires the function to be continuous. Related to the binary search.

Bit: An abbreviation for binary digit. The smallest unit of storage and manipulation, it can only be in two states, usually called 0 and 1.

Bootstrap: The program that starts a computer operating when it is powered up. It is usually located in some specific place in memory and the computer is hardwired to start from there when it is turned on.

Bubble sort: A sorting algorithm that switches adjacent elements in a list if they are out of order. On each pass, the largest unsorted element "bubbles" its way to the end of the unsorted part of the list. Other values also move closer to their final positions. Its worst initial case is a list sorted in reverse order.

Bugs: A term used by programmers to refer to errors that they have made in designing algorithms and programs. Usually means an error in the logic, as opposed to syntax and semantic errors, which can often be caught by the compiler.

Built-in functions: Functions that are provided as part of the language environment. Such functions are usually kept in a library. When the compiler detects the use of a function not defined by the programmer, it checks the library. If a matching built-in function exists, then it is used.

Byte: Usually the smallest addressable unit of memory. Often the size of a piece of memory sufficient to hold the encoding of a single character as well. On most machines, a byte is 8 bits long, but this may vary from machine to machine.

Carriage control: The method of controlling the vertical positioning of output on a line printer. Line printers use the first character of each line to determine how they should move vertically. Output statements can place an appropriate character in the first position of an output line to control printer action. Most compilers allow this feature to be turned off for terminal output, but it is good practice to assume that the first character of any output line will not be displayed.

Catenation: The process or operation of taking two strings and joining them together, one after the other, to make a new string.

Central processing unit: The part of a computer that is responsible for overseeing the operation of the other parts. It usually contains two parts: the control unit, which handles the sequencing and interpretation of instructions and the arithmetic/logic unit, which handles computation.

Comment: A statement in a program that is intended for human readers rather than as part of the program itself. Comments are ignored by the compiler except for reproducing them in the listing file. Some compilers use flags in comments to turn nonstandard compiler features on or off.

Comparison operators: These allow the values of expressions to be compared. The value of an expression involving a comparison operator is either true or false.

Compiler: A program that reads other programs as input and produces translations into machine language. A compiler also produces a listing file and sometimes other information as well.

Complexity: A measure of the number of operations that have to be done to carry out an algorithm. Complexity is usually expressed as a function of the program size.

Computer architecture: Used to refer to the components that make up a computer and their arrangement and interaction with each other. The architecture presented in this book is called a von Neumann architecture and is, at the moment, by far the most common computer architecture.

Correctness: The requirement that an algorithm or program conform to its specification or design. This means that it should produce the expected output for all legal input values. Correctness can be partially established by reasoning about programs and by testing.

Debugging: The process of finding the "bugs" in a program. It involves running test cases, inserting extra output statements into the program, tracing, and reasoning carefully about the program statements.

Definite iteration: Repetition of a statement or group of statements in which the number of repetitions is known before they begin.

Digital computers: Computers that use components with a finite number of states. Contrast with *analog computers*.

Directory: A part of a file system that collects a group of files. Plays the same role in a file system that a file drawer plays in an ordinary filing system.

Documentation: The description of a program and how to use it, written for both the user and maintainer. Usually divided into two parts—internal documentation within a program itself (see *comment*) and external documentation in the form of manuals.

Dynamic simulation: A simulation in which the notion of passage of time is important. Often has to do with analyzing the queuing behavior of systems.

EBCDIC: Stands for Extended Binary Coded Decimal Interchange Code. An encoding of characters used by IBM. See *ASCII*.

Editor: A program that allows a user to manipulate the contents of files. Usually allows sophisticated pattern matching and string manipulation.

Eigenvector: Any vector whose direction is not affected by the operation of some given matrix. Important because such vectors correspond to stable states under some transformation.

Euler's method: A method of approximating the shape of a function, given a differential equation describing its rate of change in space. Requires an initial point to select which member of the function family is being followed. Not very accurate except over small ranges.

Executable statements: Statements that correspond to machine language statements when a program is translated. Contrast with nonexecutable statements, which are usually instructions to the compiler.

Expressions: Collections of operators, variables, and constants that can be computed. An expression, when evaluated, has a value of a particular type. The type depends on what kinds of variables, operators, and constants were used.

File names: Names that are used to identify related pieces of data such as programs, text, and numeric data. File names are mapped to the actual location of the data by the file system.

File system: Part of the operating system that handles files, translates names to physical locations, and generally manages the data on the system.

Fixed point representation: The representation used for integer values. Contrast with *floating point representation*.

Flat file system: A file system in which all of a user's files are collected together into one logical grouping. There is no way to divide files used for different purposes into different groupings. There are no directories. Contrast with *hierarchic file system*.

Floating point representation: Representation used for real values, that is, those with a decimal part. The representation used consists of a fractional part, or mantissa, and an exponent. Contrast with *fixed point representation*.

Flow of control: The order in which the instructions are carried out when the program executes. The flow of control is sequential except when selection, iteration, or subprogram invocation occur.

Formal parameters: The placeholder variables that appear in a subprogram header. These placeholders must match, in number and type, the actual parameters, or arguments, that are used on invocation.

Formatter: A program that takes text as input and arranges it pleasingly. This usually involves justifying lines and inserting white space where necessary.

Functions: Subprograms that produce a particular distinguished value. They may be passed values in the usual way and may return values (though it is considered bad style). The function name must be assigned a value during the execution of the subprogram.

Gaussian elimination: A method of solving simultaneous linear equations that involves two steps: getting the coefficient matrix in upper triangular form by elementary row operations, and using back substitution to solve the new matrix equation.

Heuristic: A problem-solving approach that relies on making sensible decisions at each stage, but which is not guaranteed to provide an optimal solution to any problem.

Hierarchic file system: File system in which the files are organized into a hierarchy of directories, each of which contains logically related files. Contrast with *flat file system*.

High-level language: A computer language that is intended to be written by humans. Such a language hides many of the details of the computer on which programs execute and presents a logical model of computation to the programmer.

Identifier: A name that is given to a variable or subprogram. Most languages have rules about how an identifier may be formed.

Ill-conditioning: Applies to problems for which the algorithms are numerically unstable. This is usually because the problems are abnormally sensitive to small changes in values, which the representation of floating point values does not handle well. Such problems can result in widely differing answers with very small changes in problem parameters.

Implicit type conversion: The conversion between one internal representation for a value and another equivalent representation. This usually happens on assignment. Some language systems force the programmer to call an intrinsic function to do it explicitly.

Indefinite iteration: Repetition in which the number of times the repetition will occur is not known when the iteration begins. This is usually because the number of iterations depends on a computation that is done within the loop body.

Index of an array: The part of an array reference that specifies which element of the array is being referenced. It can be either an expression or a variable.

Infinite loop: A loop that will never end because the condition that would terminate it never becomes true. This is usually caused by failing to alter any part of the expression that is used to control indefinite iteration. Can be avoided by always making termination arguments about loops.

Input statement: Any statement that allows a program to obtain data from the outside world. May be a request for data from either the user or from a file.

Input/output devices: Devices that allow a computer to communicate with the outside world. These include terminal keyboards and screens, tape drives, card punches, and printers.

Intrinsic functions: See *built-in functions*.

Linear congruential generator: A recurrence that is used to calculate a pseudorandom number from a previous one. It involves multiplying by a constant, adding another constant, and reducing the result modulo some large number.

Linking: The process of collecting separately compiled program pieces together and making them into one executable program. This step follows compilation and usually involves incorporating library subprograms.

Local variables: Those variables that are declared within the body of a subprogram. Such variables vanish between invocations of the subprogram; they have no permanent existence.

Machine language: The language that is directly executable by a computer. The language consists of strings of binary digits that direct the computer. Humans never use machine language directly.

Maintenance: The work involved in altering programs to meet changing real world requirements or fixing errors that have been discovered in use. This is a significant component of much programming work.

Monte Carlo integration: A method of estimating integrals by sampling points randomly in the plane and calculating the number of points above and below the function being integrated. This can be used to estimate the area under the function as a fraction of the total area.

Newton-Raphson method: A method of finding the roots of an equation by following tangents to a curve until they intersect the x-axis. Requires the function to be continuous and differentiable. Produces a sequence of points that may converge to the root. Fast when it works but susceptible to being misled at or near turning points.

Numerical integration: A set of methods for estimating areas by approximating the function concerned by some other easy-to-integrate function. Methods include the trapezoid rule and Simpson's rule. Also called *quadrature*.

Overflow: Occurs when the result of some computation is too large to represent in the computer's floating or fixed point encoding. Some computers will produce an execution time error when this happens. Others won't.

Parameters: See *formal parameters*.

Power method: A method of finding the eigenvectors and eigenvalues of a transformation represented by a matrix. Depends on starting with an arbitrary vector and multiplying it by the matrix, reducing it, and repeating the multiplication. This tends to converge to an eigenvector rapidly.

Precedence: The rules that define the order in which operations are carried out when an expression contains more than one operator. The precedence is defined to agree with the ordinary rules of mathematics (more or less) and to make it easy to write expressions without parentheses.

Predictor-corrector method: A method of finding values of a function, given a differential equation describing its derivative. Uses Euler's method with an improvement.

Program synthesizer: A sophisticated form of text editor that will only allow syntactically correct language statements to be entered.

Programming languages: Languages that are designed to implement algorithms to be executed by computers. Such languages are unambiguous and allow the fundamental set of operations (sequencing, selection, repetition, and abstraction) to be expressed.

Prompting for input: Displaying a message indicating what is required before expecting a user of a program to type input values.

Pseudocode: A form of notation for writing algorithms. It is usually close to the programming language in which the program will eventually be written, but without having the formal syntax of the programming language.

Random numbers: Numbers that are distributed across some range of values according to some distribution function (usually uniform — all the values are equally likely) and which otherwise have no strong patterns. They are used in simulation. Truly random numbers are difficult to get so we usually use pseudo-random numbers.

Refinement: The method of developing an abstract solution to a problem first, and then developing each part of the abstract solution into a more detailed form.

Round-off error: The error that occurs in calculations because of the finite number of significant figures that can be maintained in the internal representation of floating point numbers. It becomes significant whenever a large number of dependent calculations are carried out. Programs should always be written to reduce round-off errors as much as possible.

Run time error: An error which occurs while the program is running. It is usually caused by some operation that is not legal but which the compiler was unable to detect or did not check for.

Scientific notation: The form of notation for real numbers in the form of a fractional part and an exponent. See *floating point representation*.

Scope rules: The rules that determine which variables can be used in each part of a program.

Semantic error: An error that occurs from writing something that is syntactically correct, but meaningless. The English sentence "Ideas are pink." is correctly formed, but semantically erroneous.

Sentinel: A special value that is used to mark something, usually the end of a list of input values.

Sieve of Eratosthenes: An algorithm to find the prime numbers in some range of integers. It depends on crossing off all multiples of numbers and so is computationally quite efficient.

Simpson's rule: A method of numerical integration that fits parabolas to sets of three successive function points in the interval of integration. The area under the function is approximated by the area under these parabolas.

Simulation: Methods of manipulating models of physical processes that are kept as data within the memory of a computer. Can be a cheap and effective way to investigate the behavior of a system.

Subprograms: Any program unit that must be invoked from somewhere else in order to execute. This is the language feature that implements abstraction.

Syntax error: An error caused by writing something that does not conform to the rules of the language.

Testing: The process of trying out various input sets and seeing how a program behaves. Can be used to raise the confidence that a program is correct.

Top-down development: See *refinement*.

Tracing execution: An important technique for debugging. Involves playing computer and recording everything that happens as a program executes.

Trapezoid rule: A method of numerical integration that uses straight line segments to approximate the function being integrated.

Variable declaration: The part of a program that tells the compiler how much memory space to allocate for the program variables. Involves telling the compiler the names, types, and sizes of variables.

Variable parameters: Parameters that pass values into a subprogram and also receive values back from the subprogram when it completes execution.

Variable: A symbolic name for a location used by a program to store values.

Word processing: The operations involved in editing and text formatting, especially when they are applied to preparing documents.

Zero of function: The point at which a function is zero. Such points are important because two functions are equal when their difference is zero and because a function has a turning point when its derivative is zero.

Index

A

Abstraction, 3
Aimag function, 442
Algorithm, 2
ALU, 25
Analog computers, 26
Area, calculating 125
Area, calculating, 163
   using parabolas, 166
   using rectangles, 162
   using trapezoids, 163
Argument, 90
Arithmetic if statement, 468
Arithmetic/logic unit, 25
Arrays, 241
   as arguments, 253
   as tables, 246
   assignment to, 242
   declaration, 242
   formatted output, 248
   in subprograms, 254
   indexes, 250
   indexing, 243
   referencing, 243
   using part of, 244
ASCII, 74
Assembly language, 6
Assigned goto statement, 468
Assignment statement, 86
Average case analysis, 266

B

Back substitution, 351
Backspace statement, 471
Bases,
   converting binary fractions to decimal, 29
   converting binary to decimal, 28
   external to internal, 298
   internal to external, 298
Binary search, 268
Binary, 5, 21
Bisection method, 178
Bit, 22
Block data subprograms, 463
Boolean data,
   See logical data
Bootstrap, 24
Boundary conditions, 37
Bubble sort, 273
Bugs, 37
Building a custom resistor, 287
Built-in functions, 145
Bytes, 74
Byte, 22, 74

C

Calculating force of impact, 201
Call statement, 58
Carriage control, 211
Catenation, 297
Central processing unit, 25
Character data, 22

Char, 251
Close statement, 223, 471
Closing a traverse, 386
Cmplx function, 442
Column numbers, 56
Comments, 37, 76
Comment, 57
Common, 461
Comparison operators, 109, 123
Compiler, 42
Complex data, 441
Complexity, 11
Computed goto statement, 468
Computer architecture, 20
Concatenation, 297
Constants, 80
Continuation character, 56
Continue statement, 104
Control unit, 25
Coordinates, 252
Correctness, 12
CPU, 25
Cramer's rule, 358

D

Data statement, 459
Data, 20
Debugger, 43
Debugging, 37, 43
Definite iteration, 3, 104
Detecting transmission errors, 430
Differential equations, 171
Digital computers, 26
Dimension statement, 463
Directory, 40
Do statement, 104, 107
Documentation, 38
Double precision, 440
Dynamic simulation, 397

E

EBCDIC, 74
Editor, 40
Eigenvalue, 368
Eigenvector, 368
Electron emission, 157
Else clause, 119
End statement, 58

Endfile statement, 470
Engineering problems,
    building a custom resistor, 287
    calculating force of impact, 201
    closing a traverse, 386
    detecting transmission errors, 430
    electron emission, 157
    escape velocity of space vehicle, 192
    failure of electronic components, 195
    file compression, 339
    finding sample points, 199
    fractal geometry, 446
    heat transfer through windows, 159
    motion of particles, 289
    neutron scattering, 437
    path of electron beam, 237
    placing chips onto boards, 432
    separation of components, 291
    simulated annealing, 435
    text compression, 389
    time between failures, 197
Enhancement, 39
Entry statement, 464
Equivalence statement, 463
Escape character, 339
Escape velocity of space vehicle, 192
Euler's method, 172, 193
Executable statements, 57
Expressions, 78
External statement, 176

F

Failure of electronic components, 195
File compression, 339
File designators, 220
File names, 40
File systems, 39
Finding sample points, 199
Fitting a straight line, 375
Flat file system, 40
Floating point data, 21, 77
Flow of control, 58
Formal parameters, 90
Format codes,
    A 206
    carriage control 211
    E 208
    F 207
    G 210

I   206
    literal string   207
    repetition of   213
    reusing   214
    T   209
    TL   210
    TR   210
    X   209
Formatter,   45
Fortran names,   56
Fortran,
    history,   7
Fractal geometry,   446
Fractal sets,   446
Full-screen editor,   41
Function subprograms,   142
Functions,   142
    abs,   171
    built-in,   145
    header,   142
    intrinsic,   145

G

Gaussian elimination,   350
General purpose computers,   26
Generic functions,   146
Goto statement,   129, 466, 477
Graphs, plotting,   224, 225, 371
Greedy algorithm,   432, 433
Group counts,   213

H

Heat transfer through windows,   159
Heuristic,   2
Hierarchic file system,   40
High-level language,   6
Horner's rule,   262

I

Ichar,   251
Identifier,   56
If statement,   118, 119
    arithmetic if   468
    logical if   467
    nesting,   121
Ill-conditioning,   357
Implicit statement,   465

Implied do loop,   215, 249
Indefinite iteration,   3, 108
Indenting,   36, 105, 119
Index function,   297
Index of an array,   243, 250
Index variable,   104
Infinite loop,   117
Input statement,   218
Input to programs,   87
Input/output devices,   25
Inquire statement,   469
Instruction execution sequence,   24
Instruction execution,   24
Integer data,   21
Integration,
    Monte Carlo,   403
    multi-dimensional,   406
    Simpson's rule,   166
    using rectangles,   126, 162
    using trapezoids,   164
Intrinsic functions,   145, 155, 251
IO,   25
Iteration,   3

K

Key,   266

L

Line editor,   309
Line layout,   56
Linear congruential generator,   400
Linear equations,   350
Linear regression,   375
Linear search,   266
Linking,   43
Local variables,   91
Locality,   23
Logical data,   22, 123
Logical if statement,   467
Logical operators,   123
Loop structure,   112

M

Machine language,   5
Maintenance,   39
Mapping characters to integers,   251
Mapping integers to characters,   251

Markup, 45
Matrices,
    determinant of, 354
    eigenvalues of, 367
    inverse, 350, 357
    lower triangular, 348, 361
    operations, 346
    upper triangular, 348, 361
Mean time between failures, 197
Mean, 113, 114, 119, 244
Merging data streams, 219
Methodology, 52
Mixed-mode, 84
Models, 396
    mathematical analysis of, 396
    simulation, 396
Monte Carlo integration, 403
Motion of particles, 289

N

Names,
    rules for constructing, 56
Nested if statement, 121
Nested loops, 117
Neutron scattering, 437
Newton-Raphson method, 184, 263
Numerical integration, 125

O

Open statement, 222, 471
Operators, 78
    order of evaluation, 82
Output, 59
Overflow in assignment, 86, 139
O(...) notation, 11

P

Paradigms, 34
Parameter statement, 81
Parameters, 90
Partial pivoting, 355
Path of electron beam, 237
Pivoting, 355
Placing chips onto boards, 432
Polynomials, 260, 375
Power method, 368
Precedence, 82, 458

Predictor-corrector method, 174
Prime numbers, 275
Print statement, 59
Printing,
    apostrophes, 64
    blank lines, 64
    text, 59
Problem requirements, 33
Program libraries, 146
Program name, 55
Program requirements, 32
Program structure, 55
Program synthesizer, 43
Programming languages, 5
Programming, 5
Programs, 20
    add list of numbers, 112
    add up positive numbers, 119
    adjust a line of text, 322
    assignments 88
    back substitution, 352
    binary search, 270
    bisection method root finding, 180, 181, 183
    blackjack, 422
    bubble sort, 274
    Caesar cipher for any alphabet, 331
    Caesar cipher, 281
    calculate areas of triangles, 46
    calculate standard deviation using arrays, 245
    calculate vector length, 255, 256
    change a constant, 130
    common example, 279
    compute powers of numbers, 118
    counting frequencies of characters, 252
    counting words and blanks in text, 321
    counting words in text using ends, 319
    counting words in text, 318
    differences between files, 221
    differentiating polynomials, 261
    dot product variations, 257
    double precision series for exponential function, 443
    end of file, 217, 218
    evaluate a function, 132
    evaluating expressions, 85

evaluating polynomials, 262
extended linear search, 267
finding eigenvectors, 369
finding numbers within 10 of largest, 240
finding the inverse of a matrix, 358
format codes: A, 207
format codes: carriage control, 212
format codes: E, 208
format codes: F, 208
format codes: G, 210
format codes: I, 206
format codes: literal string, 207
format codes: repetition, 214
format codes: reusing, 214
format codes: TR and TL, 210
format codes: T, 209
format codes: X, 209
format codes: /, 213
forward substitution, 364
frequency counting using arrays, 248
function use, 143, 144
Gaussian elimination with full pivoting, 379
Gaussian elimination with pivoting, 355
Gaussian elimination, 353
implied do loop, 215
improved sieve of eratosthenes, 277
integer division, 91
integration with Simpson's rule and multiple intervals, 187
linear regression, 377
linear search, 265
local variables, 92
logical variable, 124
LU decomposition, 362
matrix multiplication of any size, 344
matrix multiplication, 343
maximum of list of numbers, 121
mean of list of numbers with zero check, 120
mean of list of numbers, 113
merge two files, 219, 223
misuse of local variables, 92
Monte Carlo integration of volume, 406
Monte Carlo integration using sample points, 405
multiplying a vector by a matrix, 343

multiplying polynomials, 261
Newton-Raphson method, 263
Newton-Raphson root finding, 185
number base conversion, 299
odd numbers only sieve, 278
opening files, 223
output with subroutines invoking others, 61
output with subroutines, 61
permuting columns of a matrix, 348
permuting rows of a matrix, 347
plot function values, 226
powers of a constant, 94
print a grid, 64
print Dilly song, 62
print function values, 224
print reciprocal, 104, 106
print reverse successive integers, 107
print several reciprocals, 111
print square of number, 89
print successive fractions, 108
print successive integers, 107
printing numbers within 10 of largest using array, 243
roots of a quadratic, 147
selection sort, 272
series for exponential function, 169, 170
shell range, 151
shell trajectory, graphical display, 229
show contours of surface, 372
sieve of Eratosthenes, 276
simple frequency counting, 246
simple integration, rectangle rule, 127
simple integration, rectangle rule, with function, 162, 163
simple Monte Carlo integration, 403
simple nested loops, 117
simple ordinary differential equation, 173, 175
simple output with subroutines, 60
simple output, 60
simple procedure, 100, 101
Simpson's rule integration with intrinsic, 178
Simpson's rule integration, 167, 177
simulation of queue, 408
snakes and ladders, 411, 413, 415
solitaire game, 419

solving using LU decomposition, 365
square a number with a
   subprogram, 129
standard deviation of list of
   numbers, 115
static simulation, fish, 398
stream editor, 308
text editor, 311
text formatting, 325
trapezoid rule integration, 165
Triangulation, 351
   upper triangulation, 349
   using the parameter statement, 82
Prompting for input, 88
Properties of lists, 241
Prototypes, 396
Pseudocode, 34

Q

Queuing theory, 407

R

Random function with seed, 402
Random function, 402
Random numbers, 399
   functions to produce, 399
   generating, 400
   seed, 401
Read statement, 87, 217, 472
Real data, 21
Real function, 442
Refinement, 34
Regression, 375
Return statement, 58, 142
Rewind statement, 469
Root of equation,
   package to find, 263
Roots of equations,
   imaginary, 441
Roots of equation,
   finding by bisection, 178
   finding using Newton-Raphson, 184
Round-off error, 128, 170
Run time error, 42

S

Save statement, 465
Scientific notation, 21, 77
Scope rules, 92
Searching,
   binary, 268
   linear, 266
Selection sort, 271
Selection, 3, 118
Semantic error, 42
Sentinel, 111
Separation of components, 291
Sequencing, 3
Series evaluation, 168, 170
Setsd subroutine, 401
Sieve of Eratosthenes, 276, 286
Simpson's rule, 166
Simulated annealing, 435
Simulation, 396
   dynamic, 397, 407
   static, 397
Solving differential equations,
   Euler's method, 172
   predictor-corrector method, 174
Sorting,
   bubble, 273
   selection, 271
Standard deviation, 244
Statement functions, 464
Static simulation, 397
Stop statement, 57
Storage device hierarchy, 23
Stored program computers, 26
Subprogram parameters, 90
Subprograms, 57
   arrays as parameters, 254
   parameters, 90
   placement of, 59
   variables in, 91
Subroutine, 58
Substring, 296
Symbolic debugger, 43
Syntax error, 42
Synthesizer, 43

T

Termination, 9
Testing, 37

Text compression, 389
Text editor, 40
Then clause, 119
Time between failures, 197
Top-down development, 34
Tracing execution, 149
Trapezoid rule, 164
Truncation in division, 79
Type compatibility for assignment, 86
Types,
    character, 75
    complex, 441
    double precision, 440
    integer, 76
    logical, 123
    real, 78

U

Unit descriptors, 219
Utility program, 20

V

Variable declaration, 75
Variable parameters, 130
Variable, 75
Vector spaces, 252
Vectors, 252
    dot product, 256
    operations, 253
Very high-level language, 8

W

Walking bisection, 183
While statement, 108
Wolf fence, 44
Word processing, 45
Wysiwyg, 45

Z

Zero of function, 178, 184